THE ECONOMY AS AN EVOLVING COMPLEX SYSTEM

THE ECONOMY AS AN EVOLVING COMPLEX SYSTEM

THE PROCEEDINGS OF THE EVOLUTIONARY PATHS OF THE GLOBAL ECONOMY WORKSHOP, HELD SEPTEMBER, 1987 IN SANTA FE, NEW MEXICO

Edited by

Philip W. Anderson
Princeton University

Kenneth J. Arrow
Stanford University

David Pines
University of Illinois, Urbana

Volume V

SANTA FE INSTITUTE
STUDIES IN THE SCIENCES OF COMPLEXITY

Addison-Wesley Publishing Company, Inc.
The Advanced Book Program
Redwood City, California • Menlo Park, California
Reading, Massachusetts • New York
Don Mills, Ontario • Wokingham, U.K. • Amsterdam • Bonn
Sydney • Singapore • Tokyo • Madrid • San Juan

Publisher: *Allan M. Wylde*
Production Administrator: *Karen L. Garrison*
Editorial Coordinator: *Alexandra McDowell*
Electronic Production Consultant: *Mona Zeftel*
Promotions Manager: *Celina Gonzales*

Director of Publications, Santa Fe Institute: *Ronda K. Butler-Villa*

Library of Congress Cataloging-in-Publication Data

The economy as an evolving complex system / edited by Philip W. Anderson,
Kenneth J. Arrow, and David Pines.
p. cm.—(Santa Fe Institute studies in the sciences of complexity; v. 5)
Proceedings of a workshop sponsored by Sante Fe Institute.
Bibliography; p.
Includes index.
1. Economics—Congresses. I. Anderson, Philip W.,1923–
II. Arrow, Kenneth Joseph. III. Pines, David, 1924–
IV. Santa Fe Institute (Santa Fe, N.M.) V. Series.
HB21.E265 1988 88-22156
330—dc19 CIP

This volume was typeset using TEXtures on a Macintosh II computer. Camera-ready output from an Apple LaserWriter Plus Printer.

ABCDEFGHIJ-MA-898

About the Santa Fe Institute

The *Santa Fè Institute* (SFI) is a multidisciplinary graduate research and teaching institution formed to nuture research on complex systems and their simpler elements. A private, independent institution, SFI was founded in 1984. Its primary concern is to focus the tools of traditional scientific disciplines and emerging new computer resources on the problems and opportunities that are involved in the multidisciplinary study of complex systems -- those fundamental processes that shape almost every aspect of human life. Understanding complex systems is critical to realizing the full potential of science, and may be expected to yield enormous intellectual and practical benefits.

All titles from the *Santa Fe Institute Studies in the Sciences of Complexity* series will carry this imprint which is based on a Mimbres pottery design (circa A.D. 950-1150), drawn by Betsy Jones.

Santa Fe Institute Studies in the Sciences of Complexity

VOLUME	EDITOR	TITLE
I	David Pines	Emerging Syntheses in Science, 1987
II	Alan S. Perelson	Theoretical Immunology, Part One, 1988
III	Alan S. Perelson	Theoretical Immunology, Part Two, 1988
IV	Gary D. Doolen et al.	Lattice Gas Methods of Partial Differential Equations
V	Philip W. Anderson et al.	The Economy as an Evolving Complex System
VI	Christopher G. Langton	Artificial Life: Proceedings of an Interdisciplinary Workshop on the Synthesis and Simulation of Living Systems

Contributors to This Volume

Dr. Robert McCormick Adams
Secretary, Smithsonian Institute

Professor Philip W. Anderson
Department of Physics, Princeton University

Professor Kenneth J. Arrow
Department of Economics, Stanford University

Professor W. Brian Arthur
Food Research Institute, Stanford University

Dr. Eric B. Baum
Jet Propulsion Laboratory, California Institute of Technology

Professor Michele Boldrin
University of California, Los Angeles

Professor William A. Brock
Department of Economics, The University of Wisconsin

Dr. George A. Cowan
President, Santa Fe Institute and
Senior Fellow, Los Alamos National Laboratory

Professor J.-P. Eckmann
Départment de Physique Théorique, Université de Genève

Dr. J. Doyne Farmer
Theoretical Division and Center for Nonlinear Studies,
Los Alamos National Laboratory

Professor John H. Holland
Division of Computer Science & Engineering,
University of Michigan

Professor S. Oliffson Kamphorts
Départment de Physique Théorique, Université de Genève

Professor Stuart Kauffman
School of Medicine, University of Pennsylvania

Professor Timothy J. Kehoe
Department of Economics, University of Minnesota

Professor Norman H. Packard
Center for Complex Systems Research and
Department of Physics, University of Illinois

Professor Richard Palmer
Department of Physics, Duke University

Professor David Pines
Department of Physics, University of Illinois

continued

Contributors *continued*

Professor David Ruelle
I.H.E.S., France

Professor José A. Scheinkman
Goldman, Sachs & Co., New York and
University of Chicago

Professor John J. Sidorowich
Physics Department, University of California, Santa Cruz

Professor Mario Henrique Simonsen
Brazil Institute of Economics

P. W. ANDERSON, KENNETH J. ARROW, & DAVID PINES
March, 1988

Foreword

The Santa Fe Institute sponsored a workshop on "Evolutionary Paths of the Global Economy" from September 8–18, 1987 at the Institute campus in Santa Fe. The purpose of the workshop was to explore the potential usefulness of a broadly transdisciplinary research program on the dynamics of the global economic system, by bringing together a group of economists and a group of natural scientists who have developed techniques for studying nonlinear dynamical systems and adaptive paths in evolutionary systems. The workshop was convened by David Pines, Co-Chair of the Santa Fe Institute Science Board, who served as Moderator, and chaired by Philip W. Anderson and Kenneth J. Arrow, who selected the participants and arranged the program.

Following introductory remarks by the organizers, the workshop began with a series of pedagogical lectures, intended to introduce the natural scientists to the basic relevant work in economic analysis and economists to basic techniques for dealing with complex dynamical systems developed in the natural sciences. Notes prepared by the lecturers are found in Part II of these proceedings. Following the first three days of lectures, three working groups were formed to enable participants to discuss, in smaller groupings, some general questions, and to devise homework problems which focused on specific aspects of the economy as a complex evolving system. The reports of the working groups are given in Part III. Part IV contains a report on a plenary discussion which formed the last session of the conference, while Part V contains the workshop summaries prepared by the co-chairmen. One outcome of the workshop was the initiation of several transdisciplinary research collaborations. In Part VI the reader will find the first fruits of their collaborations. The Appendix contains notes on the exploratory meeting which led to the present workshop. We believe the reader might find them of interest both for historical reasons, and for the additional perspectives presented there.

At the final plenary session, the participants agreed that a more appropriate title for the workshop, and future Santa Fe Institute research activities in this area, would be "The Economy as an Evolving Complex System," and that is the title we have selected for these workshop proceedings.

It gives us pleasure, on behalf of the Santa Fe Institute, to thank our fellow participants for contributing to this volume, and Citicorp and the Russell Sage Foundation for their financial support, which has made possible both the workshops and the publication of this volume. Special thanks go to Ronda K. Butler-Villa for her assistance in preparing this volume for publication.

Table of Contents

Foreword
P. W. Anderson, Kenneth J. Arrow, and David Pines i

I. Introduction and Overview 1

Introduction and Overview
David Pines 3

II. Lectures and Perspectives 7

Self-Reinforcing Mechanisms in Economics
W. Brian Arthur 9

Neural Nets for Economists
Eric B. Baum 33

Persistent Oscillations and Chaos in Economic Models: Notes for a Survey
Michele Boldrin 49

Nonlinearity and Complex Dynamics in Economics and Finance
William A. Brock 77

Can New Approaches to Nonlinear Modeling Improve Economic Forecasts?
J. Doyne Farmer and John J. Sidorowich 99

The Global Economy as an Adaptive Process
John H. Holland 117

The Evolution of Economic Webs
Stuart A. Kauffman 125

Computation and Multiplicity of Economic Equilibria
Timothy J. Kehoe 147

A Simple Model for Dynamics away from Attractors
Norman Packard 169

Statistical Mechanics Approaches to Complex Optimization Problems
Richard Palmer 177

Can Nonlinear Dynamics Help Economists?
David Ruelle 195

Rational Expectations, Game Theory and Inflationary Inertia
Mario Henrique Simonsen 205

III. Working Group Summaries 243

Working Group A: Techniques and Webs
J. Guenther, J. Holland, S. Kauffman, T. Kehoe T. Sargent, and E. Singer 245

Working Group B: Economic Cycles
K. Arrow, W. Brock, D. Farmer, D. Pines, D. Ruelle, J. Scheinkman, M. Simonsen, and L. Summers 247

Working Group C: Patterns
P. W. Anderson, W. B. Arthur, E. Baum, M. Boldrin, N. Packard, J. Scheinkman, and L. Summers 249

IV. Final Plenary Discussion 255

Final Plenary Discussion
Richard Palmer 257

V. Summaries and Perspectives 263

A Physicist Looks at Economics: An Overview of the Workshop
P. W. Anderson 265

Workshop on the Economy as an Evolving Complex System: Summary
Kenneth J. Arrow 275

VI. Research Papers 283

Learning-By-Doing, International Trade and Growth: A Note
Michele Boldrin and José A. Scheinkman 285

Lyapunov Exponents for Stock Returns
J.-P. Eckmann, S. Oliffson Kamphorst, D. Ruelle and J. Scheinkman 301

VII. Appendix 305

Summary of Meeting on "International Finance as a Complex System" at the Rancho Encantado, Tesuque, New Mexico, August 6–7, 1986
George A. Cowan and Robert McCormick Adams 307

Index 313

I. Introduction and Overview

DAVID PINES
Department of Physics, University of Illinois, 1100 West Green Street, Urbana, Illinois 61801

The Economy as an Evolving Complex System: An Introduction to the Workshop

Any attempt, such as that in which we are presently engaged, at opening a serious dialogue between disparate, and at times hostile, communities of scholars must be regarded as a high-risk investment of resources and time. Its comparatively slim chances for success are increased if the scholars in question are equipped for the dialogue with those qualities we all strive to possess: confidence (built upon a record of accomplishment in one's own field); an open mind (could a quite different approach to familiar problems yield significant results? could techniques developed for one field find useful application in a very different area?); a willingness to learn coupled with an ability to teach; a sense of adventure; and a sense of humor. The Santa Fe Institute, as a new transdisciplinary graduate research and educational institution, has sought the involvement and advice of just such a group of scholars in defining its research agenda. In one of the early meetings of the Institute's scientific advisors, Robert McCormick Adams, Secretary of the Smithsonian Institution and Vice-Chairman of the Santa Fe Institute (SFI) Board of Trustees, told us of a suggestion made to him by John Reed, the Citicorp Board Chairman—that one examine the potential application of recent developments in the natural sciences and computer technology to the understanding of world capital flow and debt. This suggestion led SFI to sponsor, in August 1986, a small one-day exploratory meeting, organized by Adams and George Cowan, the SFI President, on "International

Finance as a Complex System." The discussions at that meeting (which are summarized in the Appendix) persuaded the Institute advisors present (Adams, Cowan, Philip W. Anderson, Carl Kaysen, John Holland, Larry Smarr, and myself) that SFI should take the next step in opening up a dialogue between economists and natural scientists by sponsoring a workshop on a topic designed to achieve that purpose.

I agreed to convene such a workshop, and was most fortunate in persuading Phil Anderson and Kenneth J. Arrow to serve as its co-chairmen. We chose as our topic, "The Evolution of the Global Economy," and decided that ten was our magic number: a ten-day intensive workshop, involving ten economists and ten natural scientists. From the start, Professors Anderson and Arrow, who selected the participants and arranged the program, held one very clear view in common. They were interested in expanding the horizons of conventional economic theory so that it might eventually be able to deal with such complex macroeconomic problems as the global economy, rather than in applying pre-existing economic theories to this problem and so becoming a forum for conflicting views of causation or cure, based on manifestly incomplete theories. The presentations at the workshop on macroeconomic problems, including the summaries of the business cycle by Larry Summers and of the present debt crisis by Mario Simonsen, were essentially masterly critical reviews and analyses of what actually has happened and is happening, and of failed theories and panaceas, *not* attempts to sell any particular set of prescriptions.

It immediately became evident that both groups of scientists were more than eager to interact, and to absorb as well as emit new ideas and new vocabularies. The largest portion of the sessions consisted of a single speaker plus extended, interleaved discussion, so that a scheduled one-hour talk typically took two-and-a-half hours to complete. After an initial six talks, the organizers scheduled a session to structure the field into topics as well as to decide upon further presentations in a more or less democratic fashion. The list of topics for working groups does more than any other single-page description to impart the basic intellectual content of the meeting:

1. Learning by the economic agents about the economy (noisy observation). Noisy optimization. Can complex adaptive learning models (e.g., Holland, Kauffman) be applied? Adaptation in sequential games (e.g., prisoner's dilemma). Is biology the same as economics or does the latter differ by emphasizing foresight? (Hicks-Arrow-Debreu trick of evading history.) How to model different types of interrelationships in biology and economics, such as predation, symbiosis, infrastructure, other links on graphs.

2. Innovations and coevolution. Ontogeny as a constraint on change. Finiteness or openness of niche space, use space, resources.

3. Economic development as a complex system, as a pattern problem. Are there many equivalent states? Can one model the diffusion of devlopment as a reaction–diffusion front? Political input as damping/enhancing factor in development.

4. Regional economy differences. Is there a tendency to polarization? Do reaction models offer a parallel? Is this paralleled by international economic differences?
5. Review of possibilities for lock-in. Do such methods as annealing suggest policies or spontaneous actions of the economic system for removing harmful lock-in?
6. Relation of deterministic to stochastic models. Role of foresight in removing positive Lyapunov coefficients (stability, chaos, etc.). Meaning of predictability in deterministic systems. Can there be deterministic models without perfect foresight?
7. Securities markets: economic theory, martingales, excess volatility, explanation of trading volumes. Does deterministic nonlinear analysis help? Psychological components (overconfidence; non-Bayesian expectation formation).
8. The global economy. Links among nations as well as over time. Monetary and real magnitudes in international flows. Changes in regime (gold standard, fixed exchange rates, floating rates).
9. Business cycles: stochastic (monetary or real) models versus deterministic nonlinear; large numbers of variables with leads and lags. Are there significant changes in regime, e.g., two world wars, Great Depression, emergence of business cycles.

These topics were apportioned among working groups in the general scheme of "Cycles" (predictive theories of dynamic behavior, such as nonlinear deterministic systems), "Webs" (theories of large numbers of interacting units with evolution as the paradigm), and "Patterns" (theories of inhomogeneity and of self-maintaining differences, with theories of growth and reaction fronts as physical analogues).

From then on, working group meetings were interspersed almost evenly with plenary sessions; some of the working groups evolved into collaborations on model problems related to the topics, while some remained primarily in the mode of continuing a general and very lively discussion of the topics in terms of model problems or of general philosophical approach. In both cases, discussions often continued far beyond the scheduled time. The working group reports in response to these topics are given in Part III of these proceedings.

Quite generally, the economists at the workshop were eager to learn as much as possible about the limits of applicability of the various kits of possibly applicable complex systems tools provided by the non-economists present, while the natural and biological scientists took every opportunity to inquire about the possible time-dependence of models of the economy. While most economists were familiar with the idea of chaos arising out of a simple dynamical system, and the possibility that time series relevant to the economy might exhibit chaotic behavior, the use of statistical physics (spin glass) methods as well as genetic and learning algorithms to describe complexity represented for many a new approach; for the non-economists, the role that foresight plays in the development of models of the economy was a quite new concept. This process of continued mutual learning lent an air of excitement to the workshop and contributed to the reluctance of so many participants to conclude the discussions and go home. There emerged a consensus that by working together over a period of time, economists and natural scientists skilled in the treatment of nonlinear processes could develop a broad interdisciplinary research program on the dynamics of the global economic system which would lead to important new results and perhaps a radically new approach to understanding the economy as an evolving complex system.

Problems anticipated prior to the workshop never materialized or were effectively minimized. These included difficulties in communicating across disciplines, tensions between experienced economists and eager amateurs, and a possible lack of focus due to the breadth of the subject matter. In fact, the meetings were intensely and mutually informative and collegial. The workshop fully achieved its principal purpose, to provide an initial basis for supportive research across several disciplines on nonlinear dynamic processes in economics.

To foster mutually supportive research in this field, the Santa Fe Institute has formed a research network and is planning a resident research program in Santa Fe on "The Economy as an Evolving Complex System." Appointments to visiting fellows and postdoctoral associates are now being made to nurture collaborations and coordinated inquiry across a number of disparate disciplines in economics. The program may well be unique, both in its comprehensive embrace of many contributing sciences and in the quality of the scientific participants.

The contribution of such a program to the further development of the science of economics will depend on the continuing commitment of the workshop participants and of other scientists who will be approached to take part in an expanded effort and on the availability of adequate resources to pursue it. Given the success of this workshop, I am optimistic on both counts, and look forward to future developments in this field.

ACKNOWLEDGMENT

I thank Philip Anderson for his insights and for the use of his summary notes on the workshop, both of which were very helpful in preparing this introduction.

II. Lectures and Perspectives

W. BRIAN ARTHUR
Dean and Virginia Morrison Professor of Population Studies and Economics, Food Research Institute, Stanford University, Stanford, CA 94305

Self-Reinforcing Mechanisms in Economics

1. PRELIMINARIES

Dynamical systems of the self-reinforcing or autocatalytic type—systems with local positive feedbacks—in physics, chemical kinetics, and theoretical biology tend to possess a multiplicity of asymptotic states or possible "emergent structures." The initial starting state combined with early random events or fluctuations acts to push the dynamics into the domain of one of these asymptotic states and thus to "select" the structure that the system eventually "locks into."

My purpose in this paper is to look at corresponding dynamical systems in economics. The literature covers a wide range of topics; therefore, I will not claim to give a complete survey. My aim is to show that the presence of self-reinforcing mechanisms, in very different problems drawn from different sub-fields of economics, gives rise to common themes, common features. I will note analogies with physical and biological systems; and I will illustrate with theory wherever it is available.

Conventional economic theory is built largely on the assumption of diminishing returns on the margin (local negative feedbacks); and so it may seem that positive feedback, increasing-returns-on-the-margin mechanisms ought to be rare. Yet there is a sizeable literature on such mechanisms, much of it dating back to the 1920's and 1930's, particularly in international trade theory, industrial organization, regional economics, and economic development. Self-reinforcement goes under different labels in these different parts of economics: increasing returns; cumulative causation; deviation-amplifying mutual causal processes; virtuous and vicious circles; threshold effects; and non-convexity. The sources vary. But usually self-reinforcing mechanisms are variants of or derive from four generic sources: large set-up or fixed costs (which give the advantage of falling unit costs to increased output); learning effects (which act to improve products or lower their cost as their prevalence increases) (Arrow, 1962; Rosenberg, 1982); coordination effects (which confer advantages to "going along" with other economic agents taking similar action); and adaptive expectations (where increased prevalence on the market enhances beliefs of further prevalence).

To fix ideas, consider a concrete example. The video technology Sony Betamax exhibits market self-reinforcement in the sense that increased prevalence on the market encourages video outlets to stock more film titles in Betamax; there are coordination benefits to new purchasers of Betamax that increase with its market share. If Betamax and its rival VHS compete, a small lead in market share gained by one of the technologies may enhance its competitive position and help it further increase its lead. There is positive feedback. If both systems start out at the same time, market shares may fluctuate at the outset, as external circumstances and "luck" change, and as backers maneuver for advantage. And if the self-reinforcing mechanism is strong enough, eventually one of the two technologies may accumulate enough advantage to take 100% of the market. Notice however we cannot say in advance *which* one this will be.

Of course, we would need a precise formulation of this example to be able to show that one technology or the other *must* eventually take all of the market. But for the moment, let us accept that conditions exist under which this happens. Notice four properties:

MULTIPLE EQUILIBRIA. In this problem two quite different asymptotic market-share "solutions" are possible. The outcome is indeterminate; it is not unique and predictable.

POSSIBLE INEFFICIENCY. If one technology is inherently "better" than the other (under some measure of economic welfare), but has "bad luck" in gaining early adherents, the eventual outcome may not be of maximum possible benefit. (In fact, industry specialists claim that the actual loser in the video contest, Betamax, is technically superior to VHS.)

LOCK-IN. Once a "solution" is reached, it is difficult to exit from. In the video case, the dominant system's accrued advantage makes it difficult for the loser to break into the market again.

PATH-DEPENDENCE. The early history of market shares—in part the consequence of small events and chance circumstances—can determine which solution prevails. The market-share dynamics are non-ergodic.

Economists have known that increasing returns can cause multiple equilibria and possible inefficiency at least since Marshall (1891; Book IV, Ch.13 and Appendix H). Modern approaches to multiple equilibria and inefficiency can be found in Arrow and Hahn, 1971, Ch.9; Brown and Heal, 1979; Kehoe, 1985; Scarf, 1981; among others. In this paper I will concentrate on the less familiar properties of lock-in and path-dependence.

Besides these four properties, we might note other analogies with physical and biological systems. The market starts out even and symmetric, yet it ends up asymmetric: there is "symmetry breaking." An "order" or pattern in market-shares "emerges" through initial market "fluctuations." The two technologies compete to occupy one "niche" and the one that gets ahead exercises "competitive exclusion" on its rival. And if one technology is inherently superior and appeals to a larger proportion of purchasers, it is more likely to persist: it possesses "selectional advantage."

Why should self-reinforcement (or increasing returns or non-convexity) cause multiple equilibria? If self-reinforcement is not offset by countervailing forces, local positive feedbacks are present. In turn, these imply that deviations from certain states are amplified. These states are therefore unstable. If the vector field associated with the system is smooth and if its critical points—its "equilibria"—lie in the interior of some manifold, standard Poincaré-index topological arguments (see Dierker, 1972; Varian, 1975, 1981; Kehoe, 1985) imply the existence of other critical points or cycles that *are* stable, or attractors. In this case multiple equilibria must occur. Of course, there is no reason that the number of these should be small. Schelling (1978) gives the practical example of people seating themselves in an auditorium, each with the desire to sit beside others. Here the number of steady-states or "equilibria" would be combinatorial.

Many sub-fields of economics that allow self-reinforcing mechanisms recognize the possibility of multiple "solutions" or equilibria, and are able to locate these analytically. Here are three examples:

1. *International Trade Theory*. The standard example here is that of two countries that each can undertake production in two possible industries (say, aircraft and cars) with large set-up costs or some other source of increasing returns. Statically, a (local minimum) least-cost arrangement would be for one country to produce all of one commodity, the other to produce all of the other. The countries could then trade to arrive at their preferred consumption mixture. But there are two such arrangements. *Which* commodity is produced in which country is indeterminate. (When the countries differ in size, these two "solutions" or equilibria can have different welfare consequences.) Statically, increasing returns trade theory has demonstrated and located multiple equilibria under a variety of increasing returns mechanisms (for example, Graham, 1923; Ohlin,

1933; Matthews, 1949; Helpman and Krugman, 1985). But as yet it does not address the question of how a *particular* trade pattern comes to be selected.

2. *Spatial Economics*. In 1909 Alfred Weber showed that where firms benefit from the local presence of other firms or agglomerations of industry, several configurations of industry can be local-minimum-cost "solutions" to the optimal location problem. Engländer (1926), Palander (1935), and Ritschl (1927) used this to argue that observed industry location patterns might not necessarily be the unique "solution" to a problem of spatial economic equilibrium, but rather the outcome of a process that is subject in part to "historical accident." One "solution"—not necessarily the optimal one—might be "selected" dynamically, in part by the historical sequence of early settlement. Thus, one region might draw ahead in concentration of industry at the expense of the others. Kaldor (1970) argued that regional *prosperity* can also be self-reinforcing, if a region's productivity is tied to its economic growth (as in Verdoorn's "law"). Similarly endowed regions can diverge in income over the long-run. Because of their intuitive appeal, these verbal arguments have had some influence, but until recently (Allen and Sanglier, 1981; Faini, 1984; Arthur, 1986) there have been few attempts to formalize them.

3. *Industrial Organization*. As one example from several possibilities here, Katz and Shapiro (1985, 1986) showed that a combination of "network externalities" (coordination effects) and expectations could lead to multiple market-share equilibria. If, *ex ante*, sufficient numbers of consumers believe that a product—the IBM personal computer, say—will have a large share of the market, and if there are advantages to belonging to a prevalent product's "network" of users, then they will be willing to purchase this product, and the producer will be induced to put a large quantity on the market, fulfilling their beliefs. But the same statement could be made for a competing product. There are therefore multiple "fulfilled-expectation equilibria" that is, multiple sets of eventual (Cournot-equilibrium) market shares that fulfill prior expectations. Here expectations are given and fixed; this is a static model with multiple solutions.

In these three sub-fields the absence of an accepted dynamics means that what I call the *selection problem*—the question of *how* a particular equilibrium comes to be selected from the multiplicity of candidates—is left unresolved.

Other areas of economics that admit self-reinforcing mechanisms do have an accepted, if simple, dynamics that shows how equilibria or steady-states are reached. An example is the neoclassical growth theory of the 1960's. Self-reinforcement occurs here for example when a threshold capital-labor ratio exists, above which sufficient savings are generated that the ratio rises, below which it falls (Solow, 1956). There are multiple equilibria. Similarly, recent attempts at an aggregate macro-dynamics (Heal, 1986), with Walrasian dynamics and unit costs falling with increased output, exhibit two long-run outcomes. One—corresponding to economic health—has high output coupled with high demand and low prices. The other—corresponding to stagflation—has low output coupled with low demand and high prices. In these parts of economics, the presence of multiple steady states is typically analyzed by standard phase-plane methods, often with demonstrations of local stability. Attractors are typically point attractors; only recently has economics begun to analyze richer possibilities.

2. LOCK-IN

When a nonlinear physical system finds itself occupying a local minimum of a potential function, "exit" to a neighboring minimum requires sufficient influx of energy to overcome the "potential barrier" that separates the minima. There are parallels to such phase-locking, and to the difficulties of exit, in self-reinforcing economic systems. Self-reinforcement, almost by definition, means that a particular outcome or equilibrium possesses or has accumulated an economic advantage. This advantage forms a potential barrier. We can say that the particular equilibrium is *locked in* to a degree measurable by the minimum cost to effect changeover to an alternative equilibrium.

In many economic systems, lock-in happens dynamically, as sequential decisions "groove" out an advantage that the system finds it hard to escape from. Here is a simple case that recurs in many forms.

Suppose an economic agent—the research and development department of a firm perhaps—can choose each time period to undertake one of N possible activities or projects, A_1, A_2, $A_3, \ldots$, A_N. Suppose activities improve or worsen the more they have been undertaken. A_j pays $\Pi_i(n)$ where n is the number of times it has been previously chosen. Future payoffs are discounted at the rate β.

THEOREM If the payoffs $\Pi_i(n)$ increase monotonically with n, then the activity A_j that is chosen at the outset is chosen at each time thereafter.

PROOF An inductive proof is straightforward.

Thus if activities increase in payoff the more they are undertaken (perhaps because of learning effects), the activity that is chosen first, which depends of course on the discount rate, will continue to be chosen thereafter. The decision sequence "grooves out" a self-reinforcing advantage to the activity chosen initially that keeps it locked in to this choice.

Notice that at each stage, an optimal choice is made under conditions of certainty; and so there can be no conventional economic inefficiency here. But there may exist *regret*. Consider the case of a person who has the choice of practising medicine or law each year. Each activity pays more, the more previous experience has been accumulated. Suppose the rewards to practising law rise rapidly with experience but then flatten out; and those to practising medicine are small initially but eventually surpass those of law. According to the theorem, whichever activity the person chooses, he will continue it to choose thereafter. If he has a high discount rate, he will choose law. And this choice will at all stages continue to be rational and superior to the alternative of first-year payoff as a doctor. Yet there may exist *regret*, in the sense that after N years in the law, an equivalent time served in medicine would have paid more at each time into the future. Self-reinforcement can lock a single, rational economic agent in to one activity, but not necessarily the one with the best long-run potential.

Sequential-choice lock-in occurs in the economics of technology (Arthur, 1983, 1984). When a new engineering or economic possibility comes along, often several technologies are available to carry it through. Nuclear power, for example, can be generated by light-water, or gas-cooled, or heavy-water, or sodium-cooled reactors. Usually technologies improve with adoption and use. In the case of several initially undeveloped technologies available for adoption, each with the potential of improvement with use, the sequence in which adoptions occur may decide which technologies improve. This time choice is by *different* agents each time—adopters acting in their own individual interest. Where adopters are exactly alike in their preferences, the outcome is trivial. If one of the embryo technologies offers a tiny advantage over the others, it is chosen by the first adopter-developer. It is used and improved. It is therefore chosen by the next developer; it is further improved. And so on. Thus, the adopter sequence can trivially lock in to the development of a technology that shows initial success in partially developed form, but that later turns out inferior in the sense that an equally developed alternative might have eventually delivered more.

2.1 LOCK-IN IN A STOCHASTIC MODEL

It might be objected that this scenario is unrealistic in that lock-in is completely predetermined and razor-edged. The technology with the slightest initial advantage becomes the one that dominates. More reasonably, we might suppose that each technology stands some chance of early development and that "luck" in the form of "random events" can influence the outcome.

As a simple example of this approach (see Arthur, 1983), consider two technologies, A and B, available to a large "market" of potential adopters, both of which improve with adoption. Assume two types of adopters, R and S, each type equally prevalent, with "natural" preferences for A and B respectively. For each adopter type, payoffs-to-adoption of A or B increase linearly with the number of previous adoptions of A or B. We can inject randomness by assuming that the order of arrival of R and S types is unknown, and that it is equally likely that an R or an S will arrive next to make his choice. Once an adopter has chosen, he holds his choice.

Initially at least, if an R-agent arrives at the "adoption window" to make his choice, he will adopt A; if an S-agent arrives, he will adopt B. Thus the difference-in-adoptions between A and B moves up or down by one unit depending on whether

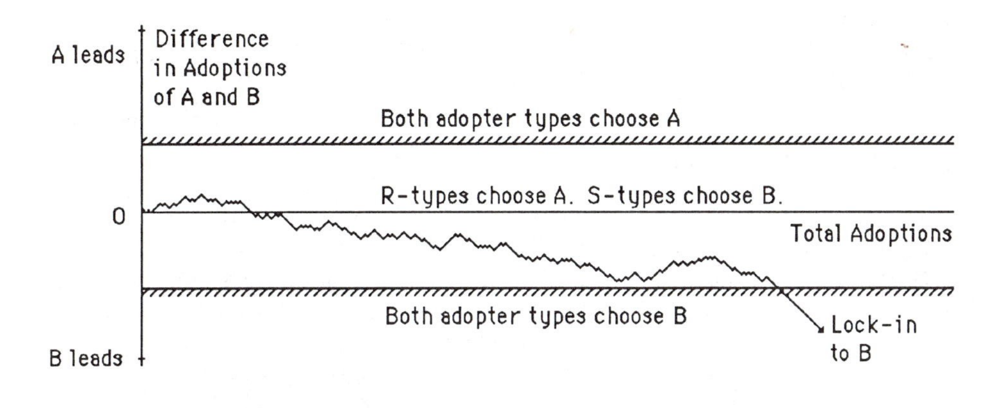

FIGURE 1 Stochastic Lock-In in a Random Walk.

the next adopter is an R or an S; that is, it moves up or down with probability one-half.

However, if by "chance" a sufficient number of R-types cumulates in the line of choosers, A will be improved in payoff over B enough to cause future S-types choosing to switch to A. From then on, both R- and S-types will adopt A, and only A. The adoption process is then locked-in to technology A. Similarly, if a sufficient number of S-types by "chance" arrives to adopt B over A, B will improve sufficiently to cause future R-types to switch to B. The process instead locks-in to B (Figure 1.) These dynamics form a random walk with absorbing barriers on each side, the barriers corresponding to the lead in adoption it takes for each agent-type to switch its choice.

In a random walk with absorbing barriers, absorption occurs eventually with probability one. Therefore, in the model I have described, the adoption process *must* lock-in to monopoly of one of the two technologies, A or B. But with the presence of random events, *which* technology is not predictable in advance. The order of arrival of agent types—the source of randomness here—"decides" the eventual outcome.

This lock-in-by-fluctuation to one pattern or structure out of several possible has parallels in thermodynamics, ferromagnetism, laser theory, and chemical kinetics, and in allele fixation through genetic drift. (For example, see Nicolis and Prigogine, 1976; Haken, 1978; and Roughgarden, 1979.)

Does the economy sometimes lock-in to a technology that is inferior in development potential, because of small, random events in history? It appears that it does. Cowan (1987) argues that a series of minor circumstances acted to favor light-water nuclear reactors in the mid-1950s over potentially superior competing alternatives, with the result that learning and construction experience gained early on locked the nuclear reactor market in to light water. David, who noted in 1975 "that marked divergences between ultimate outcomes may flow from seemingly negligible differences in remote beginnings," examines (1985) the historical circumstances under which the (probably inferior) QWERTY typewriter keyboard became locked-in.

2.2 EXIT FROM LOCK-IN

If an economic system is locked-in to an inferior local equilibrium, is "exit" or escape into a superior one possible? There is rarely in economics any mechanism corresponding to "annealing" (injections of outside energy that "shake" the system into new configurations so that it finds its way randomly into a lower cost one). Exit from an inferior equilibrium in economics depends very much on the source of the self-reinforcing mechanism. It depends on the degree to which the advantages accrued by the inferior "equilibrium" are reversible or transferable to an alternative one.

Where learning effects and specialized fixed costs are the source of reinforcement, usually advantages are not reversible and not transferable to an alternative equilibrium. Repositioning the system is then difficult. For example, in most countries road and rail are to some degree substitutes as alternative modes of transportation. Each mode is self-reinforcing in that the more heavily it is used, the more funds become available for investment in capital improvements that attract further users. Therefore, one mode may achieve dominance at the expense of the other. But reversing this or trying to assure a balance may require a significant subsidy to the weaker mode to bring it level with the advantage accumulated by the dominant mode. Capital assets—the source of advantage here—are not transferable or easily reversed, and here repositioning is costly.

Where coordination effects are the source of lock-in, often advantages *are* transferable. For example, users of a particular technological standard may agree that an alternative would be superior, provided everybody "switched." If the current standard is not embodied in specialized equipment and its advantage-in-use is mainly that of commonality of convention (like the rule against turning right on a red light in many states), then a negotiated or a mandated changeover to a superior collective choice can provide exit into the new "equilibrium" at negligible cost. Coordination *is* transferable—and possible by fiat. Even if there is no outside "standard setting" agency, and users are restricted to non-cooperative behavior, Farrell and Saloner (1985, 1986) show that as long as each user has certainty that the others also prefer the alternative, each will decide independently to "switch." But where users are uncertain of others' preferences and intentions, there can be "excess inertia": each user would benefit from switching to the other standard, as long as the others go along, but individually none dare change in case others do not follow. To use Kauffman's terminology (1987), an "adaptive walk to an adjacent peak" by nearest neighbor transitions (single-agent switches) is not possible here.

The theme of "exit" by coordination from an inferior low-level equilibrium—a locked-in position—runs through the economic development literature. Examining

Eastern Europe's post-war development prospects in 1943, Rosenstein-Rodin argued that, because of increasing returns caused by indivisibilities and complementarities in demand, industries or the firms within them may not find it profitable to expand separately; but if all could expand together via a coordinated effort, the expansion might be profitable to each. Hence development calls for coordinated expansion and investment—for what Rosenstein-Rodin called a "Big Push" on the part of government. These ideas, stimulated by Young's analysis of 1928, influenced a generation of development economists, in particular Hirshman (1958) who further examined "synergistic" or linkage effects between industries; Chenery (1959) who formalized this argument using linear programming; and Myrdal (1957) who pointed to further mechanisms of cumulative causation. A remarkably similar set of ideas has independently emerged in the recent macroeconomic literature (Weitzman, 1982). In the presence of increasing returns, large suppliers cannot create sufficient demand to make independent expansion profitable. The economy tends to "stick" below full capacity. Economy-wide coordinated stimulation is needed for exit to full employment.

3. PATH-DEPENDENCE: ALLOCATION PROCESSES

So far I have given examples of multiple equilibria and the dynamics of lock-in under very particular self-reinforcing mechanisms. Is there a general analytical framework that encompasses problems like these? The answer of course is no. But we might be able to design broad classes of analytical systems that encompass large numbers of examples. In thinking what such systems should look like, we might usefully be guided by three considerations that emerge from the discussion so far:

1. To be able to examine and track how one particular "equilibrium" or "solution" comes to be "selected" from a multiplicity of alternatives requires a dynamic approach. Therefore, we need to allow for the sequence in which actions occur or economic choices are made.

2. Many of the problems that interest us can be cast as problems of allocation or sequential choice between alternatives, where these allocations or choices are affected by the numbers or proportions of each alternative present at the time of choice.

3. Self-reinforcing systems often start off in a "balanced" but unstable position, so that their end-states may be determined by small events outside the model, as well as by initial conditions. If we treat these small events as perturbations, we need to allow for well-defined sources of randomness. In practice this means that the "state" of the system may not determine the next economic action, but rather the *probability* of the next economic action.

3.1 ALLOCATION PROCESSES

Let us consider one general class of dynamical systems that allows for these considerations. I will call these *allocation processes*.

Suppose a unit addition or allocation is made to one of K categories at each time, with probabilities that are a function of the proportion of units currently in the K categories. Time here is event time, not clock time. (In practice we might be considering the build-up of market shares by observing "allocations" of adopters, one at a time, to K technologies; or consumers to K product brands; or, in regional economics, firms to K locations.) Thus, the next unit is added to category i with probability $p_i(x)$ where x is the vector of current proportions or market shares. (The vector of probabilities $p = (p_1(x), p_2(x), \ldots, p_K(x))$ is a function—the *allocation function*—that maps the unit simplex S^K of proportions into the unit simplex of probabilities). In practice p could be obtained, at least implicitly, from the particular mechanism under consideration. Figure 2 shows two illustrative allocation functions in the case where $K = 2$.

We are interested in how market shares, or equivalently, the numbers or proportions in each category, build up. Our question is what happens to the long-run proportions in such a system. What limiting steady-states can emerge? The standard probability-theory tool for this type of problem is the Borel Strong Law of Large Numbers which makes statements about long-run proportions in processes where increments are added at successive times. But we cannot use the standard

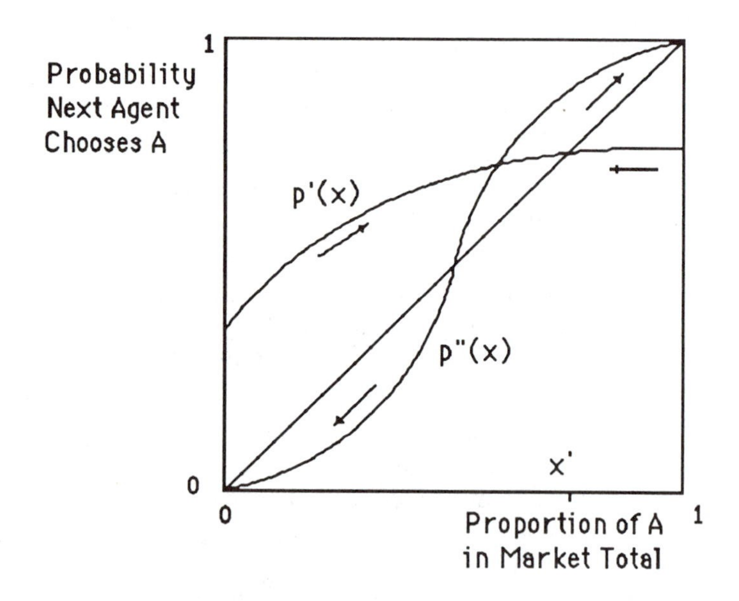

FIGURE 2 Two Illustrative Allocation Functions for dimension $K = 2$.

Strong Law in our process—we do not have *independent* increments. Instead we have increments—unit allocations to 1 through K—that occur with probabilities influenced by past increments. We have something like a multidimensional coin-tossing problem, where the probability of a unit addition to the category Heads changes with the proportion of Heads tossed previously.

To anticipate results somewhat, we can show that under non-restrictive technical conditions proportions *must* converge to one of the fixed points of the allocation function p. And where there are self-reinforcing mechanisms, typically there are multiple fixed points of p. We therefore have a useful way to identify the possible "structures" that can "emerge."

Let us look at the dynamics of this process. The process starts at time 1, with an initial vector of allocations $y = (y_1, y_2, \ldots, y_K)$, and initial total $w = \sum y_i$. Let Y_n be a vector that describes the number of units in categories 1 through K at time n (when $w + (n-1)$ units in total have been allocated). Then

$$Y_{n+1} = Y_n + b(X_n); \qquad Y_1 = y; \tag{1}$$

where b is the jth unit vector with probability $p_j(X_n)$.

Dividing Eq. (1) by the total units $w+n$, the vector of proportions in categories 1 through K, or shares, evolves as

$$X_{n+1} = X_n + \frac{1}{n+w}\left(b(X_n) - X_n\right); \qquad X_1 = y/w. \tag{2}$$

We can rewrite Eq. (2) in the form

$$\begin{aligned} X_{n+1} &= X_n + \frac{1}{n+w}\left(p(X_n) - X_n\right) + \frac{1}{n+w}\xi(X_n); \\ X_1 &= y/w. \end{aligned} \tag{3}$$

where ξ is defined as the random vector

$$\xi(X_n) = b(X_n) - p(X_n).$$

Eq. (3) is the description of the dynamics of shares that we want.

Notice that the conditional expectation of ξ with respect to the current state X_n is zero; hence, we can derive the *expected motion* of the shares or proportions as

$$E(X_{n+1} \mid X_n) - X_n = \frac{1}{n+w}\left(p(X_n) - X_n\right). \tag{4}$$

We will call

$$X_{n+1} = X_n + \frac{1}{n+w}\left(p(X_n) - X_n\right) \tag{5}$$

the *equivalent deterministic system* corresponding to our stochastic process. We see that, if the probability $p_j(X_n)$ of an addition to category j is greater than the current proportion $x_j(n)$ in category J, this category's share should increase—at least on an expected basis. Conversely, if the probability is less than the current proportion, it should decrease. Eq. (3) therefore tells us that shares are driven by the equivalent deterministic system (the first and second term on the right), together with a perturbation effect (the third term).

Depending upon the function p—and ultimately on the distribution of preferences and possibilities that defines p—there may be several points x (several market share patterns) at which deterministic motion is zero. These are the fixed points of p, where $p(x) = x$. Some of these are attractor or stable points (defined in the usual way), some are repellant or unstable points. We have the following Strong Law:

THEOREM

i. Suppose $p : S^K \rightarrow S^K$ is continuous, and that the equivalent deterministic system possesses a Lyapunov function v whose motion is negative outside a neighborhood of the set fixed points of p, $B = \{x : p(x) = x\}$. Suppose B has a finite number of connected components. Then the vector of proportions $\{X_n\}$ converges, with probability one, to a point z in the set of fixed points B.

ii. Suppose p maps the interior of the unit simplex into itself and that z is a stable point. Then the process has limit point z with positive probability.

iii. Suppose z is a non-vertex unstable point of p. Then the process cannot converge to z with positive probability.

PROOF See Arthur, Ermoliev, and Kaniovski (1983,1986). (For the case $K = 2$ with p stationary, see Hill, Lane, and Sudderth, 1980).

In other words, if allocation processes converge, they must converge to a vector of proportions (shares) represented by one of the attracting fixed points of the mapping from proportions into the probabilities of allocation. Thus in Figure 2 the process corresponding to p' converges to x', with probability one; the process corresponding to p'' converges to 0 or to 1, with probability one. Where $p = \underline{p}$, a constant function, the process converges to $\underline{p}$; therefore, the standard strong law (with unit increments added under fixed probabilities) is a special case. Where the allocation function varies with "time" n, this dependent-increment strong-law theorem still goes through, providing that the sequence $\{p_n\}$ converges to a limiting function p faster than $1/n$ converges to zero. The process then converges to one of the stable limit points of p. The theorem can be extended to non-continuous functions p and to non-unit increments to the allocations (Arthur, Ermoliev and Kaniovski, 1987b).

Allocation processes have a useful property. Fluctuations dominate motions at the outset; hence, they make limit points reachable from any initial conditions. But they die away, leaving the process directed by the equivalent deterministic system and hence convergent to identifiable attractors.

For a given problem, to identify steady-states, we do not need to compute p directly, only its fixed points. A number of studies now use this technique.

3.2 EXAMPLE: WHEN DOES SELF-REINFORCEMENT IMPLY MONOPOLY?

In many models we want to identify "competitive exclusion" or monopoly conditions under which one product, or one technology eventually must eventually gain enough advantage to take 100% of the market. In the context of our theorem, we would need to show the existence of stable fixed points only at the vertices of the unit simplex.

As an example, consider a more general version of the "adoption market" model above, in which there is now a continuum of agent types rather than just two. Assume agents—potential adopters—choose between K technologies. Suppose that if n_j previous adopters have chosen technology j previously, the next agent's payoff to adopting j is $\Pi_j(n_j) = a_j + g(n_j)$, where a_j represents the agent's "natural preference" for technology j and the monotonically increasing function g represents the technological improvement that comes with previous adoptions.

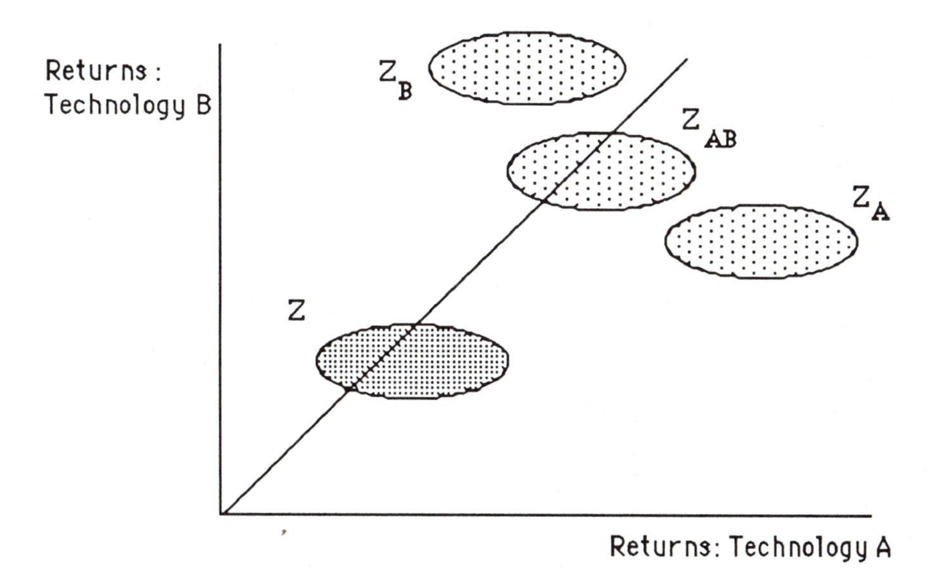

FIGURE 3 Distribution of Payoffs to Adoption of A and B under a Continuum of Adopter Types.

Each adopter has a vector of natural preferences $a = (a_1, a_2, \ldots, a_K)$ for the k alternatives, and we can think of the continuum of agents as a distribution of points a (with bounded support Z) on the positive orthant. We assume an adopter is drawn at random from this probability distribution each time a choice is made. (See Figure 3.) The distribution itself shifts either to the right or upward as returns to A or B increase with an adoption of either A or B respectively. Monopoly—in this case lock-in of a single technology—corresponds to positive probability of the distribution of payoffs being driven over the 45° line as the number of adoptions becomes large.

The path-dependent strong law allows us to derive two useful theorems (Arthur 1986). Where the improvement function g increases without upper bound as n_j increases, then the process has stable limit points only at simplex vertices, with probability one. The distribution *must* be driven over the 45° line (as with Z_A and Z_B in Figure 3) and monopoly by one technology must eventually occur. Where returns to adoption increase but are bounded (as when improvement effects become exhausted) monopoly is no longer inevitable. Here certain sequences of adopter types could bid improvements for two or more alternatives upward more or less in concert. These could then reach their increasing returns bounds together, with the adopter payoff distribution still straddled across the 45° line (as with Z_{AB} in Figure 3), and thus with the market shared from then on. The market remains shared from then on. In this case, some "event histories" dynamically lead to a shared market; other event histories lead to monopoly. Competitive exclusion no longer occurs with probability one.

3.3 EXAMPLE: LOCATION BY SPIN-OFF

Consider a particular mechanism in industry location (Arthur, 1987b). Assume an industry builds up firm by firm, and firms may locate in one of K possible regions, starting with some set of initial firms, one in each region, say. Assume that new firms are added by "spinning off" from parent firms one at a time; that each new firm stays in its parent region; and that any existing firm is as likely to spin off a new firm as any other. (David Cohen, 1984, shows that such spin-offs have been the dominant "birth mechanism" in the U.S.electronics industry.) In this case firms are added incrementally to regions with probabilities exactly equal to the proportions of firms in each region at that time. This random process is therefore a Polya process—a well-known process (see Joel Cohen, 1976) that is a special case in our class of allocation processes, one with $p(x) \equiv x$. Any point in the simplex of regions' shares of the industry is now a fixed point; regions' shares can therefore converge to *any* point. There is a continuum of "equilibrium points." From Polya theory we can say more. The random limit vector to which the process must converge has a uniform distribution on the unit simplex. (We could think of a representative outcome—regions' shares of the industry—as the result of placing $N-1$ points on the unit interval at random, and partitioning the interval at these points to obtain N "shares.") In this model "chance" dominates completely. "Historical accident" in the shape of the early random sequence of spin-offs becomes the sole determining factor of the limiting regional pattern of industry.

4. PATH-DEPENDENCE: RECONTRACTING PROCESSES

Allocation processes, as a class of models, might be appropriate for studying how an allocative pattern forms—providing choices made are irreversible. A different, useful class of model would allow recontracting within the market once it has formed. These would assume an already formed "market" or total allocation of fixed size T, divided among K categories. (For example, T hectares of land divided among K landowners; T firms divided among K regions; T voters divided among K candidates.) Transitions of units between categories are possible, with probabilities that depend in general on the market shares, or numbers in each category. Thus, self-reinforcement (or self-inhibition) are once again possible.

We have the advantage that models of this fixed-size, Markov-transition kind are standard in genetics (Ewens, 1979), epidemiology, and in parts of physics (Haken, 1978). Consider a special, but useful case, where $K = 2$, and transitions can be made only one unit at a time. (Here I follow Weidlich and Haag, 1983.) To keep matters concrete suppose a total "market" or "population" of $T = 2N$ voters with state variable m, where $N + m$ voters prefer candidate A, and $N - m$ voters prefer candidate B. Let $p_{AB}(m)$ denote the probability that a voter changes preference from A to B, and $p_{BA}(m)$ the probability that a voter changes preference from B to A, in unit time. Then the probability $P(m,t)$ of finding the system at state m at time t (strictly speaking a measure on an ensemble of such systems) evolves as:

$$P(m,t+1) = P(m,t)\big(1 - p_{AB}(m) - p_{BA}(m)\big) + P(m+1,t)p_{BA}(m+1) + P(m-1,t)p_{AB}(m-1) \tag{6}$$

which yields the Master Equation

$$\frac{dP(m,t)}{dt} = \big[P(m+1,t)p_{BA}(m+1) - P(m,t)p_{BA}(m)\big] + \big[P(m-1,t)p_{AB}(m-1) - P(m,t)p_{AB}(m)\big] \tag{7}$$

Normalizing to variable x in the continuous interval (-1,1), by setting

$$x = m/N; \quad \varepsilon = 1/N; \quad P(x,t) = NP(m,t);$$

$$R(x) = \big[p_{AB}(m) - p_{BA}(m)\big]/N; \quad \text{and} \quad Q(x) = \big[p_{AB}(m) + p_{BA}(m)\big]/N;$$

we can rewrite Eq. (7) in the form of a one-dimensional Fokker-Planck diffusion equation

$$\frac{\partial P(x,t)}{\partial t} = -\frac{\partial}{\partial x} R(x)P(x,t) + \frac{\varepsilon}{2}\frac{\partial^2}{\partial x^2} Q(x)P(x,t). \tag{8}$$

We can substitute particular drift and diffusion functions, R and Q corresponding to particular transition mechanisms and study the evolution of P over time, and its stationary, limiting distribution. In many cases it is possible to solve explicitly for the stationary distribution $P(x)$.

The important difference from allocation process models is that in this type of process transitions in market share remain of constant order of magnitude (rather than fall off at rate $1/n$). "Recontracting processes" therefore show convergence in distribution rather than strong convergence to a point. Unless they have absorbing states, permanent lock-in to one market position is not possible. Instead they exhibit "punctuated equilibria" in the shape of sojourns in the neighborhood of local maxima and transitions between them.

4.1 EXAMPLE: MARKET SHARES WITH A CONFORMITY EFFECT

Suppose a luxury car market (of size $2N$) is divided between American and German cars, denoted A and B. Suppose consumers change their preference occasionally, according to

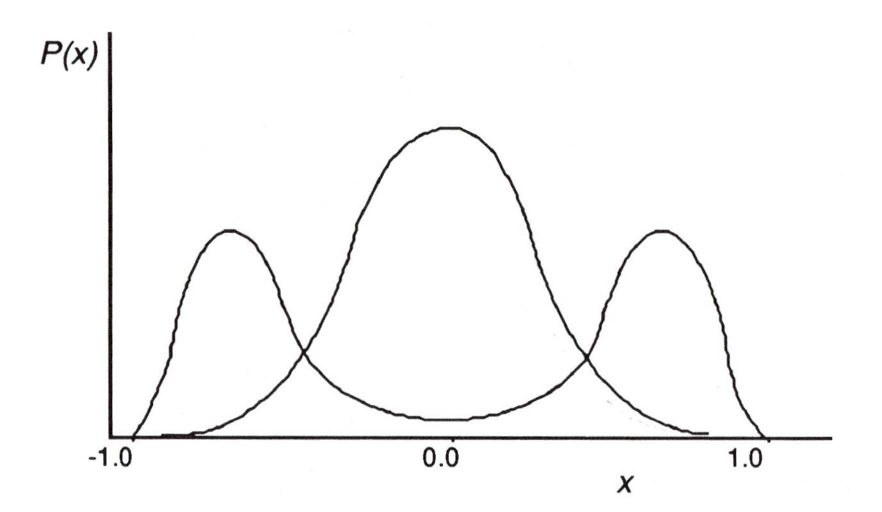

FIGURE 4 Stationary Distributions of $P(x)$ for $\kappa = 0.5$ and $\kappa = 1.3$.

$$p_{AB}(m) = \nu \exp(\delta + \kappa m)(N - m) \quad \text{and} \quad p_{BA}(m) = \nu \exp -(\delta + \kappa m)(N + m)$$

where ν denotes frequency of "switches;" δ allows for a preference bias; and κ corresponds to a fashion or conformity effect (Weidlich and Haag, 1983). In the absence of conformity, a larger population of one car type increases the chance of "switches" to the other. Hence there is a centralizing tendency. But this is offset by the conformity effect, which reinforces a concentration of one type. When κ is small, centralization dominates and the stationary distribution is unimodal (Figure 4). But as κ increases the distribution bifurcates and becomes bimodal, with maxima corresponding to the relative prevalence of one type of car over the other. In this case the market lingers at prevalence of one type of car with intermittent transitions to prevalence of the other.

5. SOME OTHER QUESTIONS

5.1 STRATEGIC ACTION

So far, most of the theory and examples I have given do not deal explicitly with the possibility of strategic action—the possibility that economic agents adjust their actions to take into account the possible responses of other agents. Strategic action does not negate the methods or frameworks discussed above. But it can add an overlay of difficulty to the derivation of allocation functions, transition probabilities, and the like.

There are now a number of studies that examine market structure in the presence of strategic action and self-reinforcement. Flaherty (1980) looks at the case where firms can strategically invest to reduce their costs. Multiple (open-loop, non-cooperative) equilibria result. Under reasonable conditions, only those equilibria that are asymmetric—where single firms become dominant in market share—are stable. Fudenberg and Tirole (1983) use learning effects as the source of reinforcement and investigate closed-loop, perfect equilibria. Again they find multiple equilibria. (The analysis here, strangely, concentrates on the symmetric equilibrium—the only one likely to be unstable.) These studies are (deterministically) dynamic in the sense that they derive trajectories that fulfill the usual Nash condition. To examine how one trajectory from the multiplicity of candidates is "selected" however, we need to push the analysis one stage further back and ask how a trajectory itself comes to be chosen or "fallen into." Spence (1981) goes part way toward this, by showing in *ad-hoc* fashion that the order in which firms enter the market makes a difference.

Hanson (1985) allows randomness to solve the "selection" problem, in the case of self-reinforcing duopoly dynamics. In a version of the stochastic adopter-sequence model above, he assumes that the two technologies that improve with adoption are *proprietary* (like VHS and Betamax)—"sponsored" by firms that can manipulate their price. Firms price low early on, possibly taking losses in exchange for market share. If both firms are evenly enough matched to stay in the market, sooner or later the cumulation of "chance events" may allow one firm sufficient adoption advantage to gradually tip the market in its favor. It can then raise its price and take monopoly profits, while keeping the other firm on the contestable margin of the market. Hanson details conditions under which monopoly by a single firm occurs with probability one. It is clear, however, that conditions can also be constructed where markets end up shared. For example, when firms discount heavily so that they are mainly interested in present sales, neither firm may have sufficient incentive to price low early on. Neither eventually wins the "natural customers" of the other with positive probability and the result is a shared market.

5.2 EXPECTATIONS

In the Katz and Shapiro model mentioned earlier, expectations are rational and fulfilled, but static—fixed at the outset. We might assume more generally that agents modify their expectations as market shares change. Suppose agents form beliefs or expectations in the shape of probabilities on the future states of the market process—probabilities that are conditioned on market share or absolute numbers of competing alternatives. We now have a *rational-expectations equilibrium* if the *actual* adoption process that results from agents acting on these beliefs turns out to have conditional probabilities that are identical to the *believed* process. Under increasing returns, if one product gets ahead by "chance," its increased probability of doing well in the market will further enhance expectations of its success. In general, expectations may interact with self-reinforcing mechanisms to further destabilize an already unstable situation. Little work has been done here yet, but it appears that the presence of dynamic, rational expectations leads more easily to monopoly outcomes (Arthur, 1983).

5.3 POLICY

Where there are multiple equilibria and a central authority favors a particular one, it can of course attempt to "tilt" the market toward this outcome. Timing is crucial here: in David's phrase (1986) there are only "narrow windows" in which policy is effective.

In many cases, however, especially where learning is the source of self-reinforcement, it is not clear in advance which equilibrium outcome has most potential promise. A central authority then faces the problem of choosing which improving products, or processes, or technologies to subsidize. This yields a version of the multi-arm bandit problem. Cowan (1988) shows that, where central authorities subsidize increasing-return technologies on the basis of their current estimates of future potential, locking into inferior technologies is less likely than in the uncontrolled adoption case. But it is still possible. An early run of bad luck with a potentially superior technology may cause the central authority, perfectly rationally, to abandon it. Even with central control, escape from inferior outcomes is not guaranteed.

It may sometimes be desirable to keep a multiple-equilibrium market "balanced" to avoid monopoly problems or to retain "requisite variety" as a hedge against future shifts in the economic environment. The question of using well-timed subsidies to prevent the "tipping" and lock-out has not yet been looked at. But its structure—that of artificially stabilizing a naturally unstable dynamical process—is a standard one in stochastic feedback control theory.

5.4 SPATIAL MECHANISMS

Economic agents may be influenced by neighboring agents' choices. Puffert (1988) examines historical competitions between railroad gauges where rail companies found it advantageous to adopt a gauge that neighboring railroads were using. Spatial mechanisms have parallels with Ising models and renormalization theory in physics (Holly, 1974) and with voter models in probability theory (Liggett, 1979; Föllmer, 1979). In economics very little work has yet been done on spatial self-reinforcing mechanisms.

REFERENCES

1. Allen, P., and M. Sanglier (1981), "Urban Evolution, Self-Organization, and Decision Making," *Environment and Planning A* **13**, 167–183.
2. Arrow, K., and F. Hahn (1971), *General Competitive Analysis* (New York: Holden-Day).
3. Arrow, K. (1962), "The Economic Implications of Learning by Doing," *Rev. Econ. Stud.* **29**, 155–173.
4. Arthur, W. B. (1983), "Competing Technologies and Lock-In by Historical Events: The Dynamics of Allocation under Increasing Returns," *I.I.A.S.A. Paper WP-83-90, Laxenburg, Austria.* Revised as *C.E.P.R. Paper 43, Stanford University.*
5. Arthur, W. B. (1984), "Competing Technologies and Economic Prediction," *Options* (Laxenburg, Austria: I.I.A.S.A.).
6. Arthur, W. B. (1986), *Industry Location and the Importance of History. Center for Economic Policy Research, Paper 84, Stanford University.*
7. Arthur, W. B. (1987a), "Competing Technologies: An Overview," *Technical Change and Economic Theory*, Eds. G. Dosi, C. Freeman, R. Nelson, G. Silverberg, and L. Soete (London: Pinter), forthcoming.
8. Arthur, W. B. (1987b), "Urban Systems and Historical Path-Dependence," *Urban Systems and Infrastructure*, Eds. R. Herman and J. Ausubel (NAS/NAE), forthcoming.
9. Arthur, W. B., Yu. M. Ermoliev, and Yu. M. Kaniovski (1983), "On Generalized Urn Schemes of the Polya Kind," *Kibernetika* **19**, 49–56. English trans. *Cybernetics* **19**, 61–71.
10. Arthur, W. B., Yu. M. Ermoliev, and Yu. M. Kaniovski (1986), "Strong Laws for a Class of Path-Dependent Urn Processes," *Proc. International Conf. on Stochastic Optimization, Kiev 1984. Lect. Notes in Control and Info. Sciences 81*, Eds. V. Arkin, A. Shiryayev, and R. Wets (New York: Springer).
11. Arthur, W. B., Yu. M. Ermoliev, and Yu. M. Kaniovski (1987a), "Path-Dependent Processes and the Emergence of Macro-Structure," *European J. Operational Research* **30**, 294–303.
12. Arthur, W. B., Yu. M. Ermoliev, and Yu. M. Kaniovski (1987b), "Non-Linear Urn Processes: Asymptotic Behavior and Applications," *Kibernetika*, forthcoming.
13. Brown, D., and G. Heal (1979), "Equity, Efficiency and Increasing Returns," *Rev. Econ. Stud.* **46**, 571–585.
14. Chenery, H. (1959), "The Interdependence of Investment Decisions," *Allocation of Economic Resources*, Ed. M. Abramovitz (Stanford: Stanford University Press), 82–120.
15. Cohen, D. L. (1984), "Locational Patterns in the Electronics Industry: A Survey," *Mimeo, Stanford University.*
16. Cohen, J. (1976), "Irreproducible Results and the Breeding of Pigs," *Bioscience* **26**, 391–394.

17. Cowan, R. (1987), *Backing the Wrong Horse: Sequential Technology Choice under Increasing Returns, Ph.D. Dissertation, Stanford University.*
18. David, P. (1975), *Technical Choice Innovation and Economic Growth* (London: Cambridge University Press).
19. David, P. (1985), "Clio and the Economics of QWERTY," *Amer. Econ. Rev. Proc.* **75**, 332–337.
20. David, P. (1986), "Some New Standards for the Economics of Standardization in the Information Age," *Paper 79, Center for Economic Policy Research, Stanford University.*
21. Dierker, E. (1972), "Two Remarks on the Number of Equilibria of an Economy," *Econometrica* **40**, 951–953.
22. Engländer, O. (1926), "Kritisches und Positives zu einer allgemeinen reinen Lehre vom Standort," *Zeitschrift für Volkswirtschaft und Sozialpolitik* Neue Folge **5**.
23. Ewens, W. J. (1979), *Mathematical Population Genetics* (New York: Springer).
24. Faini, R. (1984), "Increasing Returns, Non-Traded Inputs and Regional Development," *Econ. J.* **94**, 308–323.
25. Farrell, J., and G. Saloner (1985), "Standardization, Compatibility, and Innovation," *Rand J. Econ.* **16**, 70–83.
26. Farrell, J., and G. Saloner (1986), "Installed Base and Compatibility," *Amer. Econ. Rev.* **76**, 940–955.
27. Flaherty, M. T. (1980), "Industry Structure and Cost-Reducing Investment," *Econometrica* **48**, 1187–1209.
28. Föllmer, H. (1979), "Local Interactions with a Global Signal: A Voter Model," *Lecture Notes in Biomath 38* (New York: Springer).
29. Fudenberg, D., and J. Tirole (1983), "Learning by Doing and Market Performance," *Bell J. Econ.* **14**, 522–530.
30. Graham, F. D. (1923), "Some Aspects of Protection Further Considered," *Quart. J. Econ.* **37**, 199–227.
31. Haken, H. (1978), *Synergetics* (New York: Springer Verlag).
32. Hanson, W. A. (1985), *Bandwagons and Orphans: Dynamic Pricing of Competing Systems Subject to Decreasing Costs, Ph.D. Dissertation, Stanford University.*
33. Heal, G. (1986), "Macrodynamics and Returns to Scale," *Econ. J.* **96**, 191–198.
34. Helpman, E., and P. Krugman (1985), *Market Structure and Foreign Trade* (Cambridge: MIT Press).
35. Hill, B., D. Lane, and W. Sudderth (1980), "Strong Convergence for a Class of Urn Schemes," *Annals Prob.* **8**, 214–226.
36. Hirshman, A. (1958), *The Strategy of Economic Development* (New Haven: Yale University Press).
37. Holly, R. (1974), "Recent Results on the Stochastic Ising Model," *Rocky Mtn. J. Math.* **4**, 479–496.

38. Kaldor, N. (1970), "The Case for Regional Policies," *Scottish J. Pol. Econ.* **17**, 337–348.
39. Katz, M., and C. Shapiro (1985), "Network Externalities, Competition, and Compatibility," *Amer. Econ. Rev.* **75**, 424–440.
40. Katz, M., and C. Shapiro (1986), "Technology Adoption in the Presence of Network Externalities," *J. Pol. Econ.* **94**, 822–841.
41. Kauffman, S. (1987), "Towards a General Theory of Adaptive Walks on Rugged Landscapes," *J. Theor. Biol.* **128**, 11–45.
42. Kehoe, T. J. (1985), "Multiplicity of Equilibria and Comparative Statics," *Quart. J. Econ.* **100**, 119–147.
43. Liggett, T. (1979), "Interacting Markov Processes," *Lect. Notes in Biomath 38* (New York: Springer).
44. Marshall, A. (1891), *Principles of Economics* (London: Macmillan), 8th ed.
45. Matthews, R. C. O. (1949), "Reciprocal Demand and Increasing Returns," *Rev. Econ. Stud.*, 149–158.
46. Maruyama, M. (1963), "The Second Cybernetics: Deviation-Amplifying Mutual Causal Processes," *Amer. Scien.* **57**, 164–179.
47. Myrdal, G. (1957), *Economic Theory and Underdeveloped Regions* (London: Duckworth).
48. Nicolis, G., and I. Prigogine (1976), *Self-Organization in Nonequilibrium Systems: From Dissipative Structures to Order through Fluctuations* (New York: John Wiley and Sons).
49. Ohlin, B. (1933), *Interregional and International Trade* (Cambridge: Harvard University Press).
50. Palander, T. (1935), *Beiträge zur Standortstheorie* (Stockholm: Almqvist and Wicksell).
51. Puffert, D. (1988), *Network Externalities and Technological Preference in the Selection of Railway Gauges, Ph. D. Dissertation, Stanford University,* forthcoming.
52. Ritschl, H. (1927), "Reine und historische Dynamik des Standortes der Erzeugungszweige," *Schmollers Jahrbuch* **51**, 813–70.
53. Rosenberg, N. (1982), *Inside the Black Box: Technology and Economics* (Cambridge: Cambridge University Press).
54. Rosenstein-Roden, P.N. (1943), "Problems of Industrialization of Eastern and South-Eastern Europe," *Econ. J.* **55**, 202–211.
55. Roughgarden, J. (1979), *Theory of Population Genetics and Evolutionary Ecology* (New York: Macmillan).
56. Scarf, H. (1981), "Indivisibilities: Part I," *Econometrica* **49**, 1–32.
57. Schelling, T. (1978), *Micromotives and Macrobehavior* (New York: Norton).
58. Solow, R. (1956), "A Contribution to the Theory of Economic Growth," *Quart. J. Econ.* **70**, 65–94.
59. Spence, M. (1981), "The Learning Curve and Competition," *Bell. J. Econ.* **12**, 49–70.
60. Varian, H. (1975), "A Third Remark on the Number of Equilibria in an Exchange Economy," *Econometrica* **43**, 985–986.

61. Varian, H. (1981), "Dynamical Systems with Applications to Economics," *Handbook of Mathematical Economics*, Eds. K. Arrow and M. Intriligator (New York: North Holland), vol. 1.
62. Weber, A. (1909), *Theory of the Location of Industries* (Chicago: University of Chicago Press).
63. Weidlich, W., and G. Haag (1983), *Concepts and Models of a Quantitative Sociology* (New York: Springer).
64. Weitzman, M. (1982), "Increasing Returns and the Foundations of Unemployment Theory," *Econ. J.* **92**, 787–804.
65. Young, A. (1928), "Increasing Returns and Economic Progress," *Econ. J.* **38**, 527–542.

ERIC B. BAUM
Jet Propulsion Laboratory, California Institute of Technology, Pasadena, CA 91109

Neural Nets for Economists

ABSTRACT

I review the Hopfield model, feedforward models for associative memory, and the back-propagation learning algorithm for supervised learning.

This paper will review two aspects of recent neural networks research which may be of interest for economic modeling. First I will describe the original Hopfield model (Hopfield, 1982, 1984). This has no direct relevance to economics, but is an example of one approach to modeling a complicated, poorly understood dynamical system, namely the brain. While the Hopfield model has a number of interesting features as a dynamical system, I will also remark that other 'neural' circuits may be designed (Baum et al., 1988) which better fulfill its function of 'associative memory.' My second subject will be the 'back-propagation' learning algorithm (Rumelhart et al., 1986; Werbos, 1974). As I will describe, this algorithm has potential as an economic predictor.

Interest in neural networks has exploded in the last five years for several worthwhile reasons. The incredible advance over the last three decades of about six orders of magnitude in complexity of integrated circuits has made it expedient to look for parallel processing applications of large, regular arrays of transistors. The brain is easily the best parallel processor we are aware of, and thus it is natural to try to build brain-like circuits in silicon (or in optics). A noteworthy example of this is the research of Carver Meade (1988) into analog VLSI, for example his design of a "Silicon retina," an integrated circuit modeled explicitly on the retina. A second driving force behind neural net research is the quantity of data which has been amassed regarding the neurophysiology and anatomy of the brain. It seems that to make sense of the brain it will be necessary to study simple models designed out of similar circuitry to perform similar tasks. Without an approach like this it would be very difficult, even if supplied with a circuit diagram, to understand the function of a personal computer; and no one expects we will soon have anything like a detailed circuit diagram for the human brain! A third reason for the surge of interest in neural nets, particularly among physicists, was the perspicuous PNAS papers and lectures of John Hopfield. I urge you to read these papers; because they are so clear I will be brief in my summary of the Hopfield model.

The models I will describe begin with a simple abstracted model of a neuron first proposed (I believe) by McCulloch and Pitts (1943). They wanted to show brain-like circuits could serve as universal computers so it was natural for them to use neurons which were two-state devices. This is, in fact, a reasonable first abstraction of many real neurons which have a quiescent state, in which they fire rarely and a firing or excited state in which they fire at a maximum rate. To a first approximation, these neurons are in the quiescent state when their membrane potential is low and in the firing state when their membrane potential is high, and the membrane potential is a sum of contributions from other neurons weighted by the synaptic efficiency, which is a signed quantity.

The Hopfield model thus has each neuron described by a binary variable $v_i = \pm 1$, where v_i is called the output of the ith neuron. A matrix w_{ij} gives the weight of the connection between the output of the jth neuron and the input of the ith. Thus, the input to the ith neuron is $u_i = \sum_{j=1,N} w_{ij} v_j$, where N is the number of neurons. The output of each neuron will be decided by simply thresholding its input, so we will set $v_i = \theta(u_i)$, where θ is the Heaviside step function, $\theta(x) = +1$ for $x \geq 0$, $\theta(x) = -1$ otherwise. The dynamics is that we start with some initial assignment of values to each v_i and then randomly choose neurons (with replacement) and decide, based on the threshold rule, whether their output should change or remain the same.

Note that we have allowed every neuron to be connected to every other neuron. (Of course, if some of the w_{ij} were 0, this would correspond to no connection.) A principle difference in complexity between a pigeon brain, which has say 10^8 neurons, and a supercomputer, which has abcut 10^8 transistors, is that each transistor in the supercomputer is connected to two or three other transistors, whereas each neuron in a brain is connected to 10^3-10^4 other neurons.

Once we specify the synaptic weights w_{ij} and the initial neuronal outputs v_i, we have a dynamical system which will evolve in some complicated way and which

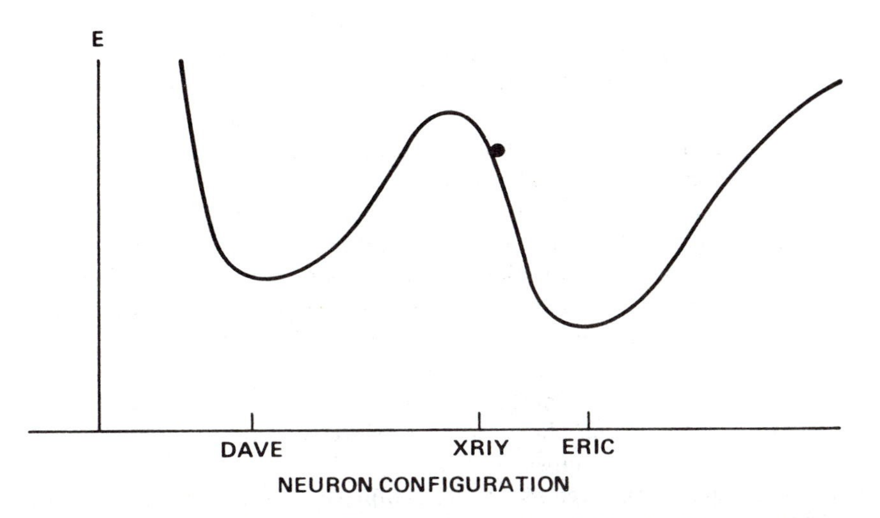

FIGURE 1 An energy landscape. The horizontal axis (which should be multidimensional) represents the N-bit word giving the value of each neuron in a particular configuration. The N-bit word for the left valley might spell 'DAVE,' for the right valley 'ERIC,' and at the dot (representing the current neuronal configuration) 'XRIY.'

we have argued is a loose but reasonable first abstraction of a brain. We now desire to specify synaptic weights so that the system performs a useful function analogous to a function of the brain. Hopfield proved an important theorem which is useful in this regard.

THEOREM (Hopfield): If the synaptic weights are symmetric, i.e., $w_{ij} = w_{ji}$ for all i and j, then the neural net always converges to a fixed point.

This follows as the 'energy' $E = -1/2 \sum w_{ij} v_i v_j$ is monotonically non-increasing as the net evolves; that is, E is a Lyapunov function. To see this, compute the change in E when neuron j, say, changes its output: $\delta E = -\sum_i w_{ij} v_i \delta v_j$. However, the threshold rule implies that $sign(\delta v_j) = sign(\sum_i w_{ij} v_i)$, so that δE is always zero or negative. E is bounded below, so the system must approach a fixed point.

We may now understand the behavior of the net by drawing its 'energy landscape.' See Figure 1. On the horizontal axis (which should really be N-dimensional), each point represents the coordinates of one neuronal configuration, thus each point represents the N-bit word giving the value of each of the N neurons. On the vertical axis is the energy of that configuration. If we start the system out at any particular configuration, it rolls downhill until it reaches a local minimum.[1]

Unlike standard computer memories, to which one must supply a specific address in order to retrieve any stored word, human memories are content addressable. Any one of a huge set of possible facts or pictures or descriptions will evoke in your mind an image of your Mom. Even if somebody trying to describe your Mom made many incorrect statements, you would have no difficulty in understanding that your Mom was being evoked and in remembering a wealth of facts about her. This content addressable, associative quality of human memory is seemingly integral to the processing power of the human mind. The difference between owning a calculus text, and thus having access through the index to all the information in it, and understanding the calculus is perhaps largely that in studying the text you form numerous associations which allow you to retrieve the information in a useful way and even to prove new theorems.

If we could choose the weights w_{ij} to sculpt the energy landscape and put the local minima wherever we wanted, we could use the neural net as an associative memory. Say we put one minima at 'ERIC,' i.e., the N-bit word specifying the neuronal configuration at this point spelled out 'ERIC.' We might also put another minima at 'DAVE.' Now if all we remembered about the word we were trying to retrieve was some of its content, namely that the second letter was R and the third I, we might start out net out at XRIY and let it evolve. It would roll downhill to the nearest minimum, which might hopefully be 'ERIC' since we presumably started out closer to 'ERIC' than to 'DAVE.' We could now read out the values of the neurons and retrieve the full stored word given as input cue only a fraction of its content.

How can we choose the w_{ij} to put minima at desired locations? Let us try to store a set of words ξ_i^μ where $\mu = 1, \ldots, M$ labels the word and $i = 1, \ldots, N$ labels the components of each word (ξ_3^2 is the third bit of the second word.) We will assume the words are random and uncorrelated, so that $\xi_i^\mu = +1$ with probability 1/2 and -1 with probability 1/2 independent of the other words or components. Then Hopfield (1982, 1984) showed that taking $w_{ij} = \sum_\mu \xi_i^\mu \xi_j^\mu$ will store the words as desired, so long as M is not too big. To see this we must at least see that each of the words are stable, that is $\xi_i^\mu = sign(\sum_j w_{ij} \xi_j^\mu)$ for all μ and i. Consider without loss of generality the first word, $\mu = 1$. Substitute in our expression for w_{ij}. We have

$$\sum_j w_{ij}\xi_j^1 = \sum_j \sum_\mu \xi_i^\mu \xi_j^\mu \xi_j^1 = \xi_i^1 \sum_j \xi_j^1 \xi_j^1 + \sum_{\mu \neq 1} \xi_i^\mu \sum_j \xi_j^\mu \xi_j^1 \tag{1}$$

[1]Given that we choose probabilistically which neuron will next decide its output, the downhill descent might proceed along different paths if we repeat the experiment and conceivably even reach different local minima. In practice this is exceedingly rare, but the possibility is precluded entirely if we consider the deterministic, continuous Hopfield model, which I will describe briefly later.

where I have broken up the sum over μ into the $\mu = 1$ term and the sum over the rest of the terms. The first term contains the inner product of ξ^1 with itself and thus is $N\xi_i^1$. The inner product in the second term between ξ^μ and ξ^1 is a number of mean zero and standard deviation $\sqrt{N}$. Similarly the sum over μ in the second term is typically of order $\sqrt{M}$. So long as we choose $N \gg M$ the first term, $N\xi_i^1$ will dominate the second term $O(\sqrt{N}\sqrt{M})$ and the memories will in fact be stable.

More detailed analysis (McEliece et al., 1987; Amit et al., 1987) indicates that we expect all the memories to be stable if $M \leq N/4ln(N)$. If we merely demand that each memory correspond to a local minima which may however have up to 5% of its bits wrong, then we may choose M as large as $.14N$.

This is an interesting and unusual model of computation in that it is statistical rather than deterministic. It is collective in the sense that the properties of a large system are radically different than the properties of a small one; this model will not work for small N. One very desirable feature which emerges is robustness. If several synapses die, performance of the memory will not noticeably degrade unless we are close to saturating its capacity. If some neurons die, we will lose some bits in each word, but none of the stored words will be lost. This robustness is a crucial property for a brain model. Your brain cells are dying as you read this, yet your performance is not noticeably degraded.

In 1984, Hopfield extended his model to allow neurons with sigmoid activation functions; that is, rather than using neurons whose output $v_i = \theta(\sum_j w_{ij}v_j)$, one may use continuous-valued neurons taking value $v_i = \sigma(\lambda \sum_j w_{ij}v_j)$, where $\sigma(\lambda x)$ is a differentiable function such as hyperbolic tangent which approaches the step function as λ, the gain parameter, becomes large. See Figure 2. (To understand this model, one shows that the $\lambda \to \infty$ limit approaches the discrete model, i.e., has the same attractors.) These neurons can easily be built out

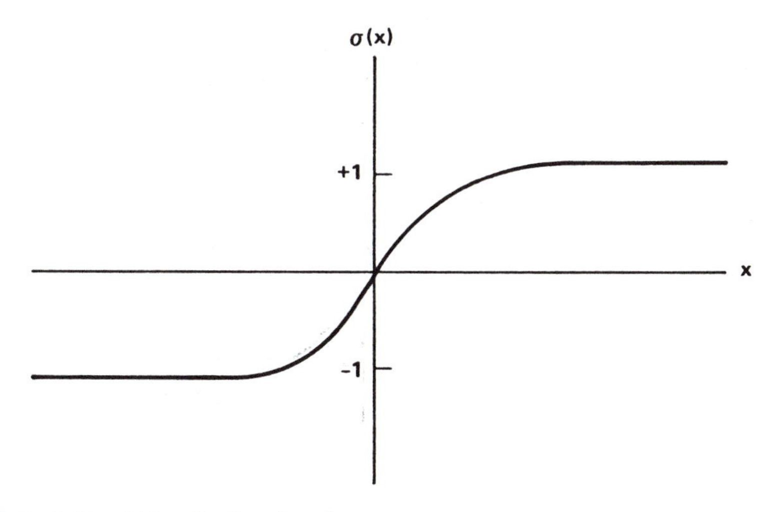

FIGURE 2 A sigmoid activation function.

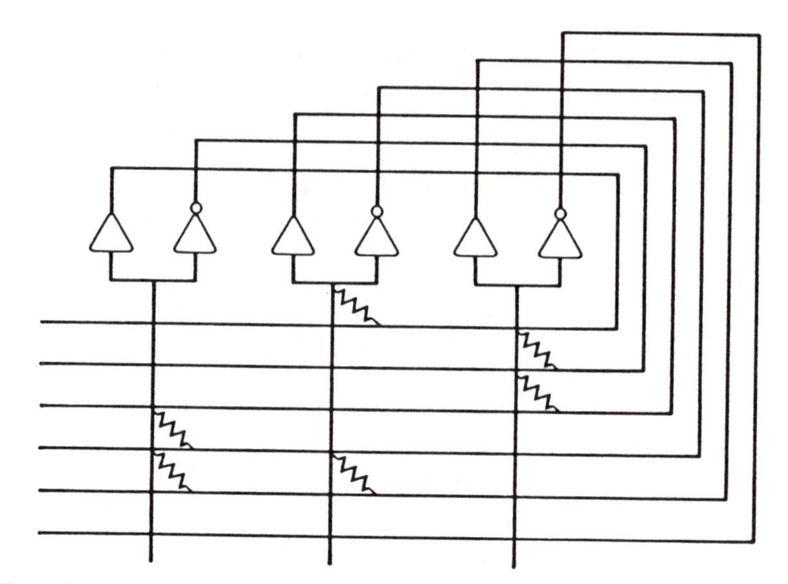

FIGURE 3 A circuit diagram for an electronic Hopfield net. The amplifiers implement a sigmoid function. Inputs are formed by current summing through connections whose conductance equals the synaptic strength, w_{ij}. Each neuron is represented as a pair of amplifiers, one inverting, so that "negative" conductances can be obtained.

of transistors, say, on CMOS chips. See Figure 3. A lot of research is underway to discover the best implementations and applications of such analog, neural network, integrated circuits.

Although the Hopfield model has aroused great interest in the field, it is worth pointing out that other designs, using the identical 'neural' building blocks, namely sigmoid neurons and resistive connections, can achieve substantially better performance for associative memory. With John Moody and Frank Wilczek, I have designed feedforward, layered nets for associative memory (Baum et al., 1988). See Figure 4. At the first layer of input neurons, one feeds an input cue, i.e., some fraction of the input neurons are set to value $+1$ or -1 according to the value of the corresponding bit in the stored word which one wishes to recall. (The input cue may also contain errors.) The other input neurons are set to zero. There are synapses between the outputs of the first-layer neurons and the inputs of the second-layer neurons. The second-layer neurons compute their outputs. What we hope is that the second-layer neurons will recover the address label or code which we have associated with the stored word most like the input cue. Having discovered the associated address, at the second level, the third-level neurons compute their values, which should give us the entire contents of the stored word, fulfilling the desired function of associative, content addressable memory. Note that this circuit does not rely on symmetric connections or attractive behavior; it is simply a forward directed, albeit highly parallel computation.

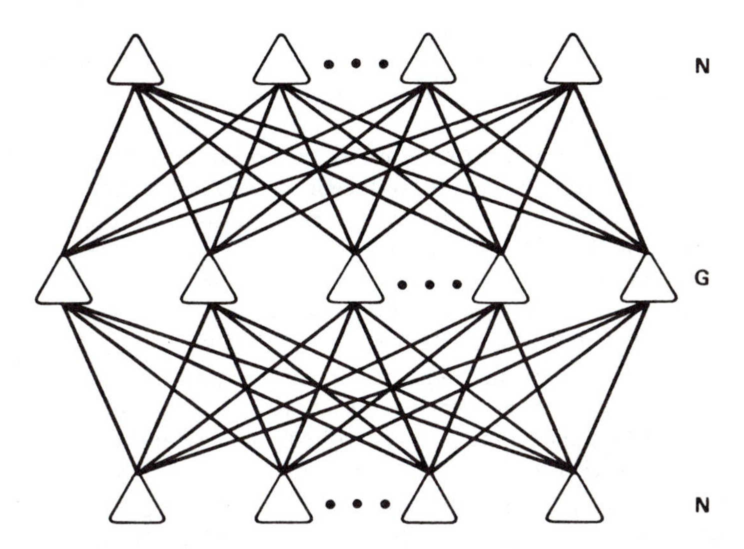

FIGURE 4 A feedforward net for associative memory. The middle level may be wired into one or several winner-take-all circuits. Computation proceeds from bottom to top, there are no feedback connections.

Since we use continuous-valued neurons, the second-layer neurons can be wired up into one or several winner-take-all circuits. The winner-take-all circuit is an important concept in neural nets, both artificial and biological. If a set of neurons are connected into a winner-take-all circuit, the one with the greatest external input turns on and suppresses all the rest. This can be imposed by using lateral, inhibitory connections, or in more effective and subtle ways (Lippmann, 1987). The use of winner-take-all circuits is not essential to the working of our feedforward associative memories, but improves their effectiveness substantially.

The choice of code words which we associate with the stored words is absolutely crucial to the functioning of the system and we have studied many interesting possibilities. The simplest, and in many ways most effective, representation is the unary or 'grandmother cell' representation.[2] In this representation the second-layer neurons are wired into one giant winner-take-all circuit and each neuron is expected to recognize one and only one stored word. When a cue is presented, the neuron

[2]The term 'grandmother cell' comes from hypothesized cells which recognize only one's grandmother. There is some biological evidence for highly specialized cells. The existence of 'grandmother cells' is, however, generally disfavored by the biological community, for reasons I do not find compelling (Baum et al., 1988).

recognizing the closest[3] stored word comes on and suppresses all the other neurons. It then single-handedly determines the values of the output neurons, configuring them to represent its stored word.

There are many advantages to the unary representation. We are able to store 1 N-bit word per neuron, where N is the number of input neurons. This gives a capacity of G N-bit words, if we have G grandmother cells, compared with a capacity of about $N/4\log(N)$ for the Hopfield model. We have the flexibility to store many relatively short words, by taking $G \gg N$, whereas in the Hopfield model the number of words stored is directly related to their length. The storage in the unary representation is independent of whether the words are random or highly correlated. Perhaps a better measure of capacity is provided by bits stored per synapse. We may construct the unary representation circuit using only two values of connection, 1 or 0. This is important as in both biology and silicon, it is difficult to design synapses with large grey scale (i.e., many possible values). Such binary valued synapses can in principle store up to 1 bit each. The unary representation saturates this information theoretic bound. (By contrast, storing $N/4\log(N)$ N-bit words on N^2 real-valued synapses, the original Hopfield model stores $1/4\log(N)$ bits per synapse.)

The unary representation provides optimum recall; that is, it always finds the stored word closest (in Hamming distance) to the input cue. Unfortunately in the Hopfield model, when the desired words are stored as local minima in the energy, many other 'spurious' minima are also unavoidably created. These severely limit retrieval as an input cue far from any stored word will likely converge to an unwanted 'spurious' minima rather than the nearest stored state.

By using a trick involving 'inverse coding,' the unary representation can be implemented so as to be extremely robust to manufacturing variation in the synaptic values (Baum et al., 1988). It can be made robust against neuron death by sacrificing some of its huge capacity for redundancy.

We have studied substantially more complex coding schemes than the simple unary representation. These may be valuable in certain applications, for instance, if one needs extreme robustness to neuron death, and so may be utilized by biology. These matters are discussed at greater length in Baum et al. (1988).

In my opinion, while the Hopfield model is much more interesting as a dynamical system and contains insights which have spurred a huge increase in neural net research and understanding, the unary representation (and other feedforward circuits) are more useful as associative memories.

[3]The μ-th grandmother cell, to store the word ξ_j^μ has connections to the inputs $w_{\mu j} = \xi_j^\mu$ and to the outputs $w_{i\mu} = \xi_i^\mu$. If I_j is the input cue, the closest stored word is the word ξ^ν for which $\sum_j \xi_j^\nu I_j$ is the largest, i.e., the closest word in Hamming distance. The neuron storing this word has, by definition, the greatest input. It takes output value +1 and suppresses all the other grandmother cells, which take value 0.

LEARNING AND GENERALIZATION

In this section I will discuss how an algorithm called 'back propagation' (Rumelhart et al., 1986; Werbos, 1974) can be used to modify the weights in feedforward circuits such as Figures 4 or 5 so as to perform 'supervised learning' or 'learning from examples.' As I will elucidate, a potential application of this algorithm is to learn to predict economic time series.

The problem of supervised learning is to model some mapping between input vectors and output vectors presented to us by some real world phenomena. To be specific, consider the question of medical diagnosis. The input vector corresponds to the symptoms of the patient; the ith component is defined to be 1 if symptom i is present and 0 if symptom i is absent. The output vector corresponds to the

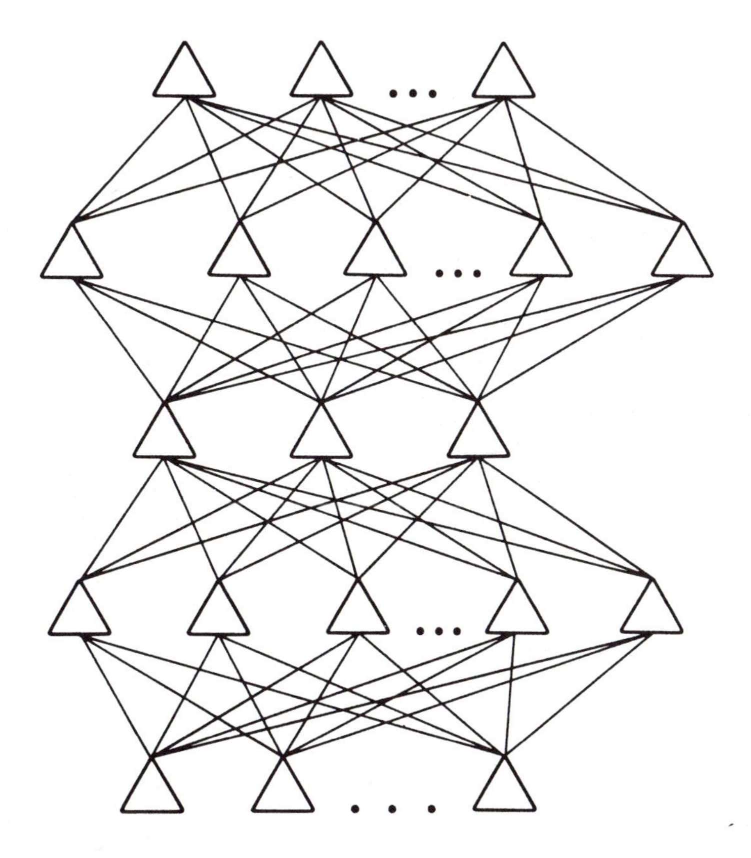

FIGURE 5 A five-layer feedforward net. Inputs are at bottom. Note the third-layer bottleneck.

illnesses, so that its jth component is 1 if the jth illness is present and 0 otherwise. Given a data base consisting of a number of diagnosed cases, the goal is to construct (learn) a mapping which accounts for these examples and can be applied to diagnose new patients in a reliable way. One could hope, for instance, that such a learning algorithm might yield an expert system to simulate the performance of doctors. Little expert advice would be required for its design, which is advantageous both because experts' time is valuable and because experts often have extraordinary difficulty in describing how they make decisions.

A feedforward neural network implements such a mapping between input vectors and output vectors. Such a network has a set of input neurons, one or several layers of intermediate neurons, and a layer of output neurons. Given a fixed set of weights $\{w_{ij}\}$, we set the input node values to equal some input vector, compute the value of the neurons layer by layer until we compute the output neurons, and so generate an output vector.

There are two important reasons why it is important, in the context of supervised learning, to use neurons with differentiable output functions.[4] First of all, the learning problem will be seen to be an optimization problem, and it has been observed by Hopfield and Tank (1985) and independently by myself (Baum, 1985, 1986) that, for a variety of arcane and little understood reasons, embedding discrete combinatorial optimization problems into continuous systems mitigates (although by no means removes) the problem of local, spurious minima. Secondly, we will be able to differentiate the values of the output neurons with respect to the weight values, and thus to adjust the weights in the appropriate direction to achieve desired output values.

Assume we are given a set of diagnosed cases $\{s_i^\mu, t_j^\mu\}$; where μ labels the case, and s_i^μ is the probability symptom i is present and t_j^μ the probability disease j is present in case μ. In our network, let us define $o_j(s^\mu, w)$, the value of the jth output neuron when vector s_i^μ is presented at the inputs and w_{ij} are the connections, as our estimate of the probability disease j is present, i.e., our estimate of t_j^μ. We desire to learn from the examples, that is to adjust the weights w_{ij} so that our estimate of the likelihood, p, of the given set of examples is as large as possible (Baum and Wilczek, 1988). Using our definitions, it is easy to write a formula for p:

$$p = \prod_\mu \left[\prod_{\{j|t_j^\mu=1\}} o_j(s^\mu, w) \prod_{\{j|t_j^\mu=0\}} (1 - o_j(s^\mu, w)) \right] \tag{2}$$

or

$$log(p) = \sum_\mu \left[\sum_{\{j|t_j^\mu=1\}} \log(o_j(s^\mu, w)) + \sum_{\{j|t_j^\mu=0\}} log(1 - o_j(s^\mu, w)) \right] \tag{3}$$

[4]In the literature the logistic function $g(x) = 1/(1+\exp(-x))$ is frequently used as it has a sigmoid shape and is easily differentiated.

The extension of Eq. (2), and thus Eq. (3) to the case where the t_j^μ are probabilities taking values in [0, 1] is straightforward and yields

$$S = \log(p) = \sum_{\mu,j} \left[t_j^\mu \log(o_j(s^\mu, w)) + (1 - t_j^\mu) \log(1 - o_j((s^\mu, w)) \right] \quad (4)$$

Expressions of this sort often arise in physics and information theory and are generally interpreted as an entropy.

It is now possible to learn in a straightforward manner. We simply start with some randomly chosen set of weights, compute the gradient of the entropy[5] with respect to the weights, and vary the weights in the gradient direction until we hit a local minimum. Although we settle for a local minimum, this seems in practical applications to perform adequately,[6] as we will discuss. Because we adjust all the weights in parallel, this algorithm converges in acceptable amounts of time for many complicated, real world problems.

In practice the descent is accomplished by presenting an example at the input, allowing the net to compute its output, moving a small distance in the gradient direction to improve the net's performance on this particular input, and then going on to the next input (Rumelhart et al., 1986; Werbos, 1974). That is, we may segregate the different cases, write $S = \sum_\mu S^\mu$, and compute independently the $\partial S^\mu / \partial w_{ij}$. By cycling through all the inputs several times, one moves approximately in the overall gradient direction.

The major reason why this algorithm is workable is that, if arranged cleverly, it takes no longer to compute the gradient, than the output for a particular input (Rumelhart et al., 1986). The forward pass through the network, and the backward pass to compute the gradient with respect to the weights, can be accomplished in about the same number of steps. To compute the gradient one first realizes $\partial S^\mu / \partial w_{ij} = \sum_k (\partial S^\mu / \partial o_k)(\partial o_k / \partial w_{ij})$. The first factor on the right-hand side is easily computed from Eq. (4). The second is computed by the chain rule, using the definition of o_k: $o_k = \sigma(\sum_j w_{kj} v_j)$. Here the sum is over the neurons in the previous layer, and the v_j themselves depend on the next layer back. o_k depends on the w_{kj} directly but on weights in previous layers through their effect on the v_j. The chain rule gives a sum over the paths in the network by which a change in a particular weight can propagate forward to effect output o_k. All the large number of paths contributing can be efficiently summed by iteratively propagating backwards on the net through each node the total sum of the ways a change in that neuron's value

[5]Most authors have simply sought to minimize a mean square error: $E = \sum_{\mu j} \left[t_j^\mu - o_j(s^\mu, w) \right]^2$. This may be more appropriate than the entropy in certain contexts such as signal processing. The notion of using the entropy is found in Baum and Wilczek (1988).

[6]This is perhaps due to the advantages of descent in continuous as opposed to discrete systems.

effects the output.[7] This 'backwards propagation' (hereafter called 'back prop') involves only about as many computations (and the same degree of parallelism) involved in the original forward pass computing the output of the net.

Economists will naturally ask whether these network techniques offer any improvement over simple curve-fitting or Bayesian reasoning. The real answer to this lies in the practical results, but before discussing these I will heuristically motivate the use of back prop. Certainly the ability in back prop to sum a complicated path integral involving many high-order terms is essential to its usefulness. This would not be true of some straightforward expansion or curve-fitting argument. Experience has shown the utility of using nets with 'hidden' or intermediate layers of neurons whose values form an 'internal representation' of the problem. In simple Bayesian reasoning (involving Gaussians), such 'hidden' units would simply be integrated out unless at least third-order correlations are included. There are too many of these for most practical problems, which involve limited computational resources and typically limited data sets.

Indeed, it is well-known folklore in curve-fitting and pattern classification that the number of parameters must be small compared to the size of the data set if any generalization to future cases is expected (LeCun, 1987).

In feedforward nets, the question takes a different form as there are other bottlenecks to information flow. If the neuronal activation functions are chosen so that the values of each of the intermediate nodes tend towards either 1 or 0,[8] then a layer with I neurons can take on no more than 2^I patterns. Thus, no matter how many input nodes, output nodes, or free parameters there are in the net, the output will be constrained to take on no more than 2^I different patterns. If I is small, some sort of 'generalization' must occur even if the number of weights is large. The image at the bottleneck layer, and more generally on all internal layers, forms an 'internal representation' of the problem. Back prop is a heuristic for finding good internal representations. One plausible reason for the success of back prop in adequately solving tasks, in spite of the fact that it finds only local minima, is its ability to vary a large number of parameters, relative to its true number of independent degrees of freedom. This leeway may allow back prop to escape from many putative traps and to find an acceptable solution.

A good expert system, say, for medical diagnosis, should not only give a diagnosis based on the available information, but should be able to suggest, in questionable

[7] Indeed, if we define a quantity δ_j^l at the jth neuron in the lth layer back from the output iteratively as follows: $\delta_j^0 = o_j'$ ($l = 0$ for the output nodes) and $\delta_j^l = v_j' \sum_i w_{ij} \delta_i^{l-1}$ where the sum is over nodes i at the $l-1$ layer, then the chain rule gives trivially: $\partial o_k / \partial w_{mn} = \delta_m v_n$ where δ_m is of course the δ defined at node m, whichever level m is on. Evidently the backwards calculation of δ is algorithmically similar to the forward calculation of the output of the net, and all the $\partial o / \partial w$ for all outputs and weights get computed in one backwards pass.

[8] Alternatively when necessary this can be enforced by adding an energy term to the log-likelihood to constrain the parameter variation so that the neuronal values are near either 1 or 0.

cases, which lab tests might be performed to clarify matters. Actually back propagation inherently has such a capability (Baum and Wilczek, 1988). Back propagation involves calculation of $\partial S/\partial w_{ij}$. This information allows one to compute immediately $\partial S/\partial s_j$. Those input nodes for which this partial derivative is large correspond to important experiments. If instead the net had been trained to predict the GNP based on a basket of indicator time series such as the CPI and various interest rates, and if subsequently the net performed well on a test time series of untrained years, one might have confidence that it had formed some reasonable internal model of the economy. One might then compute such interesting quantities as the partial derivative of the GNP with respect to the CPI. This type of information is not available in many standard statistical techniques.

Indeed, back prop can be used as a research tool in another way. Say we feed in to our net sets of symptoms $\{s^\mu\}$, and ask for the net to produce as output, after filtering through several layers including a bottleneck layer, the same set of symptoms $\{s^\mu\}$. Then the net will be forced to find at the bottleneck layer a concise description of the data. This might naturally lead to a classification into diseases, or at least syndromes.

Back prop has now been applied to many different problems. It shows great ability to solve toy A.I. problems (Rumelhart et al., 1986) such as distinguishing symmetry. For example, one may present to 16 input nodes arranged in a 4-by-4 square pattern of 1's and 0's which is either symmetric about the horizontal (Figure 6), or the vertical and ask the net to learn to recognize which situation is present. This requires the net to form a good internal representation. There are 480 patterns with horizontal but not vertical symmetry or vice versa. After presentation of a much smaller learning set of diagnosed examples, back prop has no problem in correctly classifying subsequent, untrained examples.

Back prop has been used in practice to construct reasonable "expert systems" for medical diagnosis (LeCun, 1987), translation of text to phonemes (Sejnowski and Rosenberg, 1987), playing the game of backgammon (Tesauro and Sejnowski, 1988), prediction of conflict in Latin America (Werbos and Titus, 1978), and analysis of secondary structure in protein folding (Qian and Sejnowski, 1988). It performs as well as many human constructed expert systems or classification algorithms in each of these cases; for some of them it is the best method known (including human experts). It has the advantage in all cases of being easy to implement, and requiring little human time, although it may require considerable computer time. The learning algorithm may itself be incorporated soon into special purpose, parallel hardware. In any case the feedforward circuit given for a given problem could be built, in which case one would have an expert system on a chip, yielding answers to queries within microseconds.

A third type of test to which back propagation has been subjected is that of 'intrinsically hard' prediction of chaotic time series. Using a net with only about 150 connections, Lapedes and Farber (1987) have recently made predictions of the future behavior of the fractal dimension 2.5 and 3.5 attractors of the Mackey-Glass equation. Feeding into the net as inputs $x(t_i)$, $x(t_i - 6)$, $x(t_i - 12)$, $x(t_i - 18)$, they

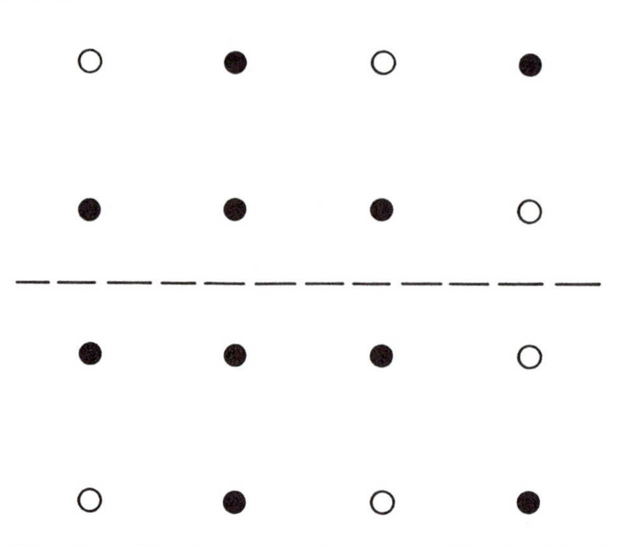

FIGURE 6 A configuration of input neuron values symmetric about the horizontal axis. Dark circles correspond to neuron value 1 and light circles to value 0.

attempted to predict 6, 12, 24, 36, 48, 60, 72, 84, and 100 timesteps into the future. Their accuracy is comparable to the best other known predictor, the local linear method of Farmer and Sidorowich (1987). Lapedes and Farber use a data set of only 500 time-series points for learning (compared to the $1-2\times10^4$ of Farmer and Sidorowich). However, they also use about ten times as much computer time.

All of this has encouraged me to attempt to use back propagation to form a predictive 'model' of the economy. Advised by W. Brock, J. Scheinkman, and L. Summers at this workshop, I hope to feed in as learning data the time series of various interrelated economic indicators, and to predict the same indicators at subsequent periods. If this works well on an untrained test set, say, the years 1983–present, it will be of particular interest to compute various derivatives and discover how these indicators effect each other. I hope to report on this work at the next Santa Fe Institute workshop.

ACKNOWLEDGMENTS

I would like to thank the Santa Fe Institute for hospitality. The research reported in this paper was performed by the Jet Propulsion Laboratory, California Institute of Technology, as part of its Innovative Space Technology Center, which is sponsored by the Strategic Defense Initiative Organization/Innovative Science and Technology through an agreement with the National Aeronautics and Space Administration (NASA).

REFERENCES

1. Amit, D. J., H. Gutfreund, and H. Sompolinsky (1987), "Statistical Mechanics of Neural Networks near Saturation," *Ann. of Phys.* **173**, 30.
2. Anderson, D. Z., ed., *Neural Information Processing Systems, Natural and Synthetic,* (New York: AIP Press), in press.
3. Baum, E. B. (1985), "Graph Orthogonalization," *MIT Tech. Report*, unpublished.
4. Baum, E. B. (1986), "Towards Practical 'Neural' Computation for Combinatorial Optimization Problems," *Neural Networks for Computing, AIP Conf. Proc. 151*, Ed. J. S. Denker (New York: AIP Press), 53–58.
5. Baum, E. B., J. Moody, and F. Wilzcek (1988),"Internal Representations for Associative Memory," to appear in *Bio. Cybernetics*.
6. Baum, E. B., and F. Wilczek (1988), "Supervised Learning of Probability Distributions by Neural Networks," to appear in *Neural Information Processing Systems, Natural and Synthetic,* Ed. D. Z. Anderson (New York: AIP Press).
7. Duda, R. O., and P. E. Hart (1973), *Pattern Classification and Scene Analysis* (New York: John Wiley and Sons).
8. Farmer, D. J., and J. J. Sidorowich(1987), "Predicting Chaotic Time Series," *Phys Rev Lett.* **59 (8)**, 845–848, (1987).
9. Hopfield, J. J. (1982), "Neural Networks and Physical Systems with Emergent Collective Computational Abilities," *Proc. Natl. Acad. Sci. USA* **79**, 2554
10. Hopfield, J. J. (1984), "Neurons with Graded Response Have Collective Properties like those of Two-State Neurons," *Proc. Natl. Acad. Sci. USA* **81**, 3088.
11. Hopfield, J. J., and D. W. Tank (1985), "'Neural' Computation of Decisions in Optimization Problems," *Bio. Cybernetics* **52**, 141–152.
12. Lapedes, A., and R. Farber (1987) "Nonlinear Signal Processing Using Neural Network: Prediction and System Modeling," *LANL preprint LA-UR-87-2662.*
13. LeCun, Y. (1987), *Address at 1987 Snowbird Conference on Neural Networks,* unpublished.
14. Lippmann, R. (1987), "An Introduction to Computing with Neural Nets," *IEEE ASSP Mag. V4* **2**, 4–22.
15. McCulloch, W. A., and W. Pitts (1943), "A Logical Calculus of the Ideas Immanent in Neural Nets," *Bull. Math. Biophys.* **5**, 115–137.
16. Meade, C. A. (1988), *Analog VLSI and Neural Systems* (Reading, MA: Addison-Wesley).
17. McEliece, R. J., E. C. Posner, E. R. Rodemich, and S. S. Venkatesh (1987), "The Capacity of the Hopfield Associative Memory," *IEEE Trans. Inf. Theory, IT33*.
18. Qian, N., and T. J. Sejnowski (1988), "Predicting the Secondary Structure of Globular Protein using Neural Network Models," *J. of Mol. Biol.*, in press.

19. Rumelhart, D. E., G. E. Hinton, and G. E. Williams (1986), "Learning Internal Representations by Error Propagation," *Parallel Distributed Processing*, Eds. D. E. Rumelhart and J. L. McClelland (Cambridge, MA: MIT Press), vol. 1.
20. Sejnowski, T. J., and C. R. Rosenberg(1987), "Parallel Networks that Learn to Pronounce English Text," *Complex Systems* **1**, 145–168.
21. Tesauro, G., and T. J. Sejnowski (1988), "A 'Neural' Network that Learns to Play Backgammon," to appear in *Neural Information Processing Systems, Natural and Synthetic,* Ed. D. Z. Anderson (New York: AIP Press).
22. Werbos, P. (1974), *Beyond Regression: New Tools for Prediction and Analysis in the Behavioral Sciences, Harvard University Dissertation.*
23. Werbos, P., and J. Titus (1978), "An Empirical Test of New Forecasting Methods Derived From a Theory of Intelligence: The Prediction of Conflict in Latin America," *IEEE Trans Sys. Man, Cyber.* **SMC-8 (9)**, 657–666.

MICHELE BOLDRIN
University of California, Los Angeles

Persistent Oscillations and Chaos in Dynamic Economic Models: Notes for a Survey

1. INTRODUCTION

It is probably not unfair to say that Nonlinear Dynamics (NLD) has not had a major impact on the development of modern economic theory. In fact, one may even be tempted to add that, until very recently, it was either an unfamiliar tool for the mathematical economist or one whose implications were often disregarded as irrelevant to the purposes of the research. Dynamical systems theory appeared for a while in the background of the studies on the stability of the tâtonnement process (see Hahn, 1982) and on optimal growth and turnpike (see McKenzie, 1986), but never really got on the stage.

During the 1950's and 1960's the only well-developed effort to use nonlinear techniques in the study of dynamic economic processes is associated with Richard Goodwin (see Goodwin, 1982, for a collection of the relevant essays). He put forward the idea of illustrating persistent, deterministic oscillations within a multiplier-accelerator setup by means of a limit cycle for a nonlinear, two-dimensional flow. His research effort toward an endogenous explanation of economic fluctuations motivated a few others' contributions within the Keynesian and Cambridge (UK) tradition, but was never able to take off and influence the whole of the profession. In fact, the late 1960's and 1970's witnessed an almost complete unanimity on the use of linear-stochastic models in order to understand business cycles. The causes of this historical process are complex and they will not concern us in this place. Similarly, I will not try to list the motivations for the sudden revival of interest in NLD which is characteristic of the last few years, but simply try to provide a (cursory) description of the problems that have been considered and of the results achieved. At the end, I will also take a timid look at the "mare magnum" of issues that are still open to investigation.

Let me make clear, at the outset, that the revival of interest in NLD to which I refer is not widespread among economists and that, as a matter of fact, the great majority of applied and theoretical researchers still look at it with doubtful eyes and believe it will not help much in understanding what is going on in the real economic world. They may, indeed, be right.

In spite of this, a group of scholars have taken the opposite perspective and have started to look at economic fluctuations under the hypothesis that a relevant portion of them can be explained as a deterministic phenomenon, endogenously created by the interaction of market forces, technologies and preferences. In particular, it is conjectured that deterministic periodic cycles affected by small stochastic forces and/or "noisy" chaotic paths generated by dynamical systems of relatively small dimensionality, can account for a relevant portion of the observed fluctuations of most of the important macroeconomic variables.

Such a research program naturally involves two lines of inquiry: (a) finding theoretical models that predict cycles and chaos as logical outcomes of "reasonable" economic hypotheses and (b) testing the available data in order to find evidence of nonlinearities in the underlying dynamic processes. A third, and most important, step should indeed be added: comparing the data and the qualitative predictions of the theoretical models in order to understand if they are at least compatible or if instead the former reject the latter. Such a task remains, for the time being, far from being undertaken. The available models are too abstract to yield any seriously testable implication and the data-screening techniques we know of seem still too weak to place any confidence on their results. In any case, it is clear that the approach we are discussing will win or lose its bet exactly on this point; any effort along those lines is, therefore, worthwhile.

This survey (fortunately!) is limited to point (a); issues and results that pertain to (b) are illustrated in Brock (1987; see also Ruelle, 1987). I have tried to the best of my ability to make this an economic survey. Given the intrinsically technical nature of the problems and the sophisticated mathematics involved, this choice entails a major consequence: I have dispensed completely with mathematical definitions, lemmas and theorems, *taking as granted* that the reader is familiar enough with the terminology to have at least an intuition of what is going on from a mathematical point of view. I have assumed the reader knows the mathematics of chaos and wants to learn a bit about the economics of chaos. For the non-technical reader, I can only recommend a few standard references: Collet-Eckmann (1980), Devaney (1986), Guckenheimer-Holmes (1983), Iooss (1979) and Lasota-Mackey (1985).

The note follows these steps: in the second section I discuss the earlier examples of competitive models having an oscillatory or chaotic dynamics. Those of them which had a macroeconomic structure were not derived from explicit maximizing behavior on the part of the agents. As it used to be believed that intertemporal maximization would eradicate instability, we consider in the next two sections models with a complicated dynamics which are grounded on rational, maximizing procedures. Section 3 takes care of the so-called overlapping generations model (OLG) and Section 4 examines the models where agents live forever. Section 5 gives some concluding comments. The bibliography at the end is meant to include all the works on nonlinear dynamics in economics which have a theoretical nature, have been published during or after the "revival" and are known to me.

2. NON-OPTIMIZING MODELS OF ECONOMIC DYNAMICS

2.1 KEYNES-KALDOR MODELS

At the outset, there was the "Keynesian" model, in one of its many possible dynamic versions. Torre (1977) for example considered:

$$\begin{aligned} \dot{Y} &= \alpha\{I(Y,R) - S(Y,R)\} \\ \dot{R} &= \beta\{L(Y,R) - L^S\} \end{aligned} \tag{2.1}$$

where Y is real income; R is the rate of interest; I and S are the investment and saving function with $\partial I/\partial Y > 0$, $\partial I/\partial R < 0$, $\partial S/\partial Y > 0$, and $\partial S/\partial R > 0$; L is the demand for money with $\partial L/\partial Y > 0$ and $\partial L/\partial R < 0$; and L^S is the fixed money supply. Finally, α and β are two positive parameters. That such a model could exhibit a limit cycle was a kind of folk theorem to which Torre provides a formal proof by showing that an Hopf bifurcation occurs when the steady state loses stability as the bifurcation parameters α and β increase. This came as no surprise: the model was formally analogous to the standard representation of Kaldor's business-cycle model, which was known to have limit cycles as a solution since the work of Chang and Smith (1971). This is in fact written as:

$$\begin{aligned} \dot{Y} &= \alpha\{I(Y,K) - S(Y,K)\} \\ \dot{K} &= I(Y,K) - \beta K \end{aligned} \tag{2.2}$$

where K is the aggregate capital stock; β is the depreciation factor; and the functions satisfy $\partial I/\partial Y > 0$, $\partial I/\partial K < 0$, $\partial S/\partial Y > 0$, and $\partial S/\partial K \neq 0$.

Dana and Malgrange (1984) contains an interesting analysis of the model of Eq. (2.2). After proving the existence of a limit cycle for a continuous-time parametric version of the model (with parameter values specified to satisfy French quarterly data for 1960–74), the authors addressed, using both simulation techniques and analytical instruments, a discrete-time version of the same model. They showed that, by using α as a bifurcation parameter, different regimes may be obtained that go from an attracting steady state to an apparently chaotic state that they label "intermittent chaos." Even if they were unable to prove the existence of a strange attractor for their model, the two authors provided a considerable amount of evidence to this end.

Both the "Keynesian" and the "Kaldorian" models are susceptible to various criticisms. A couple of them may be handled by means of the Hopf bifurcation. The first relates to the fact that you need to make strong global assumptions on the shape of I and S to obtain a cycle in Eq. (2.2) (or Eq. (2.1)) by means of the Poincaré-Bendixson theory. In fact, Chang and Smith (1971) had to assume either I or S to have an S-shape with respect to Y, for given values of K, a strong assumption with little support. By using Hopf, you may disregard this hypothesis; all you need are local information on the degree of sensitivity of I to changes in Y and K (or R) around the steady state.

A second criticism is concerned with the fact that in both Eq. (2.1) and Eq. (2.2), an important variable that does change along the business cycle is considered fixed by the model: this is K in Eq. (2.1) and R in Eq. (2.2). This means that you would like to consider a more complete model like:

$$\begin{aligned} \dot{Y} &= \alpha\{I(Y,R,K) - S(Y,R,K)\} \\ \dot{R} &= \gamma\{L(Y,R) - L^S\} \\ \dot{K} &= I(Y,R,K) - \beta K \end{aligned} \tag{2.3}$$

where the signs of the first partial derivatives are as before. The Hopf bifurcation is especially useful here as the standard Poincaré-Bendixson technique is of no help in high dimensions. The model of Eq. (2.3) has been studied by Boldrin (1983) and Cugno-Montrucchio (1983): the existence of limit cycles is proved for large sets of values of the parameters. In fact, Cugno and Montrucchio add also a fourth variable (price expectations) and still obtain oscillatory solutions by means of the Hopf theorem. Almost no one seems to have looked for the emergence of more complicated patterns of behavior in Eq. (2.3), even if it seems obvious that a bifurcation cascade may easily be originated and, therefore, the Newhouse-Ruelle-Takens (1978) theory applied to argue the existence of a strange attractor: the thing seems, indeed, to be computationally demanding and with a very low return for our economic understanding. Lorenz (1985) has tried something along these lines for a six-dimensional version of Eq. (2.2) (three different goods and three different capital stocks), but he does not seem to get any economic insight. Along similar lines, a related work is that of Medio (1984) which considers an n-dimensional multiplier-accelerator model and proves the existence of cycles by using both the ideas of synergetics and the Hopf bifurcation theorem.

Still within a Keynesian framework, there is another attempt to explain the emergency of complicated dynamics that is worth mentioning. This is Day-Shafer

(1985). A modified version of the textbook IS-LM approach provides a consumption-income function and an investment-income function:

$$\begin{aligned} C &= G(Y, M) = C[L(Y, M), Y] \\ I &= \alpha H(Y, M) = \alpha I[L(Y, M), Y] \end{aligned} \tag{2.4}$$

where M denotes the amount of money, and $L(Y, M)$ is the LM function which expresses the value of the interest rate R that guarantees a temporary money-market equilibrium for given Y and M. The other notation is as before. A dynamic process is obtained by using what is called a Robertsonian lag (current C and I depend on past income) that yields a nonlinear version of Samuelson's multiplier-accelerator process:

$$Y_{t+1} = \theta(Y_t; \alpha, M, A) = G(Y_t; M) + \alpha H(Y_t; M) + A. \tag{2.5}$$

A is a positive constant representing exogenous expenditure. The money supply here is taken as a parameter; the dynamic is therefore described by the one-dimensional map of Eq. (2.5). As the authors wanted to use the existing theory of unimodal maps, they needed to make θ unimodal. Consumption is monotonically increasing in income (with a slope bounded away from zero and bounded above by 1); therefore, the burden of nonmonotonicity is on the investment function. Their intuition goes as follows: investment demand increases with income for a low level of income because of the accelerator mechanism and low interest rates. But as income increases, the fixed amount of money supplied requires an ever-rising interest rate to clear the money market. If investment is elastic enough at high interest rates, then it must eventually fall. Also, as α grows, the hump in θ increases and it is clear that by treating it as a bifurcation parameter, the classical route to chaos can be obtained for fairly simple specifications of G and H. In fact, the authors accomplished this by using piecewise continuous maps. Even more, they showed that the accelerator mechanism is not strictly necessary to the argument: even if I depends only on R, the negative effect of Y for high Y's can be brought in by the demand for money function that determines the market interest rate. One may notice that models of this type leave plenty of room for stabilizing (or destabilizing) monetary policies. Leaving expectations considerations aside, it is clear that a procyclical monetary policy could inject the amount of money necessary to prevent R from increasing as Y increases and, therefore, eliminate the unimodal shape of θ or, at least, reduce its steepness. A similar, but not identical point is made in Day (1984). On the nonlinear accelerator, one may also want to look at Rustichini (1983), even if I must admit that the sense of the latter escapes my understanding.

For completeness, we may also recall the work of Simonovits (1982). He adopts the framework introduced by Benassy and Malinvaud of distinguishing between "classical" and "Keynesian" equilibria in the wage-price space and tries to describe a dynamical system in those variables. The author claims the existence of cycles and chaos but nothing precise is actually proven.

2.2 CLASS STRUGGLE AND CHAOS

A role was played in the earlier debate by a simple and elegant model that Richard Goodwin elaborated to formalize Marxian or conflictual views of economic growth and income distribution Even if the huge amount of research done around this model has had little or no spillover on any other area of economic dynamics, the basic idea is very interesting by its own sake and worth mentioning. The original setup (see Goodwin, 1967) is like this: q is output, k is capital stock, w is the wage rate, $a = a_0 \exp(\alpha t)$ is labor productivity trend, $\sigma = k/q$ is a fixed ratio, $u = w/a$ is the labor share in national income, and $(1 - w/a)a = \dot{k}$ are profits that are completely invested. If the work force grows like $n = n_0 \exp(\beta t)$ and $\ell = q/a$ represents the employment level, time differentiation will give:

$$\frac{\dot{v}}{v} = \left(\frac{\ell - u}{\sigma}\right) - (\alpha + \beta) \tag{2.6}$$

where $v = \ell/n$ (employment ratio). Assume wages vary according to $\dot{w}/w = f(v) = -\gamma + \rho v$ (the bargaining rule), then time differentiation will also give:

$$\frac{\dot{u}}{u} = -(\alpha + \gamma) + \rho v\,. \tag{2.7}$$

Eqs. (2.6) and (2.7) together give a bi-dimensional dynamical system well known to mathematical biologists, i.e., Lotka-Volterra "prey-predator" model.

The solution to it is given by a continuum of closed curves around the unique stationary state. Different initial conditions will place the system on different oscillatory paths along which profits/wages and unemployment oscillate perpetually. The economic interpretation is obvious, given the premises.

Pohjola (1981) modified the model in order to use the one-dimensional setup for maps. Beside the translation of the basic relations into a discrete-time version, he modified the wage-bargaining rule assuming that the level of employment determines the wage level and not its rate of change. In our notation, this reads:

$$u_t = -\gamma + \rho v_t \tag{2.8}$$

and the state of the model is fully described by the single variable v_t, the employment ratio. The dynamical system is now:

$$v_{t+1} = v_t\left[1 + A\left\{1 - \frac{v_t}{v^*}\right\}\right] \tag{2.9}$$

where

$$A = \frac{1 - \sigma(\beta + \alpha + \alpha\beta) + \gamma}{\sigma(1 + \beta + \alpha + \alpha\beta)}$$

and

$$v^* = \frac{1 - \sigma(\beta + \alpha + \alpha\beta) + \gamma}{\rho}\,.$$

A simple change of variable will transform Eq. (2.9) in the quadratic map $x_t = a x_{t-1}(1 - x_{t-1})$, which has well-known chaotic properties.

2.3 DESCRIPTIVE GROWTH MODELS

Most of the research effort illustrated so far went unnoticed outside the boundaries of a restricted number of "aficinados." A somewhat wider attention was attracted by Day (1982). His starting point is a capital-accumulation equation of the form:

$$k_{t+1} = \frac{s(k_t) \cdot f(k_t)}{1+\lambda} \tag{2.10}$$

where s is the saving function, f the production function, and $\lambda > 0$ is the exogenous population's growth rate. This is a discrete-time version of the famous Solow's growth model. The latter had used a continuous-time specification to show that under neoclassical assumptions, any capital accumulation path will converge to a steady-state position. Day, on the reverse, exploits the well-known instability of unimodal maps to provide examples of chaotic behavior within that very same framework. As the model is not an optimizing one, i.e., the aggregate saving function is not explicitly derived from considerations of intertemporal efficiency, the author is free to pick "reasonable" shapes for $s(k_t)$ (and $f(k_t)$ obviously) in order to prove his claim. He begins with a constant saving ratio σ and a Cobb-Douglas form for f; Eq. (2.10) then becomes

$$k_{t+1} = \frac{\sigma B k_t^{\beta}}{1+\lambda} \tag{2.11}$$

which is monotonic and therefore stable. By introducing a "pollution effect" in the production function, one obtains

$$k_{t+1} = \frac{\sigma B k_t^{\beta}(m-k_t)^{\gamma}}{1+\lambda} \tag{2.12}$$

which is unimodal and has period-three for certain ranges of parameter values. Returning to the Cobb-Douglas form and allowing instead for a variable saving rate, $s(k) = a(1-b/r)k/y$, he obtains

$$k_{t+1} = \left[\frac{\phi}{1+\lambda}\right] k_t \left[1 - \frac{b}{\beta B}\, k_t^{1-\beta}\right] \tag{2.13}$$

using the fact that the rate of interest must be $r = \beta y/k$. This equation also displays topological chaos for feasible parameter values.

It is obviously very easy to question the empirical validity of this exercise, but this will not eliminate the simple fact that Day showed: small perturbations of well-established models could yield dynamic predictions that go in the direction of chaos. As chaotic dynamics can be displayed by such a simple, basic model, there are no reasons to sustain the claim that it is theoretically irrelevant.

Growth models appear to be particularly well suited to provide examples of economic chaos. Day himself has worked out the chaotic properties of certain formalizations of classical (Malthus) theories of economic growth (see Day, 1983, and

Baduri and Harris, 1987, for the Ricardian system). Stutzer (1980), in turn, had already considered the growth model of Haavelmo and translated it in a discrete-time version that, again, was homeomorphic to the quadratic map. The original model he starts from is a simple one-dimensional differential equation with a globally stable steady state. Once translated into discrete time, the very same system becomes chaotic. This fact occurs often in many of the first works on chaos in economics and casts some doubt on the relevance of the findings. Indeed, one would like the qualitative results of a model to be invariant with respect to changes from continuous to discrete time. If this does not happen, then we are entitled to question the appropriateness of the chosen formalism as well as the relevance of the result. We may claim, in fact, that the change in time units introduces hidden assumptions into the model (lags, for example) and that these should be properly clarified. The issue has been scarcely considered by the theorists working in the field and it does not seem easily solvable. In any case, it is true that we have to look with some suspicion at those results that depend almost entirely on the discreteness of the time representation and that cannot be replicated in continuous time. Because one- and two-dimensional autonomous vector fields cannot produce chaos, this would imply that a "natural translation" of the low-dimensional map into a higher dimensional flow should be possible and should also preserve the qualitative predictions of those models we want to consider seriously as an explanation for dynamic economic complexities.

A second, more direct, criticism can be used against most of the results presented so far: they are derived at an aggregated, macroeconomic level, assuming some kind of competitive behavior on the part of the market participants, but without ever spelling out the kind of objectives these agents are pursuing and the set of constraints within which they are bound to operate. In short, all of the previous macromodels lack a microeconomic foundation in terms of explicit intertemporal optimizing behavior. Clearly, one can always reject such a criticism as irrelevant by claiming that actual economic agents do not, in fact, maximize consistently over time and follow instead adaptive paths and rules-of-thumb in making their decisions. Such a position has been taken, among others, by Richard Day on more than one occasion (see Day, 1986, for all of them), and it solves the problem by removing it to a different level of analysis.

As a matter of fact, the "rationality assumption" is instead taken very seriously, for one reason or another, by the majority of economists (myself included) and is therefore worth some investigation. The criticism will turn out to be wrong but it possesses, indeed, a strong intuitive basis. Economic agents (either consumer-workers or firms) are typically assumed to maximize concave objective functions over convex feasible sets. As the concavity refers to variables indexed by time, this would suggest that cycles (and *a fortiori*, chaos) should not be optimal. Averaging over cycles will exploit the concavity of the function and therefore increase the value achieved. One may conjecture from this reasoning, and it will become clearer from the discussion of Section 4 that the critical assumption behind the argument is not rationality per se, but rationality and concavity together.

In any case, the rationality-based criticism is widely spread among economists even if very seldom it has been expressed in written form. In one case (Dechert, 1984), it appeared to be especially strong and well grounded, particularly because it was independent from the concavity assumption. Dechert's argument goes as follows: pick, for example, Day's version of the Solow growth model and ask if the saving function he is using in his example could be determined, everything else being equal, as the solution to a representative-agent, infinite-horizon maximization problem. The answer is negative. More formally, let $y_t = f(k_t)$ be total output at time t, as a function of the existing stock of capital. The consumer-producer chooses how to split it between consumption and future capital in order to maximize $\sum_{t=0}^{\infty} u(c_t)\delta^t$, where u is a concave utility function, δ is a time-discount factor, $\delta \in (0,1)$, and k_0 is given as an initial condition. It turns out that, even if the production function is not concave, the optimal program $(k_0, k_1, k_2, \ldots)$ can be expressed by a policy function $k_{t+1} = \tau(k_t)$ which is monotonic. The dynamical system induced in this way cannot therefore produce cycles or chaos. The economic prediction is that such a society will asymptotically converge to some stationary position. The latter is unique when f is concave. From this we have to conclude that the chaotic examples derived from a one-sector growth model would not pass the rationality critique. Such a critique turns out to be rather special itself, as it holds true only for the special version of the one-sector growth model considered above. This will be illustrated in the next two sections.

2.4 MISCELLANEA

A variety of different economic models have been considered in recent years in order to show that they could admit, under reasonable hypotheses, chaotic outcomes. Albin (1987) considers a disaggregated model of firms' interaction that gives rise, at the aggregated level, to the behavior considered in Day (1982); many interesting simulations are provided. Baumol and Benhabib (1987) contains an introductory survey to chaos in economics, while Baumol and Wolff (1983) prove chaos in a simple model of research and development. Benhabib and Day (1981) is an extremely interesting paper in which an axiomatic approach is taken to dynamic consumer behavior: a set of conditions is imposed on preferences and the way they depend on experience in order to produce chaotic consumption paths in an environment where income and prices are fixed. Deneckere and Judd (1986) prove that the innovation dynamic may be chaotic under certain patent rules. Jensen and Urban (1982) show chaos for the old cobweb dynamics.

In a very nice paper, Rand (1978) first introduced these arguments in dynamic games by showing how a natural duopolistic interaction (of the Cournot-Nash type) could lead to chaos. In Dana-Montrucchio (1986) another class of infinite-horizon repeated games is considered: it is shown that chaos and almost everything else are possible Markov perfect equilibria.

Finally, let us recall the collective volume edited by Jean Michel Grandmont (1987) which contains some of the most significant recent papers in this area; these were presented at a conference held in Paris in June 1985 and first published in the October 1986 issue of the *Journal of Economic Theory.* We will survey some of these works in the following sections, but a direct look at the whole volume may help the interested reader to put the material in a more proper perspective.

3. OVERLAPPING GENERATIONS MODELS

To overcome the "lack of rationality" critique, we need to place our economic agents in an environment where they have to make a concrete intertemporal consumption-investment choice in order to achieve a well-specified objective in the face of given or expected prices and resource constraints. The simplest framework is provided by a model of overlapping generations with constant population and a representative agent per generation and an exogenously specified endowment stream of the consumption good. Let's indicate with a superscript y those variables pertaining to the youngs and with o those for olds, let $t = 0, 1, 2, \ldots$ indicate calendar time. Preferences are represented by a utility function $U(c_t^y, c_{t+1}^o)$, where c_t^y is consumption when young of an agent born at t, and c_{t+1}^o is the same agent's consumption when old. Finally, let (w^y, w^o) denote the time-invariant endowment pair and p_t the price of the homogeneous good at time t, so that $\rho_t = p_t / p_{t+1}$ is the interest factor at time t. The representative agent will maximize his lifetime utility $U(c_t^y, c_{t+1}^o)$ under the budget constraint:

$$c_{t+1}^o = w^o + \rho_t \left[w^y - c_t^y \right] \tag{3.1}$$

Standard concavity assumption will give two utility maximizing consumption demands: an intertemporal competitive equilibrium will then be a sequence of vectors (ρ_t, c_t^y, c_t^o) such that utility of each generation is maximized under Eq. (3.1) and the material balance constraint:

$$\left[w^y - c_t^y \right] + \left[w^o - c_t^o \right] = 0 \tag{3.2}$$

is also satisfied.

Clearly the youngs can either save or borrow and therefore they may carry claims or debts into the second period. Assume this is done by means of a universally accepted paper asset called money (or checking account). Following Gale (1973), let's call "classical" the case in which the young are impatient and borrow, and "Samuelson" the opposite one. Which state will occur clearly depends both on the shape of the utility function U and the relative magnitudes of w^y and w^o. Note also that the no-exchange (and no-money) equilibrium is always a possible outcome and it is such that if it obtains in the first period, it will be replicated forever, as our economy is time-invariant in its fundamentals. Gale also showed that, under the natural dynamics we will introduce in a moment, such autarkic equilibrium is locally unstable in the classical cases and locally stable in the Samuelson economies.

Benhabib and Day (1982) were the first to consider these economies from the point of view of nonlinear analysis (even if Gale has already pointed out the possibility of cycles). They studied the dynamics of the classical case (in an earlier paper, Benhabib and Day, 1980, they had used the overlapping generations model with capital and production to obtain chaos, but that result was critically dependent upon a rather questionable use of a future-utility discount factor varying positively with wealth). I will briefly summarize their analysis in the following pages.

Assume all solutions are interior. Using the first-order necessary and sufficient conditions for utility maximization together with the budget constraint, one obtains the equality:

$$\frac{U_1\left(c_t^y, c_{t+1}^o\right)}{U_2\left(c_t^y, c_{t+1}^o\right)} = \frac{w^o - c_{t+1}^o}{c_t^y - w^y} \tag{3.3}$$

where U_1 and U_2 are the partial derivatives of U. Under regularity assumptions, Eq. (3.3) can be solved uniquely for c_{t+1}^o, call this function

$$c_{t+1}^o = G\left(c_t^y; w^y, w^o\right). \tag{3.4}$$

Now use Eq. (3.4) to eliminate c_{t+1}^o from the left-hand side of Eq. (3.3) which, in equilibrium, must be equal to ρ_t. Let's call this newly obtained ratio the constrained marginal rate of substitution (CMRS), it will be a function of c_t^y only and of the parameters w^y, w^o. Denote it by $V(c_t^y; w^y, w^o)$. Finally, use the latter together with the material balance of Eq. (3.2) to obtain a first-order difference equation in the youngs' consumption levels:

$$c_{t+1}^y = w^y + V\left(c_t^y; w^y, w^o\right)\left(c_t^y - w^y\right) \equiv f\left(c_t^y\right). \tag{3.5}$$

The problem is now that of providing conditions for $f(c_t^y)$ to be unimodal and with the degree of steepness sufficient to produce chaotic trajectories. The authors looked for chaos in the "topological" sense (i.e., existence of a period-three orbit), but in fact provided examples of utility functions and endowment pairs for which also the stronger form of chaos (i.e., existence of an invariant, absolutely continuous and ergodic measure) can be obtained.

Naturally such sufficient conditions depend on U, through the CMRS, and amount to saying that $V(c_t^y)$ can vary sufficiently over the interval $I = (w^y, w^y + w^o)$. These conditions are: there exists a $\hat{c} > w^y$ such that:

$$\alpha_1 = V(\hat{c}) > 1 \qquad (\text{resp.} < 1) \tag{3.6a}$$

$$\alpha_2 = V\big(\alpha_1\hat{c} + (1-\alpha_1)w^y\big) > 1 \qquad (\text{resp.} < 1) \tag{3.6b}$$

$$0 < \alpha_3 = \alpha_1\alpha_2 V\big(\alpha_1\alpha_2\hat{c} + (1-\alpha_1\alpha_2)w^y\big) \le 1 \quad (\text{resp.} \ge 1) \tag{3.6c}$$

Under Eqs. (3.6a)–(3.6c), "topological chaos" will occur for the dynamical system in Eq. (3.5).

Benhabib and Day considered other relevant economic issues pertaining to the model, such as the role of a central authority regulating the credit used by the young and the Pareto efficiency of the chaotic trajectories (they may well be such, under very general conditions). In a brief remark, they also addressed the Samuelson case, pointing out that, for the case in which cyclic or chaotic trajectories could obtain, the dynamical system of Eq. (3.5) would not be well defined, in the sense that for each c_t^y there will exist two equilibrium levels of c_{t+1}^y. In the absence of a convincing selection criterion, they saw no purpose of analyzing such a system.

This very same case (the Samuelson one) was instead taken up and worked out in all its details in a paper by Grandmont (1985). The length of this article prevents a description of all the results there obtained. I will content myself with a brief discussion of the basic technique used by the author to define a meaningful dynamical system for the Samuelson case and to prove that it can be chaotic.

The basic model is, as before, one with overlapping generations. Assume the utility function is time separable and, instead of a fixed endowment of consumption good, assume each agent has a labor-time endowment $\overline{\ell}^i$, where $i = y, o$, in each period of his life. Denote with ℓ^i the amount of $\overline{\ell}^i$ he supplies for work and assume his utility depends both on consumption c^i, where $i = y, o$, and leisure time $\overline{\ell}^i - \ell^i$, according to $U_1^y(c^y, \overline{\ell}^y - \ell^y) + U^o(c^o, \overline{\ell}^o - \ell^o)$, where the utility function U^i satisfy, for $i = y, o$, standard differentiability monotonicity and strict concavity hypotheses. As we want to consider the case in which the young lends to the old in exchange for "money," let us introduce the latter explicitly in a fixed amount M. The representative individual once again maximizes his utility subject to the budget constraints:

$$p_t\,(c_t^y - \ell_t^y) + m^d = 0 \tag{3.7a}$$

$$p_{t+1}^e\left(c_{t+1}^o - \ell_{t+1}^o\right) = m^d \tag{3.7b}$$

Here m^d denotes the nominal amount of money demanded by the young. Notice that we assume the technology to be such that a unit of labor is transformed into a unit of consumption good so that we are still facing a pure exchange economy. Note also that, for now, the perfect foresight assumption about future prices, used by Benhabib and Day has not been made and p_{t+1}^e denotes the expected future price as of time t. Such a maximization problem has, once again, a unique solution which will depend only on $\rho_t^e = p_t/p_{t+1}^e$, the expected interest factor. By modifying slightly the notation used previously, we can define an excess demand for the good $z^i(\rho^e)$, where $i = y, o$, as $z^i(\rho^e) = c^i - \ell^i$. Remember that we are considering the case in which the young lends to the old in exchange for money. This implies that the $z^y(\rho^e)$ will always be negative and such that $m^d = M = -p_t z^y(\rho_t^e) = p_{t+1}^e z^o(\rho_t^e)$ at each t, along an equilibrium path. Conversely, when old, each agent will spend all of his money stock in exchange for goods. In the equality above, M denotes the fixed amount of existing bills that must all be demanded by the young in equilibrium.

Assume now that agents have perfect foresight, i.e., $p^e_{t+1} = p_{t+1}$. From the equilibrium conditions given above and the material balance, it follows that a competitive equilibrium is a sequence of p_t that solves

$$z^y(\rho_t) + z^o(\rho_{t-1}) = 0 \tag{3.8a}$$

$$p_{t+1} z^o(\rho_t) = M. \tag{3.8b}$$

Notice that what matters for the dynamic is Eq. (3.8a); once the sequence of ρ_t is determined we will get the price level from Eq. (3.8b) as a function of the given M. Our system satisfies the Quantity Theory of Money, the latter is no obstacle to chaos. If we try to invert z^y to obtain a "forward dynamics" (i.e., ρ_t as a function of ρ_{t-1}), we get into the problem recalled above as z^y may be backward bending. This is the crucial feature of the model as well as the main source of the aperiodic behavior. An increase in ρ_t has two conflicting effects on the demand for consumption by the young. It has an intertemporal substitution effect as it makes consumption more expensive today (hence, z^y should decrease), but it also makes the agent richer (wealth effect) as today's labor is paid more. This will tend to increase his demand for present consumption (hence, z^y should increase). But this is not the case for z^o. A little thought along the same lines will convince the reader that when ρ_{t-1} goes up, both substitution and wealth effect will push up the old agent's demand for consumption. Therefore z^o may be inverted and a (fictitious) backward dynamics can be obtained from Eq. (3.8a):

$$\rho_{t-1} = \phi(\rho_t) \equiv (z^o)^{-1}(-z^y(\rho_t)). \tag{3.9}$$

Even if the "trick" behind Eq. (3.9) is not quite the full solution to our problem, it suggests the line along which a perfect foresight dynamics can be studied for an economy of the Samuelson type. Conscious of this fact, Grandmont dedicates a large part of his paper to clarify the relation between the backward and the forward dynamics, as well as to work out the implications that different expectation-formation rules have for the stability of the system (on this point, see also Grandmont and Laroque, 1986). We have to skip all this for reasons of brevity. What matters to us is that, given a periodic trajectory for the backward dynamics, one may generically define a forward dynamics in a neighborhood of such trajectory so that a stability analysis can be conducted by reversing the dynamic properties of the backward paths. By following this strategy the author gives conditions under which Eq. (3.9) defines a dynamical system as an iterated map of an interval into itself and carries on a complete bifurcation analysis of such system. He makes abundant use of the techniques and results illustrated in Collet and Eckmann (1980) in order to show that a period-doubling bifurcation cascade leading to chaos will originate for a large class of utility functions. In particular, it turns out that when z^y is not monotonic (for the reasons given above), then a large enough degree of risk aversion on the part of the old trader (i.e., a "very concave" U^o) will lead to chaos if certain technical conditions are satisfied.

A more detailed consideration of his results would show that they confirm and generalize the earlier proof given by Benhabib and Day for the classical case. Chaos originates out of a conflict between the wealth and intertemporal substitution effects created by a variation in the real interest rate if the first effect is strong enough. Finally, it is worth noting that, contrary to many of the papers considered in this and the next section, such paths are not necessarily Pareto efficient and the cycles may be dampened (or created) by appropriate monetary and fiscal policies.

It is not exaggerating to say that Grandmont's paper has had a much bigger impact on the economics profession than any other of the previous (and, for that matter, subsequent) works on chaos in economics. It is after this work that macroeconomists and economic theorists in general have started to realize that, indeed, there may be something in an endogenous theory of business cycles that cannot be captured by the prevailing linear-stochastic approach.

This new attention has made it possible to better reconsider some of the former studies I have been illustrating here as well as provide incentives for new research on different models. As far as the overlapping generations economies are concerned, some recent efforts have generalized and improved upon the older results. Reasons of space suggest brevity; therefore, I will only sketch some of these discoveries. Farmer (1986) has considered a variation of the basic model where capital is introduced both as a means of production and as an asset: it is proven that, when government debt is present to finance a deficit of fixed value, periodic orbits may be obtained for the two-dimensional discrete time system that describes the economy's evolution. The technique used here is that of Hopf bifurcation for maps on the plane. The role of production is more fully analyzed in Reichlin (1986; see also Reichlin, 1987, for further improvements along the same line). In particular, the author is able to show that, when a nontrivial technology is present, one does not need the empirically unlikely assumption made by Grandmont which requires saving to be a decreasing function of the interest rate when the latter is high enough (strong wealth effect) in order to obtain complicated dynamic behaviors. In fact, by means of simple production functions (either fixed coefficients or CES) that use labor and the invested amount of the homogeneous good to produce new output, Reichlin obtains a dynamical system for the capital stock which is represented by a map of the plane into itself. He also uses the Hopf theorem to prove the existence of a limit cycle. The result is obtained even if saving is a monotonically increasing function of the rate of interest, as long as the elasticity of substitution between factors of production is low enough. This is consistent with a result obtained by Boldrin (1986; see Section 4 below) for an economy with infinitely living agents, where an example with either a CES or a Leontief production function exhibits chaos when the elasticity of substitution for the CES is low, even if agents do not have a high discount factor for future utilities. In Reichlin (1987), the same overlapping generations economy is considered with a two-sector technology. In this case, the author is able to show the existence of chaotic trajectories. Finally, Aiyagari (1987) proves the existence of periodic orbits for an exchange economy with overlapping generations that do not live only for two periods but for finitely many ones.

4. ECONOMIES WITH A FINITE NUMBER OF INFINITELY LIVED AGENTS

There are many reasons for which one may feel unsatisfied with the type of world described in an OLG model. From the dynamic point of view, which concerns us here, these models appear rather farfetched. Each time period has to be interpreted, empirically, as equivalent to 30–40 years which makes it impossible to define observable counterparts for the variables of the model. On the other hand, each agent behaves very myopically as nobody cares for the consequences of his actions more than one period ahead. This seems intuitively at odds with the existence of many institutions (firms, *in primis*) that participate in the markets for very long periods of time and that should therefore try to forecast the implications of their choices for the far future. Finally, it is reasonable to claim that a "good government" should be one which takes into account the interests of all the generations, even of the unborn, in pursuing its policies and this fact should lead to programming over infinite horizons of time. This was, indeed, Frank Ramsey's concern when, in the 1920's, he first proposed to consider optimal programs that maximize an infinite sum of society's welfares from the initial period up to infinity.

To make a long story short, such basic intuitions have led many scholars to consider the behavior of competitive economies where a finite number of agents live forever and, being endowed with perfect foresight, try to maximize the discounted sum of their utilities over the infinite horizon. The literature on this field is enormous; the curious reader is referred to Arrow and Kurz (1970), Cass and Shell (1976), Bewley (1982) and especially McKenzie (1986, 1987) for more complete treatments. For our purposes it suffices to sketch here the basic ingredients of a very general model from which most of the adopted setups can be derived as special cases. In particular, we will consider a world with a single (representative) agent that controls both consumption and production decisions and perfectly foresees even the more distant future (see Bewley, 1982, and the literature therein for a reconciliation of this abstraction with the case of many independent consumers and producers). Also we will describe only the discrete time formalism even if, later on, we will have to use the continuous-time version of the same model; the translation should be immediate.

In every period $t = 0, 1, 2, \ldots$, an agent derives satisfaction from a "consumption" vector $c_t \in \mathbb{R}^m$, according to a utility function $u(c_t)$ which is taken increasing, concave and smooth as needed. Notice that c_t denotes a flow of goods that are consumed in period t. The state of the world is fully described by a vector $x_t \in \mathbb{R}^n_+$ of stocks and by a feasible set $F \subset \mathbb{R}^{2n}_+ \times \mathbb{R}^m$ composed of all the triples of today's stocks, today's consumptions and tomorrow's stocks that are technologically compatible, i.e., a point in F has the form (x_t, c_t, x_{t+1}). Now define

$$V(x, y) = \max_c u(c) \quad \text{s.t. } (x, c, y) \in F \tag{4.1}$$

and let $D \subset \mathbf{R}_+^{2n}$ be the projection of F along the c's coordinates. Then V, which is called the short-run or instantaneous return function, will give the maximum utility achievable at time t if the state is x and we have chosen to go into state y by tomorrow. It should be easy to see that to maximize the discounted sum $\sum_{t=0}^{\infty} u(c_t)\delta^t$ s.t. $(x_t, c_t, x_{t+1}) \in F$ is equivalent to $\max \sum_{t=0}^{\infty} V(x_t, x_{t+1})\delta^t$ s.t. $(x_t, x_{t+1}) \in D$.

The parameter δ indicates the rate at which future utilities are discounted from today's standpoint (impatience): it takes values in [0,1). For $\delta = 0$ the agent is infinitely impatient and there is a sense in which a repeated myopic optimization of this kind may represent the outcomes of an OLG model. In general δ will be greater than zero.

It is mathematically simpler to consider the problem in the latter (reduced) form. The following assumptions on V and D may be derived from the more basic hypotheses on u and F:

ASSUMPTION 1. $V : D \to \mathbf{R}$ is strictly concave and smooth (if needed). $V(x, y)$ is increasing in x and decreasing in y.

ASSUMPTION 2. $D \subset X \times X \subset \mathbf{R}_+^{2n}$ is convex, compact and with non-empty interior. X is also convex, compact and with non-empty interior.

The initial state x_0 is given. Notice that the economy we are describing is essentially time-invariant: return function and feasible set do not change over time, the latter enters the picture only through discounting, and the intrinsically intertemporal nature of the production process is summarized by D (see McKenzie, 1986, for the case in which V and D evolve exogenously with time in a fairly restricted way; very little can be said about this case).

The optimization problem we are facing can be equivalently described as one of Dynamic Programming:

$$W(x) = \max \{V(x, y) + \delta W(y), \quad \text{s.t. } (x, y) \in D\} \tag{4.2}$$

The latter is the Bellman equation and $W(x)$ is the value function for such a problem. A solution to Eq. (4.2) will be a map $\tau_\delta : X \to X$ describing the optimal sequence of states $\{x_0, x_1, x_2, \ldots\}$ as a dynamical system $x_{t+1} = \tau_\delta(x_t)$ on X. The time evolution described by τ_δ contains all the relevant information about the dynamic behavior of our model economy. In particular, the price vectors p_t of the stocks x_t that realize the optimal program as a competitive equilibrium over time follows a dynamic process that (when the solution $\{x_t\}$ is interior to X) is homeomorphic to the one for the stocks. In other words, $p_{t+1} = \theta(p_t)$ with $\theta = \delta DW \cdot \tau \cdot (DW\delta)^{-1}$, where D is the derivative operator.

The question that concerns us is: what are the predictions of the theory about the asymptotic behavior of the dynamical system τ_δ? Where should a stationary economy converge under Competitive Equilibrium and Perfect Foresight? A first, remarkable answer is given by the following:

TURNPIKE THEOREM. Under Assumptions 1 and 2, there exists a level $\overline{\delta}$ of the discount factor such that for all the δ's in the non-empty interval $[\overline{\delta}, 1)$, the function τ_δ that solves Eq. (4.2) has a unique globally attractive fixed point $x^* = \tau_\delta(x^*)$. Such an x^* is also interior to X under additional mild restrictions.

Not too bad indeed. Under a set of hypotheses as general as Assumptions 1 and 2, we are able to predict that if people are not "too impatient" relative to the given V and D, then they should move toward a stationary state where history repeats itself indefinitely and no surprises ever arise. In the form given here, the Turnpike Theorem is due to Scheinkman (1976), whereas McKenzie (1976) and Rockafellar (1976) proved it for the continuous-time version; Bewley (1982) and Yano (1984) generalized it to the many-agents case (but see McKenzie, 1986, for a more careful attribution of credits).

As remarkable as it is, the Turnpike property is also very sensitive to perturbations of its sufficient conditions. In particular, how close should $\overline{\delta}$ be to one in order to obtain convergence and what happens when δ is smaller than $\overline{\delta}$? These are important questions. It is hard to rely heavily on a property that may depend critically on such a volatile and unobservable factor as "society's average degree of impatience."

The careful reader should have realized by now that the one-sector model, we briefly introduced at the end of Section 2, and used by Dechert to prove that cycles and chaos are not optimal in that framework, is a special case of the general model we are considering here, with $V(x_t, x_{t+1}) = u[f(x_t) - x_{t+1}]$ and $D = \{(x_t, x_{t+1}) \text{ s.t. } 0 \leq x_{t+1} \leq f(x_t)\}$. For that model, the Turnpike Theorem holds independently of the discount factor as τ_δ is always montonically increasing. Unfortunately, such a nice feature does not persist even if the simplest generalization of the one-sector model is taken into account. This was proved by Benhabib and Nishimura (1985). They considered a model with two goods—consumption and capital—which are produced by two different sectors by means of capital and labor. Given the two production functions, one can define a Production Possibility Frontier (PPF) $T(x_t, x_{t+1}) = c_t$, that gives the producible amount of consumption when the aggregated capital stock is x_t (a scalar), labor is efficiently and fully employed, and the decision of having an aggregated stock x_{t+1} tomorrow has been taken. The return function is now $V(x_t, x_{t+1}) = u[T(x_t, x_{t+1})]$ and $D = \{(x_t, x_{t+1}) \text{ s.t. } 0 \leq x_{t+1} \leq F(x_t, 1)\}$ where F is the production function of the capital good sector and labor has been normalized to one. In such a case, τ_δ is not always sloping upward. If the consumption sector uses a capital/labor ratio higher than the one used by the capital sector, it will slope downwards. Let x^* be the (unique) interior fixed point (i.e., $\tau_\delta(x^*) = x^*$). This is the candidate for the Turnpike. Assume, for simplicity, that τ_δ is differentiable in a neighborhood of x^*. The derivative will be $\tau'_\delta(x^*)$ at the steady state, it is negative, and it changes as δ moves in (0,1), everything else being equal. Benhabib and Nishimura showed that it may take on the value -1 for admissible δ's, in such a way that the conditions for a flip (period-doubling) bifurcation are realized. In this case, an optimal cycle of period-two will exist which can also be attractive: no more Turnpike! One may provide examples of this phenomenon showing that such an outcome is by no means due to "pathological" technologies and preferences.

Not only this, cycles are not a special feature of the discrete-time version of our model. In fact, in a much earlier work (see Benhabib and Nishimura, 1979), the two authors had used the Hopf bifurcation theorem to prove that limit cycles can occur in the continuous-time case. Let us show very briefly how this can happen. In continuous time, we face an optimal control problem of the form:

$$\max \int_0^\infty V(x, \dot{x}) \exp(-\rho t) \quad \text{s.t. } (x, \dot{x}) \in D, \quad x(0) \text{ given.} \tag{4.3}$$

Here $x(t)$ is a vector depending on time, $\dot{x}$ is its time derivative, D again the convex feasible set, and ρ the discount factor in $[0, \infty)$ ($\rho = 0$ is equivalent to $\delta = 1$ in discrete time). Using the Maximum Principle, one defines a Hamiltonian:

$$H(x, q) = \max_{\dot{x}} \{V(x, \dot{x}) + <q, \dot{x}>, \quad \text{s.t. } (x, \dot{x}) \in D\} \tag{4.4}$$

which can be interpreted as the current value of national income evaluated at the (shadow) prices q (on this point see Cass and Shell, 1976).

The dynamical system is then

$$\begin{aligned} \dot{x} &= \frac{\partial H(x,q)}{\partial q} \\ \dot{q} &= \frac{-\partial H(x,q)}{\partial x} + \rho q\,. \end{aligned} \tag{4.5}$$

Linearization of Eq. (4.5) around the steady state will yield, after some manipulations, a Jacobian matrix J that can be written as $J = \tilde{J} + (\rho/2)I$, where I is the $2n \times 2n$ identity matrix. As $\tilde{J}$ is a Hamiltonian matrix, we may consider how its eigenvalues will change with the discount factor ρ and then add $\rho/2$ to obtain those of J. If $\rho = 0$, $\tilde{J}$ has the form

$$\tilde{J} = \begin{pmatrix} A & B \\ C & -A^T \end{pmatrix} \tag{4.6}$$

with $A = \partial H(x,q)/\partial x \partial q$, $B = \partial^2 H(x,q)/\partial^2 q$, and $C = -\partial^2 H(x,q)/\partial^2 x$. It is a result of Rockafellar (1973) that under strict concavity in x and strict convexity in q of H, the $2n$ eigenvalues of $\tilde{J}$ will split into n positive and n negative ones. The steady state will be a saddle point with a stable manifold of dimension n. As the latter is also the dimension of the control vector, the optimal program will steer the system on the stable manifold, thereby guaranteeing convergence to the Turnpike. For $\rho > 0$ this is not necessarily true: the saddle-point property may be lost as some of the negative eigenvalues become positive. The Turnpike Theorems give conditions under which such stability property is preserved for small ρ. But as Benhabib and Nishimura showed when ρ grows, a pair (or more than a pair) of eigenvalues may change the sign of their real part by crossing the imaginary axis. In such a case, they proved that (care taken for the technical details) a Hopf bifurcation is realized. The limit cycle associated with it may indeed be an attractor for the system of Eq. (4.5). Once again the Turnpike property is lost as people become "a bit more impatient" than the economists would like!

Some characteristics of the oscillatory paths so obtained need to be stressed. First of all, they are realized as "equilibrium paths," in the sense that all markets are continuously clearing at each point in time, prices adjust completely and no productive resource is left "involuntarily unemployed." Moreover, they are Pareto efficient in the sense that it is impossible to modify the allocation of resources they imply, in order to increase the welfare of some agent without making somebody else worse off. Let us make it clear that none of the authors working along these lines seem to imply, with this, that economic fluctuations are intrinsically good and that nothing can or should be done to modify and control them. The idea instead is of showing that there exist forces that are intrinsic to the competitive mechanism, and depend upon the technological structure of the economy, that can be a source of wide oscillations for output and prices. As it was the case with the discrete-time two-sector model, it is the existence of certain factor-intensity relations across sectors that make it profitable for the producers (and the consumers alike) to invest, produce (and consume) in an oscillatory form. Even if all the prices are the "right ones" (i.e., no conditions for profitable arbitrage exist), still the pure seeking of individual profits will bring about cyclic behavior.

It is opportune to admit that we do not presently have an empirical representation for this phenomena and that our ability to measure how intersectoral profitability relations affect the cycle is very small or almost nil. Nevertheless, they follow from sound economic theory and it is hard to rule them out on pure *a priori* grounds. As we will see in a moment, this very same logic can be pursued further to explain the origin of chaotic movements for the same class of model economies.

Indeed, the possibility of more complicated trajectories had already been envisaged by Benhabib and Nishimura (1979, p. 433), where they quote Ruelle and Takens' celebrated paper on turbulence, after noticing that further bifurcations may follow the Hopf one, giving rise to a torus, etc. It nevertheless took a few years before a full proof was produced, and when it came, it was more general than expected. Every dynamics turned out to be a possible solution for an economy satisfying Assumptions 1 and 2 above. This was proven in Boldrin and Montrucchio (1986b; but see also Boldrin and Monrucchio, 1984 and 1986a, and Montrucchio, 1986, for additional results). The result, formally speaking, has the following form: let $\theta : X \to X$ be a C^2-map describing a dynamical system on the compact convex set $X \subset \mathbf{R}^n$; then there exist a technological set D, a return function V, and a discount factor $\delta \in (0, 1)$ satisfying Assumptions 1 and 2, such that θ is the policy function τ_δ that solves Eq. (4.2) for the given D, V and δ. The proof given was a constructive one, so that one may effectively compute a fictitious economy for each given dynamics. This clarifies that any kind of strange dynamic behavior is fully compatible with competitive markets, perfect foresight, decreasing returns, etc. At about the same time, Deneckere and Pelikan (1986) also presented some one-dimensional examples of models satisfying our assumptions and having the quadratic map $4x(1-x)$ as their optimal policy function.

All these results were given for the discrete-time version of the model, but they are not specific to it. In Montrucchio (1987), it in fact has been proven that exactly the same results hold for the case in which time is continuous. One big question that still remains open in this area pertains to the economic logic behind these theoretical and mathematical results. What is it that makes it profitable for a competitive economy to oscillate erratically over time? As we noticed with respect to the works of Benhabib and Nishimura on limit cycles, the driving force seems to be the technological structure of the different sectors. A very similar answer is true for the chaotic motions.

Unfortunately, we do not have a full-fledged analytical explanation for the multisectoral case, but something can be said for the two-sector, two-good economy that is often used in macroeconomic applications. A theoretical analysis is provided in Boldrin (1986). There are two goods—consumption and capital—produced by means of two factors—capital itself and labor. The model is therefore the same as in Benhabib and Nishimura (1985) and the resulting dynamics in the aggregate capital stock k_t is one-dimensional. It is shown that the policy function $k_{t+1} = \tau_\delta(k_t)$ is unimodal when factor-intensity reversal occurs between the two sectors. Remember that for the case in which the consumption sector always uses a capital-labor ratio higher than the one of the capital sector, period-two cycles are possible. More often than not, it is possible to find a level, say, k^*, of the aggregate capital stock such that when k_t is in $[0, k^*)$, the capital sector has a higher capital-labor ratio, whereas the opposite happens when k_t is in $(k^*, \bar{k}]$, where $\bar{k}$ is the maximum level of capital that the economy can sustain. This technological feature provides the unimodal shape for τ_δ. Variations in the level of the discount factor δ then can produce a cascade of period-doubling bifurcations that (technicalities aside) leads to period-three orbits and even to chaos in the sense of the existence of an invariant and absolutely continuous ergodic measure.

A simple example that uses standard production functions is also provided in the same paper. The problem is taken up again in Boldrin and Deneckere (1987). The case of a two-sector economy with a Cobb-Douglas and a Leontief production function is here studied in detail. That such an economy would have chaotic trajectories was first conjectured by José Scheinkman in Scheinkman (1984). Boldrin and Deneckere provide a full proof for this assertion and also show how, depending on the various parameters, cycles of different lengths can be originated.

A by-product of these exercises is to make clear that the often-adopted criticism that asserts the irrelevance of this approach to business cycle theory because of the "too high" level of discounting required for chaos is, indeed, wrong. Clearly a level of δ too close to zero as in the earlier examples would imply annual "interest rates" of the 1000% magnitude. But this need not occur. In particular, the example provided in Boldrin (1986) shows that δ may be very close to one (say, in the range (0.6,0.8)) and chaos may still be possible for appropriate values of the technological parameters.

The open question is this: can we move from the "toy models" stage to a fully specified, disaggregated and empirically calibrated model of the same class that will reproduce the so-called "stylized facts" of observed business cycles? The question is open.

This section would be incomplete without a reference to the innovative work that Mike Woodford has conducted in this area. The best example of it is provided by Woodford (1987; but see other references to the same author therein). What he uses is the one-sector growth model we described at the end of Section 2.3. We recalled there the result of Dechert (1984), according to which only monotonically convergent orbits are possible in such a setup. Woodford asks the simple question: what happens if, for some reason, one of the agent is not free to borrow? That it to say, what if (as in the real world) there are no markets open for trade at very distant dates, or loan markets for financing investments above a certain amount are not available? This intuition is captured by postulating two different types of agents—consumers and entrepreneurs. The latter are in charge of the investment process, but they cannot borrow against future returns. They must finance their investments only out of funds generated internally to the firms. The author provides very good arguments for such a state of affairs to occur (idiosyncratic shocks and private information about returns, for example) and also shows that, in such a case, both equilibrium cycles and chaotic equilibrium dynamics may exist under very general hypotheses about the technology and the level of discounting. The latter, in particular, may be as low as one likes, without affecting the result.

Considering the appealing aspect of the hypothesis on the lack of complete markets, it seems to me that such a line of research appears as one of the most promising, and worth pursuing further.

5. CONCLUSIONS

It is difficult to conclude, when a research topic is as recent, open and chaotic as the one we have surveyed. The real issue is: where do we go from here?

It seems clear that chaotic behaviors are not rare, at least theoretically, in well-formulated economic contests. In fact, they seem to be pervasive of even the simplest, descriptive representations of economic dynamics, as we have tried to show in Section 2.

When the discipline of utility maximization and rational intertemporal choice is imposed on the behavioral rules, they do not disappear at all. The research effort conducted so far has been able to identify classes of economic factors that may explain such persistence: (a) the relative importance of wealth effects as opposed to intertemporal substitution effects; (b) the degree of factor substitutability in production and the different factor-utilization ratios across sectors; (c) the degree of people's impatience and/or the extent to which they behave myopically with respect to future events; and (d) the lack of certain markets, especially of those for borrowing-lending against expected future returns.

It is difficult to establish an order of importance among these elements and they need not exhaust the class of possible explanations. What matters is that they all make sense from the point of view of economic theory; the endogenous approach to economic fluctuations appears therefore well grounded within the established General Equilibrium paradigm.

How far we go with this, on the practical side, is not clear. We are still in the stage of very abstract, purely qualitative models: their predictions are so general and so vague that any hope of testing them directly will be easily frustrated. On the other side, the empirically oriented works surveyed by Brock in his contribution to this volume suggest that there are important results we may achieve along those lines.

What we need, therefore, is to construct models that can be parameterized by using empirical evidence and that can yield testable, even if primitive, predictions by means of computer simulations. I believe it would be most useful, at the present stage, to channel our research efforts along these lines.

ACKNOWLEDGMENTS

This note has been prepared for the proceedings of the Santa Fe Institute Workshop on "Evolutionary Paths of the Global Economy," held at the Santa Fe Institute, Santa Fe, New Mexico on September 8–17, 1987. The author is very grateful to Professor Kenneth Arrow for the kind invitation to participate and to the Institute for the warm hospitality provided.

REFERENCES

1. Aiyagari, S. R. (1987), "Stationary Deterministic Cycles in a Class of Overlapping Generations Models with Long Lived Agents," *Working Paper #319, Federal Reserve Bank of Minneapolis.*
2. Albin, P. S. (1987), "Microeconomic Foundations of Cyclical Irregularities or Chaos," *Mathematical Social Sciences* **13**, 185–214.
3. Arrow, K. J., and M. Kurz (1970), *Public Investment, the Rate of Return and Optimal Fiscal Policy* (Baltimore: Johns Hopkins University Press).
4. Baumol, W., and J. Benhabib (1987), "Chaos: Significance, Mechanism and Economic Applications," *R.R. #87-16, C.V. Starr Center for Applied Economics, New York University.*
5. Baumol, W., and R. Quandt (1985), "Chaos Models and Their Implications for Forecasting," *Eastern Economic Journal* **11**, 3–15.
6. Baumol, W., and G. Wolff (1983), "Feedback from Productivity Growth in 'R and D,'" *Scandinavian Journal of Economics* **85**, 145–157.
7. Benhabib, J., and R. Day (1980), "Erratic Accumulation," *Economic Letters* **6**, 113–117.
8. Benhabib, J., and R. Day (1981), "Rational Choice and Erratic Behavior," *Review of Economic Studies* **XLVIII**, 459–471.
9. Benhabib, J., and R. Day (1982), "A Characterization of Erratic Dynamics in the Overlapping Generations Model," *Journal of Economics Dynamics and Control* **4**, 37–55.
10. Benhabib, J., and G. Laroque (1986), "On Competitive Cycles in Productive Economies," *IMSSS Technical Report #490, Stanford University.*
11. Benhabib, J., and K. Nishimura (1979), "The Hopf Bifurcation and the Existence and Stability of Closed Orbits in Multisector Models of Optimal Economic Growth," *Journal of Economic Theory* **21**, 421–444.
12. Benhabib, J., and K. Nishimura (1985), "Competitive Equilibrium Cycles," *Journal of Economic Theory* **35**, 284–306.
13. Bewley, T. (1982), "An Integration of Equilibrium Theory and Turnpike Theory," *Journal of Mathematical Economics* **10**, 233–268.
14. Bhaduri, A., and D. J. Harris (1987), "The Complex Dynamics of the Simple Ricardian System," *Quarterly Journal of Economics* **102**, 893–902.
15. Boldrin, M. (1983), "Applying Bifurcation Theory: Some Simple Results on the Keynesian Business Cycle," *Note di Lavoro #8403, Dip. di Scienze Economiche, Univ. di Venezia.*
16. Boldrin, M. (1986), "Paths of Optimal Accumulation in Two-Sector Models," *IMSSS Tech. Rept. #502, Stanford University*; forthcoming in Barnett, W., J. Geweke and K. Shell, eds., *Economic Complexity: Chaos, Sunspots, Bubbles and Nonlinearity* (Cambridge: Cambridge University Press).
17. Boldrin, M., and R. Deneckere (1987), "Simple Macroeconomic Models with a Very Complicated Behavior," *mimeo, University of Chicago and Northwestern University.*

18. Boldrin, M. and L. Montrucchio (1984), "The Emergence of Dynamic Complexities in Models of Optimal Growth: The Role of Impatience," *Working Paper #7, Rochester Center for Economic Research, University of Rochester.*
19. Boldrin, M. and L. Montrucchio (1986a), "Cyclic and Chaotic Behavior in Intertemporal Optimization Models," *Mathematical Modelling* **8**, 627–700.
20. Boldrin, M. and L. Montrucchio (1986b), "On the Indeterminacy of Capital Accumulation Paths," *Journal of Economic Theory* **40**, 26–39.
21. Brock, W. (1987), "Nonlinearities and Complex Dynamics in Economics and Finance," this volume.
22. Cass, D., and K. Shell, eds. (1976), *The Hamiltonian Approach to Dynamic Economics* (New York: Academic Press).
23. Chang, W. W., and D. J. Smith (1971), "The Existence and Persistence of Cycles in a Nonlinear Model: Kaldor's 1940 Model Re-examined," *Review of Economic Studies* **38**, 37–44.
24. Collet, P., and J. P. Eckmann (1980), *Iterated Maps on the Interval as Dynamical Systems* (Boston: Birkhauser).
25. Cugno, F., and L. Montrucchio (1982a), "Stability and Instability in a Two-Dimensional System: A Mathematical Approach to Kaldor's Theory of the Trade Cycle," *New Quantitative Techniques for Economics*, Ed. G. P. Szego (New York: Academic Press).
26. Cugno, F., and L. Montrucchio (1982b), "Cyclical Growth and Inflation: A Qualitative Approach to Goodwin's Model with Money Prices," *Economic Notes* **3**, 93–107.
27. Cugno, F., and L. Montrucchio (1983), "Disequilibrium Dynamics in a Multidimensional Macroeconomic Model: A Bifurcation Approach," *Richerche Economiche* **XXXVII**, 3–21.
28. Dana, R.-A., and P. Malgrange (1984), "The Dynamics of a Discrete Version of a Growth Cycle Model," *Analyzing the Structure of Econometric Models*, Ed. J. P. Ancot (Amsterdam: M. Nijhoff).
29. Dana, R.-A., and L. Montrucchio (1986), "Dynamic Complexity in Duopoly Games," *Journal of Economic Theory* **40**, 40–56.
30. Day, R. (1982), "Irregular Growth Cycles," *American Economic Review* **72(3)**, 406–414.
31. Day, R. (1983), "The Emergence of Chaos from Classical Economic Growth," *Quarterly Journal of Economics* **98**, 201–213.
32. Day, R. (1984), "Intrinsic Business Fluctuations and Monetary Policy: Can the Fed Cause Chaos?," *MRG Working Paper #8419, University of Southern California.*
33. Day, R. (1986), "Disequilibrium Economic Dynamics," *The Dynamics of Market Economies*, Eds. R. Day and G. Eliasson (Amsterdam, New York: North Holland).
34. Day, R., and J. W. Shafer (1985), "Keynesian Chaos," *J. Macroeconomics* **7**, 277–295.
35. Dechert, W. D. (1984), "Does Optimal Growth Preclude Chaos? A Theorem on Monotonicity," *Zeitschrift für Nationalökonomie* **44**, 57–61.

36. Deneckere, R., and K. L. Judd (1986), "Cyclical and Chaotic Behavior in a Dynamic Equilibrium Model with Implications for Fiscal Policy," *mimeo, Northwestern University, June.*
37. Deneckere, R., and S. Pelikan (1986), "Competitive Chaos," *Journal of Economic Theory* **40**, 13–25.
38. Devaney, R. L. (1986), *Chaotic Dynamical Systems* (Menlo Park, CA: Benjamin Cummings Publishing Co.).
39. Farmer, R. A. (1986), "Deficit and Cycles," *Journal of Economic Theory* **40**, 77–88.
40. Gale, D. (1973), "Pure Exchange Equilibrium of Dynamic Economic Models," *Journal of Economic Theory* **6**, 12–36.
41. Goodwin, R. M. (1987), "A Growth Cycle," *Socialism, Capitalism and Economic Growth*, Ed. C. Feinstein (London: Cambridge University Press).
42. Goodwin, R. M. (1982), *Essays in Economic Dynamics* (London: Macmillan Press Ltd.).
43. Grandmont, J. M. (1985), "On Endogenous Competitive Business Cycles," *Econometrica* **53**, 995–1046.
44. Grandmont, J. M., ed. (1987), *Nonlinear Economic Dynamics* (New York: Academic Press).
45. Grandmont, J. M., and G. Laroque (1986), "Stability of Cycles and Expectations," *Journal of Economic Theory* **40**, 138–151.
46. Guckenheimer, J., and P. Holmes (1983), *Nonlinear Oscillations, Dynamical Systems and Bifurcation of Vector Fields* (New York: Springer Verlag).
47. Hahn, F. (1982), "Stability," *Handbook of Mathematical Economics*, Eds. K. J. Arrow and M. Intriligator (Amsterdam: North Holland), vol. II.
48. Iooss, G. (1979), *Bifurcation of Maps and Applications* (New York: North Holland).
49. Jensen, R., and R. Urban (1982), "Chaotic Price Behavior in a Nonlinear Cobweb Model," *mimeo, Yale University.*
50. Lasota, A., and M. C. Mackey (1985), *Probabilistic Properties of Deterministic Systems* (Cambridge: Cambridge University Press).
51. Lorenz, H. W. (1985), "Strange Attractors in Multisector Business Cycle Models," *Universitat Göttingen, Beitrag Nr. 24, September.*
52. McKenzie, L. W. (1976), "Turnpike Theory," *Econometrica* **44**, 841–855.
53. McKenzie, L. W. (1986), "Optimal Economic Growth, Turnpike Theorems and Comparative Dynamics," *Handbook of Mathematical Economics*, Eds. K. J. Arrow and M. D. Intriligator (Amsterdam, New York: North Holland), vol. III.
54. McKenzie, L. W. (1987), "Turnpike Theory," *The New Palgrave* (New York: Stockton Press).
55. Medio, A. (1984), "Synergetics and Dynamic Economic Models," *Non-Linear Models of Fluctuating Growth*, Eds. R. R. Goodwin, M. Kruger, and A. Vercelli (Berlin, New York: Springer-Verlag).
56. Medio, A. (1987), "Oscillations in Optimal Growth Models," *Journal of Economic Dynamics and Control* **11**, 201–206.

57. Montrucchio, L. (1982), "Some Mathematical Aspects of the Political Business Cycle," *Journal of Optimization Theory and Applications* **36**, 251–275.
58. Montrucchio, L. (1986), "Optimal Decisions over Time and Strange Attractors: An Analysis by the Bellman Principle," *Mathematical Modelling* **7**, 341–352.
59. Montrucchio, L. (1987), "Dynamical Indeterminacy in Infinite Horizon Optimal Problems: The Continuous Time Case," *mimeo, Politecnico di Torino.*
60. Newhouse, S., D. Ruelle, and F. Takens (1978), "Occurrence of Strange Axiom-A Attractors near Quasi-Periodic Flows on T^m, $m \geq 3$," *Comm. in Math. Physics* **64**, 35–40.
61. Pohjola, M. T. (1981), "Stable, Cyclic and Chaotic Growth: The Dynamics of a Discrete-Time Version of Goodwin's Growth Cycle Model," *Zeitschrift für Nationalökonomie* **41**, 27–38.
62. Rand, D. (1978), "Exotic Phenomena in Games and Duopoly Models," *Journal of Mathematical Economics* **5**, 173–184.
63. Reichlin, P. (1986), "Equilibrium Cycles in an Overlapping Generations Economy with Production," *Journal of Economic Theory* **40**, 89–102.
64. Reichlin, P. (187), "Endogenous Fluctuations in a Two-Sector Overlapping Generations Economy," *Working Paper No. 87/264, European University Institute, Florence.*
65. Rockafellar, T. R. (1973), "Saddle Points of Hamiltonian Systems in Convex Problems of Lagrange," *Journal of Optimization Theory and Applications* **12**, 367–390.
66. Rockafellar, T. R. (1976), "Saddle Points of Hamiltonian Systems in Convex Lagrange Problems Having a Non-Zero Discount Rate," *Journal of Economic Theory* **12**, 71–113.
67. Ruelle, D. (1987), "Can Non-Linear Dynamics Help Economists?," this volume.
68. Rustichini, A. (1983), "Equilibrium Points and 'Strange Trajectories,' in Keynesian Dynamic Models," *Economic Notes* **3**, 161–179.
69. Scheinkman, J. A. (1976), "On Optimal Steady State of n-Sector Growth Models when Utility is Discounted," *Journal of Economic Theory* **12**, 11–30.
70. Scheinkman, J. A. (1984), "General Equilibrium Models of Economic Fluctuations: A Survey," *mimeo, Dept. of Economics, University of Chicago, September.*
71. Simonovits, A. (1982), "Buffer Stocks and Naive Expectations in a Non-Walrasian Dynamic Macromodel: Stability, Cyclicity and Chaos," *Scandinavian Journal of Economics* **84(4)**, 571–581.
72. Stutzer, M. (1980), "Chaotic Dynamics and Bifurcation in a Macro Model," *Journal of Economic Dynamics and Control* **1**, 377–393.
73. Torre, V. (1977), "Existence of Limit Cycles and Control in Complete Keynesian Systems by Theory of Bifurcations," *Econometrica* **45**, 1457–1466.

74. Woodford, M. (1987), "Imperfect Financial Intermediation and Complex Dynamics," *mimeo, June*; forthcoming in Barnett, W., J. Geweke, and K. Shell, eds., *Economic Complexity: Chaos, Sunspots, Bubbles and Nonlinearity* (Cambridge: Cambridge University Press).
75. Yano, M. (1984), "The Turnpike of Dynamic General Equilibrium Paths and Its Insensitivity to Initial Conditions," *Journal of Mathematical Economics* **13**, 235–254.

WILLIAM A. BROCK
Department of Economics, The University of Wisconsin, Madison, WI

Nonlinearity and Complex Dynamics in Economics and Finance

PART I. MACROECONOMICS

1. INTRODUCTION

Ideas from the theory of complex dynamics are useful for economics and finance. We use the term "complex dynamics" not only to include the low-dimensional, deterministic, chaotic dynamics to be explained below, but also for deterministic dynamics that converges to an attractor set that is not a point. That is to say, the dynamics does not converge to a limit steady state. We also use the term "complex dynamics" to refer to stochastic dynamics that still are complex when the stochastic shocks are shut down.

In order to explain why the theory of complex dynamics is useful, I'll first give a "toy" version of economic/financial theory so that readers other than economists can follow. Then, in Section 3, I'll explain some basic concepts of the theory of complex dynamics and report on evidence of chaotic dynamics in macroeconomic time-series data. In Section 4, evidence consistent with 6- to 7-dimensional chaos in stock returns and 2-dimensional chaos in monthly T-bill returns is presented. Evidence of nonstationarity and seasonalities in returns data is also presented. An attempt is made to interpret the evidence from the point of view of chaos theory and financial theory.

In particular throughout the article, I try to identify possible routes to chaos and complex dynamics in economics following much the same intellectual strategy as that followed by natural scientists as reported in the survey by Eckmann and Ruelle (1985). As we will see, due to the special character of economics and economic data, this intellectual strategy had to be drastically modified. The obstacles that arose in economics and methods reported on here to deal with them may be instructive to natural scientists that must deal with data-limited situations with shifting dynamics like those commonly faced by economists. In order to move on, we must present a simple version of standard economic theory to get into a position to understand what situations in economics are likely to lead to complex dynamics.

This theory is an attempt to capture and formalize what many (but not all) economists believe are inherent stabilizing mechanisms of a smoothly functioning capitalist economy. Economists are divided in their beliefs whether these mechanisms squash complex dynamics in the real world. In order to get complex dynamics we must find the forces that frustrate these smoothing mechanisms.

A COARSE-GRAINED SKETCH OF RECURSIVE GENERAL EQUILIBRIUM THEORY

Consider the following mathematical cartoon of the theory of value. Let

$$PPF = \{(x,y) \mid y \leq f(x)\}, \qquad U = U(x,y) \tag{1.1}$$

Here *PPF* denotes "Production Possibility Frontier" for two goods 1 and 2; $f(x)$ denotes the maximum amount of 2 you can get when you are producing x units of 1. Let $U(x,y)$ be a quasi-concave utility function. That is, assume $\{(x,y) \mid U(x,y) \geq U_0\}$ is convex for each utility level U_0 and *PPF* is convex. Also assume $f(.)$ is decreasing, concave, and $U(.,.)$ is increasing. Let (x^*, y^*) solve

$$\max U \text{ s.t. } (x,y) \text{ in } PPF. \tag{1.2}$$

It is a theorem of competitive analysis that there is a price vector (p_x, p_y) such that if a representative consumer solves

$$\max U \text{ s.t. } p_x x + p_y y \leq p_x x^* + p_y y^*, \tag{1.3}$$

and a representative firm solves

$$\max p_x x + p_y y \text{ s.t. } (x,y) \text{ in } PPF, \tag{1.4}$$

then the social optimum allocation (x^*, y^*) results. This is a formalization of Adam Smith's "Invisible Hand."

The close connection between social optimization and competitive equilibrium outlined in the cartoon above has been generalized to many goods and many consumers in various formal setups. Many of these are treated in Arrow and Hahn (1971).

In the classical Arrow-Debreu (cf. Arrow and Hahn, 1971) model of a competitive economy, a finite number of economic agents solve problems like Eq. (1.3) over choices of dated goods and contracts (contingent claims) subject to the constraint that the value of expenditures is less than or equal to the value of their resources. The sequence of budget restraints faced by each agent can be collapsed into one intertemporal budget restraint under the assumption of a complete set of markets. This intertemporal budget restraint reflects the constraint that the present value of intertemporal consumption is bounded by the present value of intertemporal income. A fixed-point theorem can then be used to produce existence of equilibrium prices that balance supply and demand for all dated goods and contracts. At such a competitive equilibrium, the reduction, for each agent, of the entire sequence of budget restraints into one grand intertemporal budget restraint allows you to prove that the resource allocation across agents is Pareto optimal. That is to say, it is impossible to reallocate resources across agents in such a way to make one of them better off without making another worse off. This is the power of the complete markets assumption. This is one popular way to formalize Adam Smith's Invisible Hand in the multi-goods/multi-agent case.

Much effort in theoretical economics has gone into investigating conditions under which competitive equilibrium gives social optimum (i.e., Pareto optimum) and when it doesn't, and what can be done to fix it when it doesn't. We are not concerned with that matter here. We are interested in temporally *recursive* economic models that yield intertemporal equilibria whose dynamics can be analyzed by dynamical systems theory. We are also interested in causes of instability and emergence of complexity in such equilibria. But the abstract mathematical structure of the models that we shall look at is like that treated above.

Look at the following stripped-down version of economic growth theory. This will be turned into recursive general equilibrium theory. There is one good "shmoos" which can be eaten or put to pasture to give birth to new shmoos next period. The herd of shmoos radioactively depreciates at rate δ. The social optimum problem is

$$\max U = \sum b^{t-1}u(c_t) \text{ s.t. } c_t + x_t - x_{t-1} \leq g(x_{t-1}) - \delta x_{t-1}. \qquad (1.5)$$

Here $t = 1, 2, \ldots, T$; initial condition x_0 is given; T is the planning horizon; $b = 1/(1+r)$ is the discount factor on future utility; $u(.)$ is the utility function; $\sum$ runs from $t = 1$ to $t = T$; and $g(.)$ is the production function. Maintained assumptions are $u'(0) = \infty$, $u' > 0$, $u'' < 0$, $u'(\infty) = 0$; $g'(0) = \infty$, $g' > 0$, $g'' < 0$, $g'(\infty) = 0$; and $\delta > 0$, $0 < b < 1$.

There is an equilibrium problem that parallels the one for the two-good model above. We add a "stock market" to the equilibrium problem for future use below. Let a representative consumer with the same preferences U as in Eq. (1.5), facing the sequence $\{q, \Pi, r\}$ parametrically, with $x(0) = z_0 = 1$, choose $\{(c_t, x_t, z_t)\}$ to solve

$$\max U \text{ s.t. } c_t + x_t + q_t z_t \leq (q_t + \Pi_t)z_{t-1} + r_t x_{t-1}, \qquad (1.6)$$

a representative firm chooses x_{t-1} at each date t to solve, facing r_t parametrically,

$$\max f(x_{t-1}) - r_t x_{t-1}, \qquad (1.7)$$

where $f(x) = g(x) - (1-\delta)x$. I exposit matters using f to save notation.

Competitive equilibrium is $\{q, \Pi, r\}$ such that solutions to Eqs. (1.6) and (1.7) clear markets at all points in time. There is one perfectly divisible equity share outstanding: $z_t = 1$. Let's look at the first-order necessary conditions (sufficient by concavity) for a competitive equilibrium when $T = \infty$:

$$\text{Stocks: } q_t = G(t+1)[q_{t+1} + \Pi_{t+1}], \quad G(t+1) \equiv \frac{bu'(c_{t+1})}{u'(c_t)}, \tag{1.6}$$

$$\text{Capital: } u'(c_t) = bu'(c_{t+1})r_{t+1}, \quad r_{t+1} = f'(x_t), \tag{1.7}$$

$$\text{Balance: } c_t + x_t = f(x_{t-1}), \quad z_t = 1, \tag{1.10}$$

$$\text{Transversality for } z\text{: } q_t z_t b^{t-1} u'(c_t) \to \infty, \quad t \to \infty, \tag{1.11}$$

$$\text{Transversality for } x\text{: } x_t b^{t-1} u'(c_t) \to \infty, \quad t \to \infty. \tag{1.12}$$

Just as in the two-good case, competitive equilibrium solves Eq. (1.5) and vice versa.

The "turnpike" theorem of optimal growth asserts that for $T = \infty$, the optimal solution $\{c, x\}$ of Eq. (1.5) converges to a unique steady state (c, x) independent of initial conditions as $t \to \infty$. Hence, the market prices q_t and r_t, and profits Π_t also converge to a unique steady state.

LOW DISCOUNTING LEADS TO STABLE DYNAMICS IN GENERAL The striking thing is that this same type of convergence behavior occurs for vector versions of Eq. (1.5) as long as utility and technology display a time-separable structure like that above and are assumed to be strictly concave, and b is near enough to one. The intuition follows. For $b = 1$ optimal steady states are unique and solve a strictly concave, static, nonlinear programming problem subject to a convex constraint set. For $b = 1$ an "optimum" ("optimum" in the sense of the overtaking ordering) path must converge to the unique steady state, else it will "lose value" too much of the time and the sum (integral) will diverge to $-\infty$. For b near one, continuity of the global stable manifold in b at $b = 1$ motivates the result.

Another result is that global asymptotic stability holds when there are IID or Markov stochastic shocks imposed on tastes and technology. In this case, cumulative distribution functions of the optimal consumption and capital stocks converge stochastically to limit distributions independently of initial conditions when b is near one. This parallels the result in the deterministic case. Convergence always obtains in the one-sector model above. This, too, parallels the result in the deterministic case.

The function $V = (W'(x_t) - W'(y_t))(x_t - y_t)$ is a Lyapunov function for the optimal dynamics of a pair of optimal paths starting from initial conditions x, y respectively, provided b is close enough to one. Here $W(x)$ is the maximum sum of utilities you can get from the solution of Eq. (1.5) for $T = +\infty$ and initial capital stock x. By concavity of W, V is always negative. It rises over time along with the optimal dynamics when b is close to one. Therefore, for b near one, the optimal dynamics is globally asymptotically stable. This is true in the stochastic case as well as for many sectors.

The natural scientist may find the following slight digression illuminating. Consider any economic model where the social optimization problem boils down to a continuous-time, optimal control problem of the form

$$W(x_0) = \max \int_0^\infty e^{-bt} u(x, dx/dt) dt. \tag{1.13}$$

Here x is a vector in R^n, the utility $u(.,.)$ is jointly concave, and the optimization is taken over the set, X, of piecewise continuously differentiable functions $x(t)$ s.t. $x(0) = x_0$. Bounds are placed upon X to get existence of an optimum. Put

$$H(q, x) = \max \left(u(x, v) + qv\right), \qquad v = dx/dt. \tag{1.14}$$

H is the maximized value of the pre-Hamiltonian over the control v. H is convex in q, concave in x.

Write down the Pontryagin necessary (also sufficient in our case because of concavity of u) conditions of optimality,

$$\frac{dx}{dt} = H_q, \qquad \frac{dq}{dt} = bq - H_x, \tag{1.15}$$

$$qxe^{-bt} \to 0, \qquad t \to \infty. \tag{1.16}$$

Put $V = (dx/dt)'(dq/dt)$. Take dV/dt along solutions of Eqs. (1.15) and (1.16) to get the quadratic form,

$$\frac{dV}{dt} = \left(\frac{dx}{dt}, \frac{dq}{dt}\right)' Q \left(\frac{dx}{dt}, \frac{dq}{dt}\right), \tag{1.17}$$

where the $2n \times 2n$ matrix Q has $n \times n$ diagonal elements $-H_{xx}$, H_{qq} and $n \times n$ off-diagonal elements, $(b/2)I$, where I is the $n \times n$ identity matrix.

PROPOSITION: V is a Lyapunov function for solutions of Eq. (1.15) that satisfy Eq. (1.16) provided that Q is positive definite. The matrix Q is positive definite provided that $b^2/4 < L_q L_x$ when L_q and L_x are the smallest eigenvalues of H_{qq} and $-H_{xx}$ respectively.

Once you have a Lyapunov function, stability theorems are routine. Notice that H_{qq} and H_{xx} are positive semi-definite by convexity of H in q and concavity of H in x. So when b is near zero, a slight tightening of convexity/concavity of H gives global asymptotic stability. A version of this argument can be given for stochastic models also. Versions of this theorem exist for discrete-time models as well as continuous-time models. See Epstein (1987) and references for details. These results are useful for our understanding of the causes of instability and complex dynamics in economic systems.

They are useful for the following reasons. Turn back to competitive equilibrium theory with a finite number of long-lived agents. If markets are complete and tastes and technology have a recursive structure like that of optimal growth theory, then, since competitive equilibrium is Pareto optimal, there are positive weights, w_h, one for each agent $h = 1, 2, \ldots, H$ such that competitive equilibrium maximizes the weighted sum of utility subject to technological constraints and material balance over time. Because of the recursive structure of preferences and technology, this is a problem like that of optimal growth theory. Therefore, if the agents don't discount the future very much and their planning horizons are long, the solution will converge to a steady state independently of initial conditions. See Epstein (1987) for this result.

This is not a good breeding ground for instability and complex dynamics. Stochastic shocks are no help because points get replaced by distributions, but stochastic stability remains.

We sum up the conclusion of this section.

BASIC PRINCIPLE: Broadly speaking, if we want complex dynamics on a time scale that is short relative to agent's lifetimes and preserve recursive structure in preferences and technology, it is helpful to sever the connection between equilibrium and Pareto optimality.

HOW CAN WE GET COMPLEX DYNAMICS? To get complex dynamics and instability in dynamic economic evolution, we must do at least one of the following:

1. Introduce agents that act as if they discount the future relatively heavily.
2. Abandon the concavity assumptions on tastes and technology. The introduction of increasing returns coupled with externalities might lead us to discover possible examples of complex dynamics. Potential example: the growth and decay of cities.
3. Abandon the assumption of complete markets.
4. Abandon the assumption of price-taking agents.
5. Impose "complex preferences" and/or "complex technology." That is to say, abandon the assumption of recursive or stationary-state variable representations of preferences and technology.
6. Abandon the assumption that the system is in equilibrium. This could be done by taking an evolutionary perspective, for example. A more Austrian or Schumpeterian style of modeling might have a better chance of leading us to discover complex dynamics in economics.
7. Admit direct effects of some agents' actions upon the tastes or technologies of others (Arrow and Hahn, 1971, Chapter 6, externalities). This would allow complex dynamics of fashion to infect tastes or allow complex dynamics of technological diffusion to influence technology. These dynamics would then be transmitted to prices and quantities through the equilibration process.
8. Introduce exogenous "forcing functions." Population dynamics or the dynamics of technological change might be promising examples.

It is important to realize the joint constraints imposed on potential complex dynamical fluctuations and the operative time scale of self-interested economic agents. High-frequency fluctuations in personal income would be intertemporally smoothed by optimizing consumers (due to concavity of utility plus low discounting relative to the frequency of the oscillations). Similar smoothing mechanisms operate at the firm level. For example, Epstein (1987) uses intertemporal duality theory to show that the global asymptotic stability theorems discussed above hold if the rate of return to production by firms is not too large relative to the time scale. It is the message of the stability literature in economics that these smoothing mechanisms must be "shut down" to some extent in order to get instability and complex dynamics.

Let us discard the poorest potential paths to complex dynamics first. Population dynamics is a poor "forcing function" for the introduction of complex economic dynamics. Montroll and Badger (1974) reviews demographic literature that shows that the Pearl-Verhulst logistic law,

$$\frac{dN}{dt} = kN\left[1 - \left(\frac{N}{\theta}\right)\right], \tag{1.17}$$

where θ is the saturation population, tracks human population growth quite well for appropriate choices of $N(0)$, k and θ. This "law" seems to work well for the 18th century to the present for many countries, especially in Europe and the United States. Major deviations from Eq. (1.17) look like shocks such as wars; the U.S. Great Depression, the Irish potato famine of 1845, and the increase in size of the Netherlands by land reclamation. We give up the search for economic complex dynamics caused by complex dynamics in human population in this article.

Boldrin and Deneckere (1987) and references, especially to Montruccio and Pelikan, show for both discrete- and continuous-time versions of Eq. (1.13) that, given any complex dynamics $dx/dt = h(x(t))$, a concave utility u and a discount rate b can be found such that the optimum dynamics of Eq. (1.13) are the same as $dx/dt = h(x(t))$. While this is a strong mathematical result, neither the dynamics nor the parameters of u nor the value of b have been calibrated or estimated from data at any time scale. Therefore, the Boldrin-Deneckere-Montruccio result must stand only as a logical possibility at this stage. The discount b must be typically large to get the result. This is intuitive. The result needs "myopic" agents. A value of b near 0 together with concavity of u implies global asymptotic stability of the optimum dynamics.

Agents that plan as if the future is worth nearly as much as the present contribute to economic stability.

The survey by Baumol and Benhabib (1987) discusses recent work in economics where chaos appears as a logical possibility in economic models that are consistent with the usual axioms of economic theory. Woodford (1987) gets chaos in a model with two types of agents. One type acts as if they discount the future a lot due to restraints on borrowing. This is how Woodford gets around the stability results above. No one has calibrated a model on real data and shown that it generates chaotic dynamics. No one has adduced strong evidence that any actual economic time series (after reduction to stationarity by detrending) is chaotic. We'll get into evidence for chaos in financial data below in Section 4. Evidence for chaos and complex dynamics in macroeconomic data is discussed in Section 3.

Turn now to a brief exposition of some basic ideas in financial theory. The idea that there should be no arbitrage profits (in some set of economically relevant units) in financial equilibrium is linked up with the growth theory above to show how dynamics in the "dividend" process are transmitted through the equilibration mechanism to equilibrium asset prices.

2. FINANCE

The most important idea in financial theory is the idea that financial equilibrium means the absence of arbitrage profits and the absence of arbitrage profits implies restrictions on returns to holding financial assets. This is explained below.

Consider the following equation

$$p(i,t) = E\{G(t+1)\,[p(i,t+1) + y(i,t+1)] \mid I(t)\} \tag{2.1}$$

where $p(i,t)$ denotes the price of asset i at date t; $y(i,t+1)$ denotes the net cash flow from asset i when it is held over the time interval $(t,t+1)$; $\{G(t+1)\}$ is a discounting process; $I(t)$ is information available to traders at date t; and E denotes conditional expectation. Eq. (2.1) means this: if you buy asset i at date t at price $p(i,t)$, collect net cash flow $y(i,t+1)$ over $(t,t+1)$ (which is treated as being paid at $t+1$), and sell the asset at price $p(i,t+1)$, you can't make a profit based on information $I(t)$ after discounting at $G(t+1)$. Here the process $\{G(.)\}$ may be stochastic. Notice also that a stand must be taken on what expectations are being assumed on prices and cash flow expected to materialize at date $t+1$ when decisions to buy or sell are being taken at date t. We'll assume rational and homogeneous (across traders) expectations for simplicity in this article. But keep in mind that this is a special assumption, even though it is a very popular way of "closing" the model in finance. This forms a benchmark case for more sophisticated financial models with different kinds of traders. The process of solving Eq. (2.1) may be explained by looking at special cases.

A special case is when $\{G\}$ is constant, nonstochastic, and $G = b = 1/(1+r)$. We will assume G is constant until further notice. Drop the index "i" for now. Suppose by way of illustration that $y(t) = y$, constant and nonstochastic. Then a solution, called the "fundamental" solution of Eq. (2.1) is

$$p^* = \frac{by}{1-b} = \frac{y}{r}. \tag{2.2}$$

All solutions of Eq. (2.1) are of the form $p^*(t) + b(t)$ where

$$b(t) = E\{G(t+1)b(t+1)\}. \tag{2.3}$$

The process $\{p^*(t)\}$ is gotten by forwardly iterating Eq. (2.1) up to T and assuming the "tail" term goes to zero as $T \to \infty$ in the infinite series development. The process $\{b(t)\}$ is called a "bubble" process. In the special case $G = b = 1/(1+r)$, it grows on average at rate r. Absence of arbitrage profits between any two finite periods of time will not rule out bubbles. There are ideal market institutions that will get rid of them. "Bubbles" is an area of controversy and great interest in finance. But let us assume the absence of bubbles for now.

Look at another example. Let $\{y(t)\}$ be given by

$$y(t) = \alpha y(t-1) + e(t), \tag{2.4}$$

where $\{e(t)\}$ is independently and identically distributed (IID) with mean 0 and variance σ^2. Let $\{I(t)\}$ contain current and past y's. then it is easy to see that the fundamental solution of Eq. (2.1) is given by

$$p^*(t) = ay(t), \qquad a \equiv \frac{\alpha}{1+r-\alpha}. \tag{2.5}$$

Look at a popular example used a lot in empirical work on finance, the geometric random walk,

$$\ln\big(y(t)\big) = \mu + \ln\big(y(t-1)\big) + e(t), \tag{2.6}$$

where $\{e(t)\}$ is IID $N(0, \sigma^2)$. Let $\{I(t)\}$ contain current and past y's. Following Kleidon (1986), the fundamental solution of Eq. (2.1) is given by

$$p^*(t) = ay(t), \qquad a \equiv \frac{1+g}{r-g}, \qquad 1+g = \exp\left(\mu + \frac{\sigma^2}{2}\right). \tag{2.7}$$

In general, if $y(t)$ is a kth-order Markov process and $\{I(t)\}$ consists of current and past y's, then the fundamental $p^*(t)$ will be of the form $p^* = P(y(t), \ldots, y(t-k))$.

The Eq. (2.1) emerges naturally from general equilibrium theory. For example, look at the Eq. (1.8) above. Replace, in Eq. (1.8), q_t by $p(t)$, Π_t by $y(t)$, and $bu'(c_{t+1})/u'(c_t)$ by $G(t+1)$; you have Eq. (2.1). This illustrates an important principle.

PRINCIPLE: Solve the real side of the model by nonlinear programming as in the growth model illustrated above; then use Eq. (2.1) to find the price of the equities. The transversality condition at ∞ gets rid of bubbles. This same mechanism works for stochastic multisector models as well as deterministic models. It won't work for incomplete markets models.

We have developed enough background now to explain how financial markets, even in equilibrium, can transmit nonlinearity/linearity, or even chaos of the $\{y\}$ process into the $\{p^*\}$ process through Eq. (2.1). Turn now to a development of some basic concepts from chaos theory and empirical testing for the presence of chaos.

3. BASIC DEFINITIONS AND CONCEPTS OF CHAOS THEORY

Since most of the expository material on chaos and testing for it is explained in Brock (1986) and Eckmann and Ruelle (1985), we'll be brief here.

Consider the deterministic chaos

$$x(t+1) = 2x(t), \qquad x(t) \leq 1/2; \qquad x(t+1) = 2 - 2x(t), \qquad x(t) \geq 1/2. \quad (3.1)$$

Two basic properties of deterministic chaos are well illustrated by this simple recursion. First, this recursion generates, for almost all (in the sense of Lebesgue measure) initial conditions $x_0 \in (0,1)$, trajectories such that

$$\#\left\{\frac{x(t),\ 1 \leq t \leq T \mid x(t) \in [a,b]}{T}\right\} \longrightarrow b - a; \qquad a, b \in [0,1], \qquad a < b. \quad (3.2)$$

That is to say, the map of Eq. (1.1) has a nondegenerate invariant measure just like a stochastic growth model. Almost all trajectories converge to it.

Second, suppose you make a small error in measuring the initial state so that you only know that it lies in the interval

$$I = [a - \epsilon, a + \epsilon], \qquad \epsilon > 0 \quad (3.3)$$

Now imagine that at date 1 you must forecast $x(t)$ based on the knowledge of Eq. (3.3). The loss of precision in your forecast (the length of the interval where you know that $x(t)$ lies) grows exponentially fast as t grows in the short term until you know nothing. That is, you only know that $x(t)$ lies in $[0,1]$. This property means that you have the *potential* ability (if you could measure x_0 with infinite accuracy) to forecast $x(t)$ perfectly.

A third property of deterministic chaos (redundant given the two properties listed above but repeated anyway for emphasis) is that the time series $\{x(t)\}$ appears stochastic even though it is generated by a deterministic system, Eq. (3.1). More precisely, in the case of Eq. (3.1), the empirical spectrum and empirical autocovariance function is the same as that of white noise, i.e., the same as that generated by independently and identically distributed, uniform, $[0,1]$ random variables.

Eq. (3.1) illustrates the need for a test for stochasticity beyond spectral and autocovariance analysis. Simply plotting $x(t+1)$ against $x(t)$ will not do because examples like

$$x(t+1) = F(x(t), \ldots, x(t-q)), \qquad q \geq 1 \quad (3.4)$$

can be generated that are chaotic.

An efficient way to test for chaos is to consider the following quantity

$$C_m(e,T) = \#\left\{\frac{(t,s),\ 1 \leq t,\ s \leq T \mid \ \|x_t{}^m - x_s{}^m\| < e}{T_m{}^2}\right\} \quad (3.5)$$

where $T_m = T - (m-1)$ and $x_t{}^m = (x(t) \ldots, x(t+m-1))$.

Brock and Dechert (1986) show that

$$C_m(e,T) \to C_m(e), \qquad T \to \infty, \tag{3.6}$$

for almost all initial conditions. They prove Eq. (3.6) for noisy chaotic systems also.

Dimension is "estimated" by natural scientists (cf. the survey of Eckmann and Ruelle, 1985) by plotting $\ln(C_m(e,T))$ against $\ln(e)$ for large T and looking for constant slope zones of this plot that appear independent of m for large enough m. Here "ln" denotes the natural log.

The definition of correlation dimension in embedding dimension m is

$$d_m = \frac{\lim\lim\ln\left[C_m(e,T)\right]}{\ln(e)}, \qquad T \to \infty, \qquad e \to 0, \tag{3.7}$$

where the limit is taken w.r.t. T first, w.r.t. e second.

The correlation dimension itself is given by

$$d = \lim d_m, \qquad m \to \infty. \tag{3.8}$$

Brock and Sayers (1986) estimated various measures of dimension that were stimulated by the theoretical quantities above.[1] They are

$$SC_m(e,f,T) = \frac{\ln\left(C_m(e,T)\right) - \ln\left(C_m(f,T)\right)}{\ln(e) - \ln(f)}, \tag{3.9}$$

$$\alpha_m(e,T) = \frac{\ln\left(C_m(e,T)\right)}{\ln(e)}. \tag{3.10}$$

The quantity SC_m is an estimate of the slope of the plot of $\ln(C_m)$ against $\ln(e)$, i.e.,

$$\text{slope at } e = \frac{C_m{}'(e,T)e}{C_m(e,T)}. \tag{3.11}$$

Note that the slope at e is just the elasticity of C_m at e. It measures the percentage change in new neighbors that a typical m-history $x_t{}^m$ gets when e is increased to $e + de$. Hence, dimension is a crude measure of the level of parsimony (the minimal number of parameters) needed in a dynamic model to fit the data.

Low dimension estimates were found for many economic time series by Brock and Sayers (1986). But it is well known that many macroeconomic time series are well modeled by linear autoregressive processes with near-unit roots (Nelson and Plosser, 1982). It is easy, based on dimension estimation alone, to confuse such processes with deterministic chaos. Brock (1986) proposed a diagnostic, called the

[1] To my knowledge, these ideas were first applied in economics in papers by Brock (cf. Brock, 1986) and Scheinkman (cf. Scheinkman and LeBaron, 1986) which were presented at the Conference on Nonlinear Dynamics organized by J. M. Grandmont, Paris, France, June 1985. Some of the papers presented at that conference are in Grandmont (1986).

residual diagnostic, to get around this obstacle. Estimate a linear autoregressive time series model by your favorite technique. Estimate the dimension of the residuals of that model and compare with the original data. If you have low-dimensional deterministic chaos, the dimension of the residuals will be the same as the dimension of the original data. The estimated dimension of the residuals of the best-fit linear models of Brock and Sayers (1986) looked like the dimension of random numbers. The evidence for low-dimensional deterministic chaos was weak.

It is tricky to carry out this diagnostic in practice. In the sample size typically available to macroeconomics (100 to 400 observations), estimation errors in estimated dimension can cause the residual diagnostic to reject deterministic chaos even when it is true. Therefore, we not only used the residual diagnostic but also tested the residuals for temporal dependence using a more powerful test which will be discussed below.

SIZE AND POWER CHARACTERISTICS OF THE BDS TEST Brock, Dechert, and Scheinkman (1987), hereafter "BDS," created a family of statistics based upon the correlation integral C_m. This is calculated by first putting

$$C_m(e,T) = \#\left\{\frac{(t,s),\ 1 \leq t,\ s \leq T \mid \|x_t{}^m - x_s{}^m\| < e}{T_m^2}\right\}, \qquad T_m \equiv T - (m-1). \tag{3.12}$$

Second, the limit of Eq. (3.12) exists almost surely under modest stationarity and ergodicity assumptions on the stochastic process under scrutiny (Brock and Dechert, 1986). Call this limit $C_m(e)$.

The BDS paper considered tests based upon the statistic

$$W_m(e,T) \equiv \frac{T^{1/2} D_m(e,T)}{b_m(e,T)}, \qquad D_m \equiv C_m - [C_1]^m, \tag{3.13}$$

where b_m is an estimate of the standard deviation under the IID null. BDS showed, under the null of IID, $W_m \to N(0,1)$, as $T \to \infty$. The W statistic was shown to have a higher power against certain alternatives than the tests of independence based upon the bispectrum. The reason for this is that the bispectrum is zero for the class of processes with zero third-order cumulants. There are many dependent processes with zero third-order cumulants. The W statistic catches many of these processes.

Given an IID stochastic process $\{X_t\}$, consider the formula

$$b_m = (1, -mC_{m-1})' \sum (1, -mC_{m-1}), \tag{3.14}$$

where

$$\sum_{22} = 4(Q - C^2), \tag{3.15}$$

$$\sum_{11} = 4(Q^m - C^{2m}) + 8\sum_{j=1}^{m-1}(Q^{m-j}C^{2j} - C^{2m}) \tag{3.16}$$

$$\sum_{12} = 2(Q + Q^m + 2QC^{m-1} - (C + C^m)^2) +$$

$$4\sum_{j=1}^{m-1}(QC^{m-1} + Q^{m-j}C^{2j} - C^{1+m} - C^{2m} - .5\left(\sum_{11} + \sum_{22}\right),$$

where

$$C = E\{I_e(X_i, X_j)\} \equiv C_1(e), \tag{3.17}$$

$$Q = E\{I_e(X_i, X_j)I_e(X_j, X_k)\}. \tag{3.18}$$

Here $I_e(X, Y)$ is just the indicator function of the event $\{| X - Y |< e\}$ and $\sum_{j=1}^{m}$ denotes the sum from $j = 1$ to $j = m$.

This formula for the standard deviation, b_m, used in the W statistic (cf. Eq. (3.14)), was adapted from BDS by Scheinkman and LeBaron (1987).

Look at Table 1, the results from a revision of BS (1986). BS fitted "best" linear models (after transformation of units and detrending) in the usual manner. Recall that under the null hypothesis of correct fit, the estimated residuals are asymptotically IID. If the residuals were *actually* IID, the W statistics reported below would be asymptotically $N(0,1)$. Of course, the estimation process induces "extra variance." Based on our own computer experiments and those reported in Scheinkman and LeBaron (1987), we do not believe that the correction is very large for most cases in practice. In any event the reader is advised to keep this in mind while reading Table 1.

Evidence is strong for nonlinearity in (a) industrial production, (b) civilian employment, (c) unemployment rate, (d) pigiron production, and (e) Wolfer's sunspot numbers. BS also performed symmetry tests on many of these same series. Symmetry testing confirms the presence of nonlinearity. Let us sum up at this point.

SUMMARY FOR MACROECONOMICS

Despite the logical possibility of chaos in theoretical macroeconomic models, an empirical examination of post-WWII U.S. aggregative macroeconomic time-series data has failed to find evidence of deterministic chaos. There is evidence of nonlinearity that is robust to units changes and detrendings. The tests for chaos that were applied may have rejected low-dimensional deterministic chaos when it was true. Presentation of evidence of low correlation dimension or even positive estimated Lyapunov exponents does not make the case for chaos.

TABLE 1 W Statistics for Residuals of Linear Models

Series	Dim	#prs	W	N
ed:	2	902	-.63	147
residuals of AR(2) fit to detrended real U.S. gnp	3	272	.16	147
from B (1986) and BS (1986).	4	92	1.44	147
dsear1pd:	2	12554	5.10	433
residuals from AR(1) to fit to the first difference	3	5176	6.52	433
of log U.S. Industrial Production.[1]	4	2310	8.34	433
dsear4p:	2	12555	5.21	430
residuals from an AR(4) fit to the first difference	3	5216	6.61	430
of log U.S. Industrial Production.	4	2338	8.35	430
dsegpdi1:	2	1173	.94	147
residuals from AR(1) fit to the first difference of	3	423	1.77	147
log real U.S. quarterly gross domestic investment	4	152	1.881	147
(GPCI) (cf. BS, 1986)				
dsegpdi1:	2	1087	1.61	144
residuals from AR(4) fit to the first difference of	3	391	2.53	144
log real U.S. quarterly gross domestic investment	4	148	3.21	144
(GPCA) (cf. BS, 1986)				
dsegpdi1:	2	970	-.50	140
residuals from AR(8) fit to the first difference of	3	344	1.40	140
log real U.S. quarterly gross domestic investment	4	121	1.79	140
(GPCI) (cf. BS, 1986)				
edpig:	2	54848	15.97	715
residuals of an AR(2) fit to detrended U.S. pigriron	3	32072	20.09	715
production as in BS (1986)	4	19519	24.10	715
edsun:	2	1617	5.19	170
residuals of an AR(2) fit to the Wolfer sunspot	3	642	6.90	170
series as in BS (1986)	4	239	6.47	170
edunemp:	2	990	3.24	130
residuals of an AR(2) fit unemployment rates as	3	404	4.94	130
in BS(1986)	4	163	5.32	130
egpdid:	2	1345	1.21	147
residuals of an AR(2) fit to linearly detrended,	3	506	1.57	147
log real U.S. GPDI as in BS (1986)	4	191	1.60	147
empar2:	2	841	4.21	130
residuals of an AR(2) fit to linearly detrended,	3	316	5.94	130
log U.S. employment as in BS (1986)	4	128	7.64	130

[1] the industrial production series is U.S. post-war quarterly and is taken from Litterman and Weiss (1985).

Turn now to finance.

PART II. APPLICATION TO FINANCE

In the first three sections of this paper, we outlined a toy version of general equilibrium theory, recursive general equilibrium theory, global asymptotic stability theory, concepts from empirical nonlinear science, and looked at evidence for chaos in macroeconomic data. Below we look for evidence of chaos in financial data.

4. THE CHOICE OF TIME SCALE ON WHICH TO LOOK FOR COMPLEX DYNAMICS IN FINANCE

In serious confrontation of the theory of complex dynamics with data, the issue of appropriate time scale comes up. Returns on stock i over time period $[t, t+h]$ are defined by

$$r_{1,t,t+h} = \frac{p_{i,t+h} + y_{i,t,t+h}}{p_{i,t-1}}. \tag{4.1}$$

At the minute-to-minute level, Wood et al. (1985) have documented: (i) returns tend to be higher at the opening of the exchange, fall over the first 20–30 minutes, look unpredictable in the middle, and tend to rise in the last 20–30 minutes to the close; (ii) trading volume and price volatility follow the same pattern near the opening and the close. These patterns are small, but appear statistically significant. Details (Marsh and Rock, 1986) of market micro-structure matter a lot at this frequency. We will not spend any time analyzing dynamics at the minute-to-minute frequency because of standard economic reasoning discussed below.

DAILY/WEEKLY PRICE CHANGES ARE UNPREDICTABLE?

Sims (1984) is a nice formalization of the standard economic reasoning that price changes must be unpredictable over small time intervals in a frictionless market like stock markets in modern countries. This is formalized by:

DEFINITION (Sims, 1984): A process $\{P(t)\}$ is instantaneously unpredictable if and only if, a.s.,

$$E_t \frac{\left[\left(P(t+v) - E_t\left[P(t+v)\right]\right)^2\right]}{E_t\left[\left(P(t+v) - P(t)\right)^2\right]} \longrightarrow 1, \qquad v \to 0. \tag{4.2}$$

Here E_t is taken w.r.t. the information set I_t. ($\{I_t\}$ is an increasing sequence of sub-sigma fields of a master sigma field, I.)

In words, "for an instantaneously unpredictable process prediction error is the dominant component of changes over small intervals. Of course, for a martingale of Eq. (4.3) below with finite second moments, the ratio in Eq. (4.2) is exactly 1." Sims points out that, under Eq. (4.2), regressions of $P_{t+s} - P_t$ on any variable in I_t have $R^2 \to 0$, as $s \to 0$. Under Eq. (4.3), $R^2 = 0$. He also points out that Eq. (4.2) doesn't rule out predictability over longer time periods. He then argues that this roughly squares with empirical evidence in finance. "Short periods" for Sims is daily to weekly.

The martingale hypothesis is

$$E\left[P_{t+s} \mid I_u,\ u \leq t\right] = P_t, \qquad s > 0. \tag{4.3}$$

Notice that Eq. (4.3) is the same as Eq. (1.8) above with $G = 1$ and $y = 0$. In practice, Eq. (4.3) is "adjusted" for discounting, G, and net cash flow, y, over the period.

Now recall that chaos theory teaches us that trajectories generated by chaotic maps are potentially perfectly predictable provided that you can measure the state perfectly. But if you measure the state today with error, then forecasts of the future state become worthless at an exponential rate. Hence, nonlinear dynamicists sometimes say that chaotics dynamics are unpredictable. Yet the financial logic outlined above leads us to believe that low-dimensional chaotic deterministic generators for stock prices and returns over minute-to-minute or even daily to weekly time period should be extremely unlikely. Turn now to some evidence on this matter.

EVIDENCE ON COMPLEX DYNAMICS IN STOCK RETURNS Scheinkman and LeBaron (SLB, 1986) have estimated the correlation dimension for 1226 weekly observations on the CRSP value-weighted U.S. stock returns index for the mid-60's. They get roughly 6.

Similar results for closing prices over the mid-1970's to the mid-1980's for gold and silver on the London Exchange were reported by Frank and Stengos (1986). They get correlation dimension estimates between 6 and 7 for daily, weekly, and biweekly series.

I examine several stock returns series below: (a) VW = monthly returns on the value-weighted NYSE index for the mid-20's to the late 70's; (b) EW = monthly returns on the equal-weighted NYSE index for the same period; and (c) BLVW = weekly returns on the value-weighted CRISP index used in the Scheinkman/LeBaron (1986) study. I deliberately make life difficult for myself in estimating correlation dimension by reporting step-wise empirical derivatives of the Grassberger-Procaccia diagram. Recall that the GP diagram is $\ln(C_m(e))$ plotted against $\ln(e)$. These empirical derivatives are going to look rough because we lose the inherent smoothing involved in estimating the slope of the "constant slope zone" of the GP diagram by regression analysis or by eyeball analysis. I then compare this series of empirical derivatives for the shuffled counterpart. The shuffled counterpart refers to a series that is independent and identically distributed, and has the same stationary distribution as the series under scrutiny.

TABLE 2 Stock Returns Series
(all data was divided by the standard deviation and multiplied by .2)

e	$.9^6$	$.9^7$	$.9^8$	$.9^9$	$.9^{10}$	$.9^{11}$	$.9^{12}$	$.9^{13}$	$.9^{14}$	$.9^{15}$
EW[1]:										
SC[2]	2.9	3.4	4.0	5.0	5.6	5.9	6.3	7.1	7.0	8.0
SC_{sf}[3]	5.0	5.6	6.1	6.6	7.1	7.5	8.0	8.3	9.1	9.7
VW[4]:										
SC	4.3	5.0	5.6	5.9	7.0	7.1	7.0	7.2	8.7	6.7
SC_{sf}	5.3	6.0	6.6	7.3	7.8	8.6	9.0	9.0	9.0	10.4
BLVW [1:1600][5]:										
SC	3.7	4.1	4.2	4.4	4.8	4.9	5.4	6.0	7.1	7.4
SC_{sf}	6.5	7.3	7.6	8.6	7.3	10.7	6.1	11.9	19.7	644.2
BLVW [601:1226][6]:										
SC	5.9	6.5	7.2	7.6	7.8	8.8	8.5	8.5	9.7	14.6
SC_{sf}	6.7	7.2	7.4	7.7	7.7	8.6	10.4	10.1	7.5	645.6
DSE8TBLM[7]:										
SC	1.2	1.4	1.6	1.8	1.9	1.9	2.0	1.8	1.7	1.8
SC_{sf}	3.3	3.9	4.4	4.6	4.8	5.2	5.3	5.6	5.9	7.1
KODAK[8]:										
SC	5.5	6.1	7.1	7.5	7.9	8.9	7.8	9.1	9.2	9.5
SC_{sf}	5.0	6.6	6.9	8.0	7.3	8.3	10.9	12.7	8.7	3.8

[1] N=696, m=10 where N equals the number of observations and M equals embedding dimension (for all notes); returns on equal-weighted CRISP index.

[2] Empirical slope of the GP diagram for the original series.

[3] Empirical slope of the GP diagram for the shuffled series.

[4] N=696, m=10; returns on value-weighted CRISP index.

[5] N=600, m=10; first 600 returns on SLB's value-weighted index.

[6] N=626, m=10; next 626 returns on SLB's value-weighted index.

[7] N=628, m=10; residuals of an AR(8) fit to first difference's of T-bill returns.

[8] N=720, m=10; returns on Kodak stock.

Put

$$SC(.9^n) = \frac{\ln\left(C_m(.9^n)\right) - \ln\left(C_m(.9^{n-1})\right)}{\ln(.9)}. \tag{4.4}$$

SC is an estimate of the empirical derivative at $.9^n$. See Table 2.

We have succeeded in replicating the Scheinkman and LeBaron study. The dimension of both halves of the data set is between 6 and 7 when you take into account the roughness of empirical derivatives relative to estimated empirical slopes of GP plots. The dimension of the shuffled series is surely bigger in both halves of the data set. Note, however, that the difference in behavior between the original series and the shuffled series is smaller for the second half than for the first half.

By way of comparison, we looked at the monthly EW and VW indices. Notice that they display similar behavior to the weekly BLVW data set. This is comforting. Gennotte and Marsh (1987) calculated Grassberger-Procaccia plots for this same data set after taking out the January effect and taking out linear structure. They find significant evidence of nonlinear dependence. It is significant because they calculate BDS (1986) statistics for the spread between the GP plot for their prewhitened January adjusted data and the GP plot for the shuffled data. The BDS statistics were highly significant, especially for EW. It is striking that the significance of dependence was so high for EW because most of the January effect resides in small firms which loom relatively large in EW compared to VW. But Gennotte and Marsh took out the January effect.

Gennotte and Marsh looked at the subsample January 1942–December 1971. The BDS statistic fell to 2.45 for EW and 1.01 for VW in contrast to 7.23 for EW and 4.39 for VW for the whole sample. Since Genotte and March fitted best linear models to the data as well as taking out the January seasonality, therefore under the null hypothesis of linearity, the 5% level of the BDS statistic is around ± 2. Therefore, the size of the BDS statistic could be looked upon as a rough measure of nonlinearity in these data. It is interesting that the subsample January 1942–December 1971 appears linear.

In contrast with the above results (especially for the value-weighted indices), the behavior of the monthly returns for the individual stock, KODAK, is striking. Returns for KODAK appear IID. There is little difference in behavior between the original returns and the shuffled returns.

Contrast this with the behavior of Treasury Bills taken from Ibbotsen-Sinquefield (1977, p. 90, Exhibit B8). These data are also used by Gennotte and Marsh (1987) to measure the risk-free rate of return. We reduced this series to stationarity by taking first differences. Now "test" the null hypothesis that first differences of T-bill returns have a low-dimensional, deterministic, chaotic explanation.

We identified best-fit linear models by Box-Jenkins method to get rid of linear dependence in T-bill returns. Recalling the discussion of the residual diagnostic above this prewhitening does not change the dimension of the underlying dynamics if they are chaotic. The results for DSE8TBLM are typical. We get the low dimension of the residuals even after fitting an autoregressive process of order 8 to the first differences of T-bill returns. The dimension appears to be around 2. Furthermore, the dimension of the shuffled counterpart is much larger although it is not equal to the theoretical value which is 10.

We are puzzled by the T-bill results. There seems to be strong evidence consistent with a chaotic explanation. We do not conclude that, however, because the largest Lyapunov exponent has not been estimated and shown to be positive. Also unlike, for example, the case for the Belousov-Zhabotinski chemical reaction (Eckmann and Ruelle, 1985), the dynamics have not been "reconstructed." All we can say at this stage is that T-bill returns are worthy of further investigation.

There is one thing that the scientist must always keep in mind when interpreting this evidence. It is this. During the period under scrutiny, the government intervened in determining interest rates. Hence, T-bill rates were influenced by government intervention. There were periods where T-bill rates didn't move at all because the government was controlling interest rates. In other periods, the government was more concerned about the growth of the monetary base and less concerned about fluctuations in interest rates. Hence, the "dynamics" is subject to "regime shifts." Shifts in the government posture toward controlling ease of credit and controlling interest rates impacts on the short-run opportunity cost of funds. This influences stock returns. The market watches the central bank very carefully.

The bottom line for the scientist is that government activity in the financial markets makes it even harder to interpret the evidence.

WHAT MAY EXPLAIN THE EVIDENCE FOR NONLINEARITY ADDUCED ABOVE? It appears that some structure of correlation dimension less than or equal to six generates weekly and monthly returns on aggregate stock market indices. Potential explanations are: (i) apparent deterministic seasonalities in returns such as (a) the Monday effect (returns are lower from Friday close to Monday open than over the rest of the week), (b) the monthly effect, and (c) the January effect; (ii) linearly autoregressive conditional heteroscedasticity (current variance is a linear function of past variances and past squared deviations about the conditional mean); (iii) mean reversion at business cycle frequencies, i.e., negative autocorrelations at 3- to 5-year frequencies;and (iv) systematic movements in the tradeoff between risk and return over the business cycle. I discuss these possibilities in detail in Brock (1988). Tentative conclusions that I draw are that the January effect, linearly autoregressive conditional heteroscedasticity, and mean reversion in the sense that a near-unit root autoregressive process is added to the fundamental can be discarded as explanations for the low correlation-dimension estimates.

ACKNOWLEDGMENTS

I would like to thank the University of Wisconsin Graduate School, the National Science Foundation, and the Guggenheim Foundation for essential financial support. I thank W. D. Dechert, J. A. Scheinkman, and C. L. Sayers for years of shared research activity on nonlinear science. I wish to thank the Santa Fe Institute for hosting a most exceptional nine-day conference at which this material was presented. None of the above are responsible for errors in this paper.

REFERENCES

1. Arrow, K., and F. Hahn (1971), *General Competitive Analysis* (San Francisco: Holden Day).
2. Baumol, W., and J. Benhabib (1987), "Chaos: Significance, Mechanism, and Economic Applications," *R.R. #87-16, Department of Economics, New York University.*
3. Boldrin, M., and R. Deneckere (1987), "Simple Growth Models with Very Complicated Dynamics," *University of Rochester and Northwestern University*, unpublished.
4. Brock, W. A. (1986), "Distinguishing Random and Deterministic Systems," *Journal of Economic Theory* **40(1)**, 168–195.
5. Brock, W. A., and C. Sayers (1986), "Is the Business Cycle Characterized by Deterministic Chaos?," *SSRI W.P. #8617, Department of Economics, The University of Wisconsin*; *Journal of Monetary Economics*, forthcoming in 1988.
6. Brock, W. A., and W. D. Dechert (1986), "Theorems on Distinguishing Deterministic and Random Systems," *University of Wisconsin, Madison, and University of Houston*; Barnett, W., E. Berndt, and H. White, eds. (1988), *Dynamic Econometric Modelling* (Cambridge: Cambridge University Press), forthcoming.
7. Brock, W. A., W. D. Dechert, and J. Scheinkman (1986), "A Test for Independence Based on the Correlation Dimension," *Department of Economics, University of Wisconsin, Madison, University of Houston, and University of Chicago*, unpublished.
8. Brock, W. A. (1988), in *Differential Equations, Stability, and Chaos in Dynamics Economics*, Eds. W. Brock and A. Malliaris (Amsterdam: North Holland), chapter 10, forthcoming in 1988.
9. Eckmann, J., and D. Ruelle (1985), "Ergodic Theory of Chaos and Strange Attractors," *Reviews of Modern Physics* **57**, July, 617–656.
10. Epstein, L. (1987), "The Global Asymptotic Stability of Efficient Intertemporal Allocations," *Econometrica* **55(2)**, 329–356.
11. Frank, M., and T. Stengos (1986), "Measuring the Strangeness of Gold and Silver Rates of Return," *Department of Economics, University of Guelph, Canada*, unpublished.
12. Gennotte, G., and T. Marsh (1986), "Variations in Ex Ante Risk Premiums on Capital Assets, *University of California, Berkeley, Business School*, unpublished.
13. Grandmont, J. (1985), "On Endogenous Competitive Business Cycles," *Econometrica* **53**, 995.
14. Grandmont, J., ed. (1986), *Journal of Economic Theory Symposium on Nonlinear Economic Dynamics*, Vol. 40, #1, October, 1–196.
15. Grassberger, P. and I. Procaccia (1983a), "Measuring the Strangeness of Strange Attractors," *Physica* **9D**, 189–208.

16. Grassberger, P., and I. Procaccia (1983b), "Estimation of the Kolmogorov Entropy from a Chaotic Signal," *Physical Review A* **28**, October, 2591–2593.
17. Ibbotson, R., and R. Sinquefield (1977), *Stocks, Bonds, Bills, and Inflation: The Past (1926–1976) and the Future (1977–2000)*, (University of Virginia: Financial Analysts Research Corporation).
18. Kleidon, A. (1986), "Variance Bounds Tests and Stock Price Valuation Models," *Journal of Political Economy* **94(5)**, 953–1001.
19. Litterman, R., and L. Weiss (1985), "Money, Real Interest Rates, and Output: A Reinterpretation of Postwar U.S. Data," *Econometrica* **53**, 129–156.
20. Marsh, T., and K. Rock (1986), "The Transaction Process and Rational Stock Price Dynamics," *Berkeley: University of California, Berkeley, Business School*, unpublished.
21. Montroll, E., and W. Badger (1974), *Introduction to Quantitative Aspects of Social Phenomena* (New York: Gordon and Breach).
22. Nelson, C., and C. Plosser (1982), "Trends and Random Walks in Macroeconomic Time Series," *Journal of Monetary Economics* **10**, 129–162.
23. Scheinkman, J. and B. LeBaron (1986), "Nonlinear Dynamics and Stock Returns," *University of Chicago*; *Journal of Business*, forthcoming.
24. Scheinkman, J., and B. LeBaron (1987), "Nonlinear Dynamics and GNP Data," *University of Chicago*, forthcoming in Barnett et al.
25. Sims, C. (1984), "Martingale-Like Behavior of Prices and Interest Rates," *Department of Economics, The University of Minnesota*, unpublished.
26. Shiller, R. (1984), "Stock Prices and Social Dynamics," *Brookings Papers on Economic Activity* **2**, 457–510.
27. Wood, R., T. McInish, and J. Ord (1985), "An Investigation of Transactions Data for NYSE Stocks," *The Journal of Finance* **XL(3)**, 723–741.
28. Woodford, M. (1987), "Imperfect Financial Intermediaries and Complex Dynamics," *University of Chicago, Graduate School of Business*, unpublished.

J. DOYNE FARMER & JOHN J. SIDOROWICH†
Theoretical Division and Center for Nonlinear Studies, Los Alamos National Laboratory, Los Alamos, NM 87545 and †Physics Department, University of California, Santa Cruz, CA 95064 (permanent address)

Can New Approaches to Nonlinear Modeling Improve Economic Forecasts?

ABSTRACT

We discuss new approaches to nonlinear forecasting and their possible application to problems in economics. We first embed the time series in a state space, and then employ straightforward numerical techniques to build nonlinear dynamical models. We pick an *ad hoc* nonlinear representation and fit it to the data. For higher dimensional problems we find that breaking the domain into neighborhoods and using local approximation is usually better than using an arbitrary global representation. For some examples, such as data from a chaotic fluid convection experiment, our methods give results roughly fifty times better than those of standard linear models. We present scaling laws for our error estimates based on properties of the dynamics, such as the dimension and Lyapunov exponents, the number of characteristic time scales, the extrapolation time, and the order of approximation of the model. Our methods also provide strong self-consistency requirements on the identification of chaotic dynamics and a more accurate means of computing statistical quantities such as dimension. They also naturally lead to a new method for noise reduction. We compare our methods to neural networks and argue that the basic principles are similar, except that our methods are orders of magnitude faster to implement on conventional computers. At this point we have only applied these new methods to a few economic time series, but so far do not see improvements over conventional linear forecasting procedures.

1. INTRODUCTION

In this paper we summarize results of previous work on nonlinear modeling of chaotic dynamical systems (Farmer and Sidorowich, 1987, 1988a, 1988b), and discuss possible applications to economic time series. Our work follows in the tradition of nonlinear time-series modeling (Gabor, 1954; Weiner, 1958; Priestley, 1980, 1987; Tong and Lim, 1980), but addresses issues that naturally follow from the assumption that a random process is produced by deterministic chaotic dynamics. This leads to new forecasting models, new estimates of error scaling with variations in parameters, and a method of nonlinear smoothing for noise reduction. For other recent work along these lines (see Grassberger, 1987; Cremers and Hübler, 1987; Crutchfield and McNamara, 1987; Casdagli, 1988; Lapedes and Farber, 1987a).

2. MOTIVATION AND PHILOSOPHY

Economic time series are often cited as examples of randomness. But what does it really mean to say that something is random? In this paper we will take the practical view that randomness occurs to the extent that something cannot be predicted. Thus, the statement that economic time series are random is an empirical one: with a better understanding of the underlying dynamics, better measurements, or the computational power to process a sufficient amount of information, behavior that was previously believed random might become predictable.

What causes randomness? There are several possibilities—unpredictability can come about for many reasons. One common cause is ignorance. If we do not know the forces that cause something to change, then we cannot predict its behavior. For example, a common way to choose something randomly is to flip a coin. In the absence of any other information, heads is just as likely as tails, and the outcome is unpredictable. However, if we made precise measurements of the motion of the coin as it left our hand, we could predict the final outcome. People who are skilled in flipping a coin properly can do this. Another classic example of randomness is the game of roulette—the final resting position of the ball is uncertain, which provides an excuse for a great deal of money to change hands. However, as demonstrated by E. O. Thorp and Claude Shannon (Thorp, 1979), and independently by Norman Packard and one of us (jdf) (Bass, 1985), the motion of roulette balls can be predicted using simple physical laws, to a sufficient degree of accuracy to give a significant advantage over the house. Thus we see that when we know the dynamics and have enough information about the state of the system, some classic examples of randomness cease to be random.

Aside from ignorance, three seemingly more basic causes of randomness are quantum mechanics, complexity, and chaos. At the present moment in scientific history, quantum mechanics appears to be the most fundamental cause of randomness. There is apparently an intrinsic limit to the accuracy of measurement that we cannot go beyond, no matter how much we know or how careful we are. Although at this point in time the randomness of quantum mechanics seems inescapable, one wonders whether when we peel back the next layer of the scientific onion, and replace quantum mechanics by a new theory, this apparently fundamental randomness might be caused by something more understandable, such as chaos or complexity. Quantum mechanics seems very far from economics, and while it may ultimately have something to do with the unpredictable nature of human beings, it is probably irrelevant on any immediate level. A more likely cause of randomness in economics is complexity, which we will define here as behavior that involves many irreducible degrees of freedom. Alternatively we will measure the complexity in terms of the *dimension* of the motion in the state space, which roughly speaking should be the same as the number of irreducible degrees of freedom.

Consider the motion of gas particles. Even laying quantum mechanics aside, we can never hope to gather enough information to monitor each particle. Similarly, in an economy there may be so many different agents at work that we cannot hope to keep track of all of them, much less learn the forces that describe their interactions. But even very complex systems may have collective modes of behavior that are described by simple laws. For example, many bulk properties of a gas can be predicted through equilibrium statistical mechanics. While the behavior of the individual particles is random, many bulk properties of the gas are not.

Obviously the analogy to a gas is far too simple to have much bearing on economics. One immediate problem is the simplicity of the interaction of the components. The economy is certainly far too complex to be described by laws as simple as those of statistical mechanics. Another immediate problem is the assumption of equilibria. To someone schooled in nonlinear dynamics, economic time series look very far from equilibrium, and the emphasis of economic theories on equilibria seems rather bizarre. In fact, the use of the word equilibrium in economics appears to be much closer to the notion of attractor as it is used in dynamics rather than any notion of equilibrium as used in physics.

Although still far too simple, fluid turbulence is probably a slightly better analogy for thinking about economic problems. Fluid turbulence is the result of nonequilibrium complex behavior. While the motion is quite complex, it is not so complex that we must describe it in terms of the motion of individual molecules. A turbulent fluid has an effectively infinite number of *possible* degrees of freedom, but in reality only a finite number of them are excited, a much smaller number than one would think possible from naive arguments. Furthermore, there can be large-scale structures in fluid turbulence which persist over long periods of time, and may have predictable dynamics over short periods of time.

Turbulent fluid motion, like the economy, has long been perceived as "random." Where this randomness comes from is one of the fundamental questions in fluid mechanics, and still a matter of debate. One theory, which is sometimes attributed to Landau, is that the randomness arises from the complexity of fluid motion, through the excitation of independent degrees of freedom. Another theory, due to Lorenz (1963) and Ruelle and Takens (1973), is that at least part of the randomness comes about from *sensitive dependence on initial conditions.*

The examples of random behavior cited above, such as the coin toss or roulette, are very fundamental physical systems. At least on a macroscopic level, they can be described in terms of simple physical laws. As originally pointed out by Poincaré (1908), their randomness comes about because a small cause at one time determines a large effect at a later time. Imperceptible errors are amplified to macroscopic proportions, so that the system is random over long times, even though it is described by simple laws. Anyone can roughly predict the motion of a roulette ball by eye a second into the future, but predicting it fifteen seconds into the future is difficult even with more quantitative means. Randomness emerges from determinism, without the need for any complexity on a macroscopic scale. *Chaos* is sensitive dependence to initial conditions that occurs in a sustained way. For a deterministic dynamical system, chaos occurs when on the average nearby trajectories separate at an exponential rate. This is measured by the *Lyapunov exponents*; if any of them are positive, then there is exponential separation and the system is chaotic. Chaos can cause random behavior even in systems with only a few degrees of freedom.

Chaos and complexity by no means preclude each other; chaos also occurs in systems with many degrees of freedom. The Lorenz/Ruelle/Takens theory of turbulence does not exclude the possibility that turbulent flows are also complex—it just says that at least some of the randomness derives from chaos. However, it is also true that the motion of many complex systems asymptotically collapses onto only a few degrees of freedom, and thereby loses much of its complexity, while maintaining chaos and randomness.

For example, in Figure 1, we show a time series obtained from a convecting fluid flow, a $Helium^3$-$Helium^4$ mixture cooled from below (Haucke and Ecke, 1987). This time series shows some regularities, but it is also demonstrably aperiodic and chaotic. Although this motion describes a bulk property of a fluid with an infinite-dimensional state space, there is considerable evidence indicating that the *asymptotic* motion, once transients have died out, lies on a finite-dimensional attractor of roughly dimension three. What this means is that if this system is left unperturbed, only three degrees of freedom are needed to describe the motion. Of course, this is only an approximation—there may be other modes of oscillation that are too small to be resolved here, as well as other very small amplitude modes, such as thermal noise. If we neglect these effects, however, the motion is much simpler than what one might naively expect from an infinite-dimensional state space.

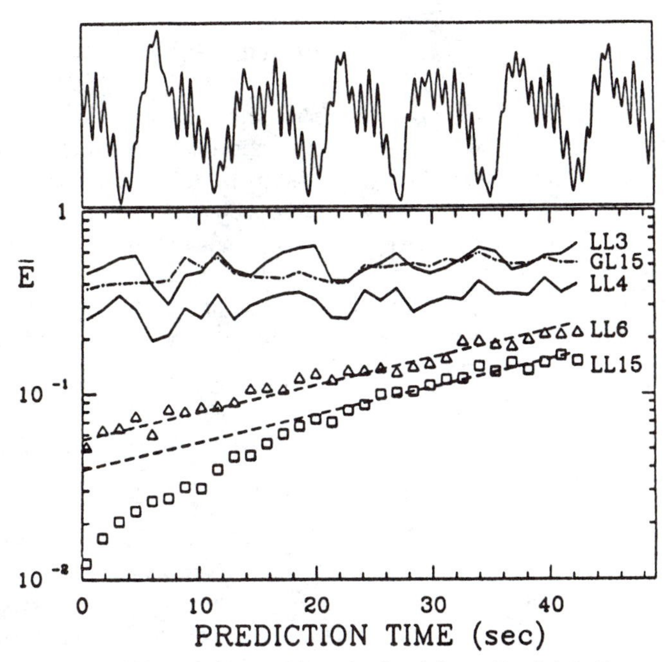

FIGURE 1 Top: An experimental time series obtained from Rayleigh-Benard convection in an He^3-He^4 mixture (Haucke and Ecke, 1987), with Rayleigh number $R/R_c = 12.24$ and dimension $D = 3.1$. Bottom: The normalized prediction error $\bar{E}$ (defined in Eq. (3)) making forecasts with the Local Linear and Global Linear (linear autoregressive) methods. Numbers following the initial indicate the embedding dimension. The dashed lines are from Eq. (6), using computed values of the metric entropy from Haucke and Ecke (1987). Our predictions were based on $N = 30,000$ data points; with a sampling time $\Delta t = 0.07$ seconds, and a delay embedding time $\tau = 10\Delta t$. Based on the mean frequency in the power spectrum the characteristic time is roughly $t_c = 1.5$ seconds, so our database contains roughly 1400 characteristic times.

This brings up another way in which chaos and complexity coexist. Sometimes there may be low-dimensional chaotic behavior, superimposed on much more complex behavior. For example, Shaw (1984) has shown that in some cases the time intervals between drips falling from a water faucet can be described to a good degree of approximation by simple chaotic dynamics. He also makes it clear that in some cases both chaos and complexity are present simultaneously. Pictures of the dynamics suggest simple maps, but they are fuzzy, due to superimposed complexity in the motion. The important point here is that we may find low-dimensional chaotic behavior mixed together with high-dimensional complex behavior.

Does the analogy to fluid flow have any validity for the economy? Is the economy so complex that we cannot hope that there are any low-dimensional attractors, even in an approximate sense? Must we model every possibility, or are their collective modes of behavior that follow deterministic dynamics, involving a tractable number of degrees of freedom?

Work by Boldrin and Scheinkman (this volume) and Brock and Sayers (1986 and Brock, this volume) suggests that some important economic time series might come from chaotic attractors. Their results give encouragement that the quest for low-dimensional chaotic behavior may be appropriate in economics. These results are all based on computations of the fractal dimension, which is currently the most commonly used tool for discovering chaos. Unfortunately, however, this procedure is notoriously unreliable as a means of detecting chaos. Although they have applied the technique quite carefully, it would be nice to have stronger proof.

3. NONLINEAR MODELING

The nonlinear modeling techniques that we discuss here give stronger tests for the presence of low-dimensional deterministic chaos than fractal-dimension computations. In addition, if chaos is present, they make it possible to take advantage of this to make short-term predictions, giving significant improvements over the usual linear techniques. For example, for fluid turbulence data such as that of Figure 2, our predictions are as much as 50 times better than those of standard linear models.

Our approach to forecasting can be summarized as follows:

1. Embed the time series in a d-dimensional state space.
2. Pick a nonlinear d-dimensional representation for the dynamics, and fit the parameters of the representation to the data.
3. Test the model by measuring the accuracy of forecasts, as well as performing other qualitative statistical tests.

EMBEDDING IN A STATE SPACE

There are many ways to do this. For example, for a time series $\{v(t_i)\}$ one can embed with delay coordinates (Yule, 1927; Packard et al., 1980; Takens, 1981), [1]

$$\begin{aligned} x_1(t) &= v(t), \\ x_2(t) &= v(t-\tau), \\ &\vdots \\ x_d(t) &= v(t-(d-1)\tau) \end{aligned} \tag{1}$$

[1] This was also suggested by David Ruelle.

If the time series is produced by nonlinear dynamics on a moderate-dimensional attractor, and the dimension d and the delay time τ are chosen properly, then the resulting vector $x(t)$ will contain enough information to uniquely determine future evolution, and hence define unique states for the system. In practice, this is not always possible with finite precision data, but in many cases it works quite well. For a more complete discussion, see Farmer and Sidorowich (1988a).

PICKING A NONLINEAR REPRESENTATION

Once the data is embedded in a proper state space, we will assume that its future evolution is deterministic, so that we can write $x(t+T) = f(x(t))$. The problem is to fit a functional form $\hat{f}$ to approximate f. We assume that we know some of the history of the system, which gives us some sample pairs $(x(t'), x(t'+T))$ where $t'+T < t$. We want to approximate f, so that given a point $x(t)$ whose future value $x(t+T)$ is unknown we can use $\hat{f}$ to estimate $x(t+T)$.

For example, we could use m^{th} degree d-dimensional polynomials to make a prediction $\hat{x}(t,T)$ at time t.

$$\hat{x}(t,T) = A_m(x(t)_1, \ldots, x(t)_d) = \sum_{i_1=0,\ldots,i_d=0} a_{i_1,\ldots,i_d} x(t)_1{}^{i_1} \cdots x(t)_d{}^{i_d}. \quad (2)$$

where $\sum_{j=1}^{d} i_j \leq m$. The free parameters $a_{i_1,\ldots,i_d}$ can be fit to minimize $||x(t'+T) - \hat{x}(t,T)||^2$ for the sample data points.

The basic problem is that there are an infinite number of nonlinear functional representations, and while in principle any complete representation is sufficient, in practice some perform much better than others. In the absence of any *a priori* information, we are forced to make an arbitrary choice. While some representations perform much better than others across a wide spectrum of problems, there are few rigorous results (Lorentz, 1966), and the selection of optimal representations is largely a black art.

For the global polynomial representation above, for example, the number of parameters $a_{i_1,\ldots,i_d}$ is $(d+m)!/(d!m!) \approx d^m$. For $m > 1$ this number gets large fast as d increases. To get stable fits, the number of parameters must be significantly less than the number of data points, but if m is too small, this representation may be unable to accommodate the variations of f. A similar problem is encountered for any global representation with a finite number of terms.

As discussed in Farmer and Sidorowich (1988a), we have experimented with several different representations, such as polynomials, ratios of polynomials, and radial basis functions (Casdagli, 1988). Lapedes and Farber (1988, 1987a) have also used neural nets, which can be viewed as an alternative representation. Some of these representations produce good approximations, particularly in lower dimensions, but none of them perform very well in higher dimensions.

An alternative approach is to use local representations, by dividing the data into neighborhoods and fitting parameters in each neighborhood separately. We call this *local approximation*. With this approach, item (2) above is expanded into three parts:

1. Pick a local representation.
2. Assign neighborhoods.
3. Find a local *chart* that maps the points in each neighborhood into their future values. To make a prediction evaluate the chart at x.

The basic idea is illustrated in Figure 2.

A simple way to assign neighborhoods is to partition the domain into disjoint sets. For example, we could use a rectangular grid. This approach is convenient, but it has the disadvantage that there is no overlap between the neighborhoods, and therefore no continuity between charts. A point near the boundary of its neighborhood may be poorly approximated. This is particularly true for representations such as polynomials that do a poor job of extrapolating outside their domain of validity.

One way to cope with this problem is to enforce matching conditions between adjacent neighborhoods. For many interpolation schemes, for example, this is an essential element in achieving accuracy. Unfortunately, this becomes a difficult problem for data in more than two dimensions, which is precisely the situation that we are most interested in here.

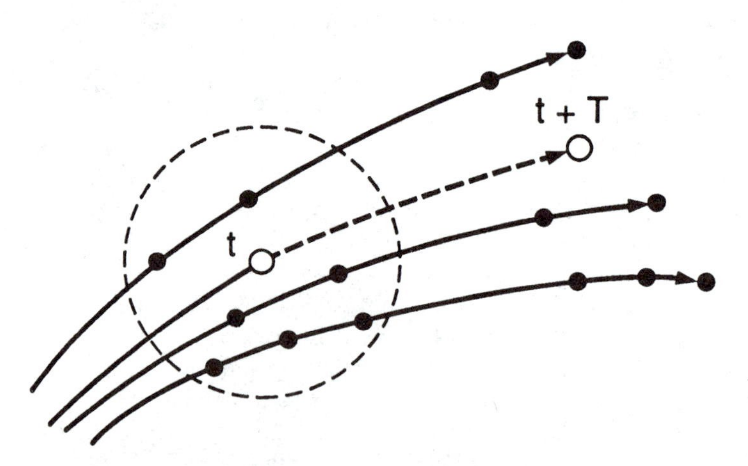

FIGURE 2 Local approximation. The current state $x(t)$ and its unknown future value $x(t+T)$ are represented by open circles. The black dots inside the dashed circle are the neighbors of $x(t)$. To make a prediction we fit a local chart with the neighbors in its domain, and the states they evolve into a time T later in its range. To make a prediction we evaluate the chart at $x(t)$.

An alternative that is more accurate than disjoint partitions and more convenient than enforcing matching conditions is to overlap the neighborhoods, so that each chart is constructed from a good set of neighbors. Let $\{y_i\}$ be the n points of the neighborhood. We want to choose $\{y_i\}$ so that the predictions are as good as possible. A simple criterion is nearness: for a given metric $\|\ \|$ and a given n we will say that $\{y_i\}$ is the *nearest neighborhood* of x if it minimizes $\sum_i \|x - y_i\|$. This criterion is not optimal, as can be seen by considering linear interpolation in two dimensions; if the triangle defined by the three nearest neighbors does not enclose x, then the interpolation may be poor. Problems such as this can be overcome by varying the neighborhood size, testing the sensitivity of the fits and using bigger neighborhoods when they are needed. In practice we find that choosing good neighborhoods makes a big difference in the quality of our predictions.

Once we have chosen neighborhoods, the next step is to fit charts to them. This is essentially the same problem as for global representations. While we anticipate that dividing the domain into neighborhoods reduces the dependence on representation, it is nonetheless the case that some representations are better than others. We are again forced to make an *ad hoc* decision and evaluate the results empirically based on performance.

Note that when we use linear charts, these methods have similarities to the threshold autoregressive model of Tong and Lim (1980) and Priestley's local linear model (Priestley, 1980). For a more complete discussion, see Farmer and Sidorowich (1988a).

To measure the accuracy of forecasts, we use the normalized error,

$$E(t,T) = \frac{(\hat{x}(t,T) - x(t+T))^{\frac{1}{2}}}{\langle (x(t) - \langle x(t) \rangle)^2 \rangle} . \tag{3}$$

In numerical experiments we take the root-mean-square average over many values of t, which we call $\bar{E}$.

Local approximation schemes can be classified according to the order of the derivatives that the errors depend on. For example, suppose our charts are polynomials of degree m. Suppose we want to approximate a function f that is not itself a polynomial of degree m. In the ideal case, the error depends on $f^{(m+1)}(x)$, the $m+1$ derivative of f. This implies that the errors are proportional to ϵ^{m+1}, where ϵ is the spacing between data points. The average spacing between N points uniformly distributed over a D-dimensional space is $\epsilon \approx N^{-1/D}$. Calling q the *order of approximation*, we find[2]

$$\bar{E} \sim N^{-\frac{q}{D}}, \tag{4}$$

where in this case $q = m + 1$.

Achieving the ideal case where $q = m+1$ is difficult for large q, since in general fitting a polynomial of degree m does not produce a fit that is accurate to order

[2] We will use the symbol "$\sim$" to mean "scales as," i.e., $z \sim y(x)$ implies $z = Cy(x)$, where C includes all dependencies on variables other than x.

$m+1$. For example, suppose the number of data points is equal to the number of free parameters. The approximation may go through each point precisely, but in between it may oscillate wildly, producing an extremely inaccurate approximation. Even when we use more neighbors this may limit the accuracy.

To avoid this confusion, we will use Eq. (4) to *define* the order of local approximation, taking the limit as $N \to \infty$, and letting D be the information dimension of the underlying measure of the data points.

$$q = \lim_{N\to\infty} \frac{D|\log \bar{E}|}{\log N} \tag{5}$$

In general q may depend on D, f, the way in which we choose the neighborhoods, and other factors.[3]

Moving to representations of higher degree involves a tradeoff—higher degree representations potentially promise more accuracy, but also require larger neighborhoods. A larger neighborhood usually implies that the variations of f in the neighborhood will be more extreme. Finding the best compromise between these two effects is a central problem in local approximation.

A trivial example of local approximation is *first-order*, or *nearest neighbor* approximation. This amounts to simply looking through the data set for the nearest neighbor, and predicting that the current state will do what the neighbor did a time T later. We approximate $x(t+T)$ by $\hat{x}(t,T) = x(t'+T)$, where $x(t')$ is the nearest neighbor of $x(t)$. For example, to predict tomorrow's weather we would search the historical record and find the weather pattern most similar to that of today, and predict that tomorrow's weather pattern will be the same as the neighboring pattern one day later.[4] First-order approximation can sometimes be improved by finding more neighbors and averaging their predictions, for example, by weighting according to distance from the current state.

An approach that is usually superior is *local linear* or *second-order* approximation. For the neighborhood $\{x(t')\}$, we simply fit a linear polynomial to the pairs $(x(t'), x(t'+T))$. When the number of nearest neighbors $M = d+1$ and the simplex formed by the neighbors encloses $x(t)$, this is equivalent to linear interpolation. If the data is noisy, the chart may be more stable when the number of neighbors is greater than the minimum value. Again, this procedure can be improved somewhat by weighting the contributions of the neighboring points according to their distance from the current state. Linear approximation has the nice property that the number of free parameters and, consequently, the neighborhood size grows slowly with the embedding dimension.

[3] Note that the order of approximation as we have defined it here is one larger than the definition we gave in Farmer and Sidorowich (1987, 1988b). We have changed our definition to correspond to common usage.

[4] This was attempted by E.N. Lorenz, who examined roughly 4000 weather maps (Lorenz, 1969). The results were not very successful because it was difficult to find good nearest neighbors, apparently because of the high dimensionality of weather.

Since the accuracy increases with the order of approximation, it is obviously desirable to make the order of approximation as large as possible. Any nonlinear representation is a candidate for higher order approximation. The criteria for a good local representation are somewhat different from those for a good global representation. On one hand, getting a good fit within a local neighborhood is easier, because the variations are less extreme. Wild variations or discontinuities can be accommodated by assigning neighborhoods properly. On the other hand, a local neighborhood necessarily has less data in it, so the representation must make efficient use of data. For reasonably low-dimensional problems, we have found that we usually easily achieve third-order approximation, for example using quadratic polynomials. In low dimensions (two or less), it is sometimes possible to do better.[5] In higher dimensions we often find that we cannot do any better than second order approximation, at least with our current techniques.

4. ERROR SCALING

The accuracy of predictions naturally decays as we attempt to extrapolate further forward in time. The rate of decay depends on the prediction method. There are two basic approaches: *direct* prediction, in which we directly approximate for each extrapolation time T, and *iterated* approximation, in which we approximate for some shorter time T_{comp} and then iterate this approximation to make an estimate for time T. As we discuss in Farmer and Sidorowich (1988a), in the limit where $\bar{E}$ is small, for direct approximation the errors scale roughly as

$$\bar{E} \sim N^{-\frac{q}{D}} e^{q\lambda_{max}T}, \tag{6}$$

whereas for iterated approximation the errors grow roughly as

$$\bar{E} \sim N^{-\frac{q}{D}} e^{\lambda_{max}T}. \tag{7}$$

λ_{max} is the largest Lyapunov exponent, which is the rate at which the logarithm of the distance between nearby trajectories separate. Clearly iterated approximation is superior. At $T = 0$ the average error is the same in both cases, but for iterated approximation it increases more slowly with time. Note that while we have stated this in terms of the number of data points, for continuous time series a more appropriate measure is the number of characteristic time scales; as the time series is sampled more finely, we reach a point of diminishing returns where more data gives very little improvement, unless we examine a longer stretch of the time series.

[5] Casdagli has independently reached the same conclusions (Casdagli, 1988). In two dimensions he apparently achieves *sixth*-order approximation in some cases using global radial basis functions, but this does not seem to carry over in more dimensions.

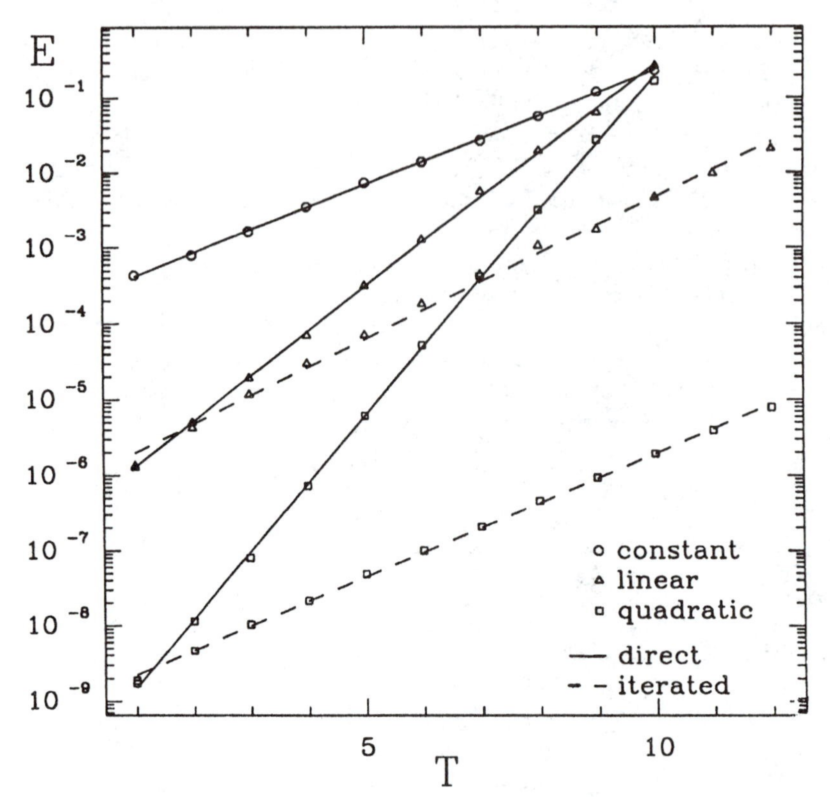

FIGURE 3 Forecasts made using local polynomial approximation, building the data base from a $5,000$-point time series generated by Eq. (8). We make 500 forecasts, and plot the average error as a function of the extrapolation time T. The logarithm of the error for direct forecasts grows roughly according to $q\lambda$ whereas for iterative forecasts it grows according to λ, where in this case $q = m + 1$. λ is the Lyapunov exponent, which in this case is one bit per iteration.

The validity of these estimates for a simple one-dimensional map,

$$x_{t+1} = sin(\pi x_t), \tag{8}$$

where $-1 < x_t < 1$, is demonstrated in Figure (3), where we show the approximation error as a function of the extrapolation time T. As predicted, the error grows roughly according to Eq. (6) for direct approximation and according to Eq. (7) for iterative approximation.

We do not always observe this scaling in practice; in fact, in some cases direct forecasting is apparently superior to iterated forecasting (Farmer and Sidorowich, 1988a). We do not understand this at present, and intend to investigate it further in the future.

4.1 EXPERIMENTAL DATA ANALYSIS

Once we have a nonlinear model, to the extent that the model accurately represents the data, all the techniques that were previously available only in numerical experiments become available in physical experiments. As we argue in Farmer and Sidorowich (1988a), if we can achieve approximation of order higher than one, we can use this to compute quantities such as the fractal dimension more accurately.

In addition, forecasting provides a much stronger test for the presence of low-dimensional chaos than the usual procedure of computing fractal dimension and Lyapunov exponents. It is very unlikely to achieve accurate forecasts unless the dynamics are really low dimensional. Furthermore, the scaling laws given in Eq. (7) provide strong self-consistency requirements.

4.2 NOISE REDUCTION

Forecasting can also be used as a means of noise reduction. When the equations of motion are known exactly, even with a very small number of data points noise reductions of ten orders of magnitude are not hard to achieve. When we must learn the equations of motion, the extent of noise reduction is limited by the accuracy of forecasts, but under favorable circumstances noise reductions of several orders of magnitude are still possible. Please see Farmer and Sidorowich (1988a) for a more detailed discussion.

5. REAL WORLD APPLICATIONS?

The error-scaling formulas of the previous section give an idea of the circumstances in which we can expect our methods to give improvements over the standard techniques. The dimension D is a measure of the complexity of the system, and if the complexity is too large, or $\lambda_{max}T$ is too large, then we expect very little improvement over standard linear methods. But if the time series has components of its behavior which are described by nonlinear dynamics on an attractor and we have enough data, we may see significant improvements over conventional linear methods.

So far we have very little experience applying these techniques to more realistic applications. We have made limited attempts to study a few financial time series, such as the Dow-Jones Industrial Average, IBM stock, and gold currency prices. In each of these cases we see very little if any advantage over conventional linear forecasting methods. However, we do not feel that we have enough evidence as yet to conclude that our techniques will not ultimately give improvements, at least in some cases. The few time series we have analyzed so far have 1000 points or less, and are probably not the most appropriate time series for our techniques. We have not had time yet to implement several of the improvements described in Farmer and Sidorowich (1988a), such as methods for dealing with nonstationary behavior, which are certain to be very important in dealing with economic time series. In the near future, we hope to apply our methods to a much broader class of time series.

6. ADAPTIVE DYNAMICS

There has been a great deal of interest recently in solving artificial intelligence problems with *adaptive networks* such as neural nets (Rummelhart and McClelland, 1986; Cowan and Sharp, 1987) and the classifier system (Holland, 1986). Although on the surface the straightforward approximation techniques that we employ here seem quite different from neural nets, the underlying principles are actually much the same. However, since our representations are more convenient numerically, fitting parameters is hundreds of times faster.

Forecasting is an example of what is often called *learning with a teacher*. The task is to predict "outputs," based only on "inputs." For forecasting, the input is the present state and the output is the future state. The record of past states provides a set of known input-output pairs which acts as a "teacher." The problem is to generalize from the teaching set and estimate unknown outputs.

We can restate the problem more formally as follows: given an input x_i and an output y_i, we want to find maps F and G of the form

$$\hat{y}_i = F(x_i, \alpha_i) \tag{9}$$

$$\alpha_{i+1} = G(x_i, \hat{y}_i, y_i, \alpha_i) \tag{10}$$

that minimize $||y_i - \hat{y}_i||$, where $\hat{y}$ is an estimate of y, and the metric $||\ ||$ provides a criterion for the estimation accuracy. α_i are parameters for F, and G is a map that changes α_i, i.e., a *learning algorithm*. x, y and i can be either continuous or discrete. For a forecasting problem, for example, i corresponds to time, x is the current state x_t, and y is a future state, $y_t = x_{t+T}$.

Neural nets correspond to a particular class of functional forms for F and G. Although this form was originally motivated by biology, there is no reason to be constrained by this in artificial intelligence problems, as reflected by many recent developments in this field. Neural nets have had success in certain problems that can be solved by learning with a teacher, for example, text to speech conversion (Sejnowski and Rosenberg, 1987) or finding gene locations on DNA sequences (Lapedes and Farber, 1987b). Lapedes and Farber have also shown that neural nets can be effective for forecasting (Lapedes and Farber, 1987a).

However, as recently pointed out by Omohundro (1987), alternative approaches that depart significantly from the usual form of neural nets may be computationally much more efficient. Our approach to forecasting provides a good example of this; our methods give equivalent or more accurate forecasts than the neural net of Lapedes and Farber (1987b), and are several orders of magnitude more efficient in terms of computer time. Furthermore, since the computations can be performed in parallel, we expect that this speed discrepancy will persist even with future parallel hardware. Omohundro has pointed out that similar methods may be employed for other problems, such as associative memory, classification, category formation, and the construction of nonlinear mappings. Many aspects of the methods that we have proposed here are applicable to this broader class of problems.

Although we have assumed that x and y are continuous, with the addition of thresholds our methods are easily converted to the discrete domain. Our work, taken together with that of Lapedes and Farber, makes it clear that the neural network solves problems by surface estimation. They show that the same is true in the discrete domain, except that answers are obtained by "rounding" the surface, truncating to a discrete value. Generalization occurs through the extrapolation of the surface to regions in which there is no data (Lapedes and Farber, 1988). There is no *a priori* reason to constrain the functional representation to those that are currently popular in the study of neural networks.

7. CONCLUSIONS

By making explicit nonlinear models using straightforward numerical techniques, we can sometimes do a very good job of forecasting and reproducing qualitative dynamics. Thinking in terms of deterministic chaos naturally leads to several improvements over previous nonlinear methods, such as error estimates, higher orders of approximation, and a new approach to noise reduction. These methods are still new, and a great deal remains to be done to improve them. However, we are rapidly reaching the stage where they can be applied to more realistic applications. Within the next year we hope to know whether or not there are circumstances in economics in which nonlinear models can make better forecasts than conventional linear techniques. The answer to this question should also give us fundamental information about the intrinsic complexity of the economy.

We are grateful for support from the Department of Energy and the Air Force Office of Scientific Research under grant AFOSR-ISSA-87-0095.

We urge the reader to use these results for peaceful purposes.

REFERENCES

1. Bass, Thomas (1985), *Eudaemonic Pie* (New York: Houghton-Mifflin).
2. Boldrin, M., and J. A. Scheinkman (1988), "Learning-By-Doing, International Trade and Growth: A Note," this volume.
3. Brock, W. A. (1988), "Nonlinearity and Complex Dynamics in Economics and Finance," this volume.
4. Brock, W. A., and C. L. Sayers (1986) "Is the Business Cycle Characterized by Deterministic Chaos?" *Technical Report 8617, Social Systems Research Institute, University of Wisconsin, Madison.*
5. Casdagli, M. (1988), "Nonlinear Prediction of Chaotic Time Series," *Technical Report, Queen Mary College, London.*
6. Cowan, J. D., and D. H. Sharp (1987), "Neural Nets," *Technical Report LA-UR-87-4098, Los Alamos National Laboratory.*
7. Cremers, J., and A. Hübler (1987) "Construction of Differential Equations from Experimental Data," *Z. Naturforsch* **42a**, 797–802.
8. Crutchfield, J. P., and B. S. McNamara (1987), "Equations of Motion from a Data Series," *Complex Systems* **1**, 417–452.
9. Farmer, J. D., and J. J. Sidorowich (1987), "Predicting Chaotic Time Series," *Physical Review Letters* **59(8)**, 845–848.
10. Farmer, J. D., and J. J. Sidorowich (1988a), "Exploiting Chaos to Predict the Future and Reduce Noise," *Los Alamos National Laboratory Preprint.*
11. Farmer, J. D., and J. J. Sidorowich (1988b), "Predicting Chaotic Dynamics," *Dynamic Patterns in Complex Systems*, Eds. J. A. S. Kelso, A. J. Mandell, and M. F. Shlesinger (Singapore: World Scientific).
12. Gabor, D. (1954), "Communication Theory and Cybernetics," *Transactions of the Institute of Radio Engineers* **CT-1(4)**, 9.
13. Grassberger, P. (1987), "Information Content and Predictability of Lumped and Distributed Dynamical Systems," *Technical Report WU-B-87-8, University of Wuppertal.*
14. Haucke, H., and R. Ecke (1987), "Mode Locking and Chaos in Rayleigh-Benard Convection," *Physica* **25D**, 307.
15. Holland, J. (1986), "Escaping Brittleness: the Possibilities of General Purpose Machine Learning Algorithms Applied to Parallel Rule-Based Systems," *Machine Learning II*, Eds. Michalski, Carbonell, and Mitchell (Kaufmann).
16. Lapedes, A. S., and R. M. Farber (1988), "How Neural Nets Work," *Technical Report, Los Alamos National Laboratory.*
17. Lapedes, A. S., and R. Farber (1987a), "Nonlinear Signal Processing Using Neural Networks: Prediction and System Modeling," *Technical Report LA-UR-87-, Los Alamos National Laboratory*; submitted to *Proc. IEEE.*
18. Lapedes, A. S., and R. M. Farber (1987b), "Neural Net Learning Algorithms and Genetic Data Analysis," in preparation.
19. Lorentz, G. G. (1966), *Approximation of Functions* (New York: Holt, Rinehart, and Winston).

20. Lorenz, E. N. (1969), "Atmospheric Predictability as Revealed by Naturally Occurring Analogues," *Journal of the Atmospheric Sciences* **26**, 636–646.
21. Lorenz, E. N. (1963), "Deterministic Nonperiodic Flow," *Journal of Atmospheric Science* **20**, 130–141.
22. Omohundro, S. (1987), "Efficient Algorithms with Neural Network Behavior," *Complex Systems* **1**, 273–347.
23. Packard, N. H., J. P. Crutchfield, J. D. Farmer, and R. S. Shaw (1980), "Geometry from a Time Series," *Physical Review Letters* **45**, 712–716.
24. Poincaré, H. (1908), *Science et Methode* (Biblioteque Scientifique); English translation by F. Maitland (1952), *Science and Method* (Dover).
25. Priestley, M. B. (1987), "New Developments in Time-Series Analysis," *New Perspectives in Theoretical and Applied Statistics*, Eds. M. L. Puri, J. P. Vilapiana, and W. Wertz (New York: John Wiley and Sons).
26. Priestley, M. B. (1980), "State Dependent Models: a General Approach to Nonlinear Time Series Analysis," *Journal of Time Series Analysis* **1**, 47–71.
27. Ruelle, D., and F. Takens (1973), "On the Nature of Turbulence," *Comm. Math. Phys.* **20**, 167.
28. Rummelhart, D., and J. McClelland (1986), *Parallel Distributed Processing* (Cambridge, MA: MIT Press), vol. 1.
29. Sejnowski, T. J., and C. R. Rosenberg (1987), "Nettalk," *Complex Systems* **1**.
30. Shaw, R.S. (1984), *The Dripping Faucet as a Model Dynamical System* (Santa Cruz, CA: Aerial Press).
31. Takens, F. (1981), "Detecting Strange Attractors in Fluid Turbulence," *Dynamical Systems and Turbulence*, Eds. D. Rand and L.-S. Young (Berlin: Springer-Verlag).
32. Thorp, E. O. (1979), "The Mathematics of Gambling: Systems for Roulette I," *Gambling Times* **January 1979**, 62.
33. Tong, H., and K. S. Lim (1980), "Threshold Autoregression, Limit Cycles and Cyclical Data," *Journal of the Royal Statistical Society B* **42(3)**, 245–292.
34. Weiner, N. (1958), *Nonlinear Problems in Random Theory* (New York: John Wiley and Sons).
35. Yule, G. U. (1927), *Philos. Trans. Roy. Soc. London A* **226**, 267.

JOHN H. HOLLAND
The University of Michigan

The Global Economy as an Adaptive Process

By way of prologue: this paper is no more than an idiosyncratic view of the problems posed by the global economy. It constitutes a kind of annotated outline with mention of relevant concepts from mathematics and the computer sciences. It is recklessly egotistical in citing five pieces of my own work that carry these ideas further (many who have contributed to this outlook are properly cited therein). Other than that, it cites only four other papers that enlarge upon particular points along the way. While these ideas are presented in my own peculiar dialect, they owe much to long-term interactions with my colleagues in the BACH group at the University of Michigan, and to intensive discussions at the Santa Fe Institute and the Los Alamos National Laboratory. [Technical comments related to the more general discussion are included in square brackets.]

The global economy has features that, *in the aggregate*, make it a difficult subject for traditional mathematics:

1. The overall direction of the economy is determined by the interaction of many dispersed units acting in parallel. The action of any given unit depends upon the state and actions of a limited number of other units.

2. There are rarely any global controls on interactions—controls are provided by mechanisms of competition and coordination between units, mediated by *standard operating procedures* (SOPs), assigned roles, and shifting associations.

3. The economy has many levels of organization and interaction. Units at any given level typically serve as "building blocks" for constructing units at the next higher level. The overall organization is more than hierarchical, with all sorts of tangling interactions (associations, channels of communication) across levels.

4. The building blocks are recombined and revised continually as the system accumulates experience—the system adapts.

5. The arena in which the economy operates is typified by many *niches* that can be exploited by particular adaptations; there is no universal *super*-competitor that can fill all niches (any more than such would be the case in a complex ecology such as a tropical forest).

6. Niches are continually created by new technologies and the very act of filling a niche provides new niches (cf. parasitism, symbiosis, competitive exclusion, etc., in ecologies). Perpetual novelty results.

7. Because the niches are various, and new niches are continually created, the economy operates far from an optimum (or global attractor). Said in another way, improvements are always possible and, indeed, occur regularly.

The global economy is an example, par excellence, of an *adaptive nonlinear network* (ANN hereafter). Other ANNs are the central nervous system, ecologies, immune systems, the developmental stages of multi-celled organisms, and the processes of evolutionary genetics. ANNs provide for a substantial extension of traditional economics. The building blocks of traditional economics are fixed rational agents that operate in a linear, static, statistically predictable environment. In contrast, ANNs allow for intensive nonlinear interactions among large numbers of changing agents. These interactions are characterized by limited rationality, adaptation (learning), and increasing returns. (Typical examples in the global economy are: entrainment of speculators in the stock market, anticipation of shortages and gluts, learning effects in high-technology, and niche creation wherein a successful innovation creates a web of supportive and augmentive economic activities.)

The usual mathematical tools, exploiting linearity, fixed points and convergence, provide only an entering wedge when it comes to understanding ANNs. Deeper understanding requires a mathematics that puts more emphasis on combinatorics coordinated with computer modeling. In particular, the understanding of ANNs require a mathematics and modeling techniques that emphasize the *discovery* of building blocks and the *emergence* of structure through the combination and interaction of these building blocks.

One major feature of ANNs is well illustrated by the global economy: ANNs do *not* act in terms of stimulus and response, they *anticipate*. In the global economy, the anticipation of an oil shortage or of a significant default of foreign loans can have profound effects upon the course of the economy, whether or not the anticipated events come to pass. Participants in the economy build up models of the rest of the economy and use them to make predictions. These models, sometimes called *internal models* are rarely explicit. They are usually more *prescriptive* (prescribing what should be done in a given situation) than *descriptive* (describing the options in a given situation).

[A prescriptive model plays a role in economics quite similar to the role of a *strategy* in the theory of games. In games that are not trivial (checkers, chess, or go, for instance), the feasible strategies are not optimal (the *min-max* strategies for these games are unknown). The strategies are typically formulated in terms of *lookahead*, anticipation of countermoves by opponents. Lookahead obviously requires models of the opponents and the game itself. The emphasis is upon progressive improvement of strategies through improvements in the models, and innovations are continually being made. In mathematical terms, operation is far from any global attractor and strategies are faced with perpetual novelty. Even the most complex games are quite simple compared to ANNs, so these considerations hold *a fortiori* for ANNs in general, and economies in particular.]

In an economy, the prescriptions of an internal model are typically procedures, SOPs, that suggest and dictate actions to be taken under specific conditions. It is important that these SOPs have been acquired (learned) from encounters with similar situations in the past. Instead of being explicitly situated in some central planning bureau, such a model is distributed, being localized in the units that act upon the SOPs. Moreover, the actions and predictions made by the SOPs at one level often combine in their effects to provide "higher-level" implicit internal models that yield more globally directed actions and predictions. As a result, the anticipations generated are much more complex, and subtle, than those generated by any single explicit model.

[Models have traditionally been defined in mathematics by using *homomorphic* maps. However, the constraints that must be satisfied in order to have a homomorphic model can almost never be met in a realistic ANN, such as an economy. Internal models are almost always imperfect, and they are being continually revised. Hierarchical, imperfect models defined in terms of SOPs can be captured mathematically as *quasi-homomorphisms* (see Holland et al., 1986). In this context, the notion of progressive refinement of a model through refinement of SOPs is well captured by the addition of levels to a quasi-homomorphism.]

In an economy, as in ANNs in general, accumulated experience provides increasingly refined SOPs and progressively more sophisticated interactions between them. This process is a familiar one. Classical SOPs for building vaulted ceilings became refined and elaborated until they culminated in the gothic cathedrals. Even modern structural analysis can justify gothic structures only in conjunction with appropriate computer models—an illustration of the way in which the ability to control and predict via SOPs can vastly surpass the ability to analyze and model explicitly.

Nevertheless, theoretical understanding, once achieved, provides increased scope, allowing the design of Wright's mile-high skyscraper and NASA's space shuttle, both of which are distant structural cousins of the gothic cathedral. Ultimately, for systems as complex as the global economy, whatever mathematical understanding is attained must be coordinated with computer models designed to test extrapolations and predictions based on data.

[The perpetual novelty engendered by the progressive refinement of internal models, and by revisions and recombinations of building blocks (technological innovation, and the like), poses an additional problem not traditionally encountered in mathematics. One way to tackle perpetual novelty mathematically is to consider a Markov process in which each of the states has a recurrence time that is large with respect to feasible observation times. The states of this process are the counterparts of the *unknown* possibilities for the overall process. The units in an ANN are constrained to observe only certain features of these states, so that states with the same observed features are indistinguishable. In effect, equivalence classes are imposed upon the states of the original process. By treating these equivalence classes as the states of a new process, we implicitly define a *derived Markov process*. If the equivalence classes are large enough, the states of the derived process have shorter recurrence times, allowing repeated observation. In effect, perpetual novelty is approached by treating certain features as irrelevant, thereby defining *events* that do recur in feasible times. The basic mathematical task is to determine conditions under which the derived Markov process provides valid predictions for the underlying process (see Holland, 1986a).]

For me, the premier theoretical question about the evolution of the global economy, and of ANNs in general, is: how is it that, in response to the selective pressures of the ANN's history, internal models emerge to exert control on the ANN's responses? Because implicit internal models are so pervasive in ANNs, and because anticipatory behavior is *the* characteristic behavior of ANNs, I believe that we will be constrained to a relatively primitive level of understanding until we can make some progress along these lines. The present situation seems quite similar to the situation in evolutionary theory prior to the development of a mathematical theory of genetical selection, and, in some respects, prior to Darwin's great synthesis.

Clearly the selective pressures placed on an ANN come from its environment. Different environments determine different evolutionary paths. Thus, to study the emergence of models, one must characterize the ANN's environment. It is traditional in economics to carry out this characterization by attaching a *utility* to the various states of the environment. The role of utility in economics is quite similar to the role of *payoff* in game theory, the *error function* in control theory, *fitness* in evolutionary genetics, and so on. It puts a worth on the states and, hence, implicitly defines goals for the system. (Such a function is sometimes called a sufficient function in that it sums up a myriad of effects contributing to the system's evolution.) Somehow the ANN must search out the valuable states in the environment. For the reasons cited earlier, this is primarily a search for improvements, not a search for optima.

Given a representation of the state space in terms of components (features or properties), the problem facing an ANN is particularly simple if the utility function is linear over the components. It becomes progressively more difficult as the utility function becomes increasingly nonlinear (the counterpart in economics of *epistatic effects* in genetics.)

[Because nonlinear is a definition by exclusion—the set of all functions that are *not* linear—it has been difficult to determine the difficulty of the environment from a specification of the utility function. However, there is now a kind of transform, the *hyperplane transform*, that provides a relation between the specification of a nonlinear function and the information provided by any biased sample of the function's argument space (see Holland, 1988). Specifically, the hyperplane transform represents the utility function in terms of its averages, under the sampling distribution, over selected hyperplanes in the environmental state space. It is easily shown that a system gathers information about the averages associated with high-dimensional hyperplanes much more rapidly than it does for progressively lower-dimensional refinements. This simple observation has strong implications for the way in which an algorithm should search for improvements in nonlinear domains. The transform makes possible the comparative study of various search operators, such as sampling by making random changes, and sampling by recombining building blocks (see the discussion of mutation and recombination below).]

My own particular approach to the study of ANNs involves parallel, message-passing, rule-based systems, called *classifier systems*, in which many rules are active simultaneously (see Holland, 1986b). The rules in a classifier system are the counterparts of the interacting SOPs in an economic system. The rules are in a condition/action form, so that a given rule takes action whenever its conditions are satisfied. The typical action is the posting of a message that brings other rules into action. The set of rules active at any instant amounts to a system's model of its current situation. Because a given rule can be active in many situations, it constitutes both a component for a variety of models, and a hypothesis to be tested by use.

[Typically, classifier systems are organized in, or evolve to, a tangled default hierarchy. Rules at the top of the hierarchy are quite general and often wrong—their conditions specify coarse, high-dimensional hyperplanes in the environmental state space. Rules at successively deeper levels are ever more specific, specifying exceptions to the higher level rules. Rules also interact "horizontally" to provide relations and associations, yielding an overall organization much like Fahlman's (1979) NETL.]

Each rule in a classifier system is assigned a strength that reflects its usefulness in the context of other active rules. When a rule's conditions are satisfied, it competes with other satisfied rules for activation. The stronger the rule, the more likely it is to be activated. This procedure assures that a rule's influence is affected by both its relevance (the satisfied condition) and its confirmation (the strength). Usually many, but not all, of the rules satisfied will be activated. It is in this sense that a rule serves as a hypothesis competing with alternative hypotheses. Because of the competition there are no internal consistency requirements on the system; the system can tolerate a multitude of models with contradictory implications.

A rule's strength is supposed to reflect the extent to which it has been confirmed as an hypothesis. This, of course, is a matter of experience, and subject to revision. In classifier systems, this revision of strength is carried out by the *bucket-brigade* credit assignment algorithm. Under the bucket-brigade algorithm, a rule actually bids a part of its strength in competing for activation. If the rule wins the competition, it must pay this bid to the rules sending the messages that satisfied its conditions (its suppliers). It thus pays for the right to post its message. The rule will regain what it has paid only if there are other rules that in turn bid and pay for its message (its consumers). In effect, each rule is a middleman in a complex economy, and it will only increase its strength if it turns a profit.

[Some classic fixed-point theorems in economics can be used to show that under certain, statistically stable conditions, this algorithm actually adjusts the strengths in an appropriate fashion.]

Because the systems being studied are complex, and the possible rule sets are astronomical in number, it is likely that the rules the system starts can be greatly improved. This is no longer a matter of confirmation. Even if the bucket-brigade credit assignment algorithm works perfectly, it can only rank the rules already present. The system must have some means to replace weak rules with new rules that are *plausible.* It is not enough to generate new rules at random; somehow the generation process must be dependent upon the system's experience as to what is reasonable and what is unreasonable. The rule-generating algorithms for classifier systems are *genetic algorithms.* In effect, a genetic algorithm carries out a subtle search for useful building blocks (*schemas*) that can be recombined to form new rules. The building blocks used are components of the more successful rules. The rules generated by the genetic algorithm replace weak rules and serve in turn as plausible hypotheses to be tested under competition.

[Classifier systems are unlike most rule-based systems in that they evolve and can be studied theoretically (several useful theorems already exist). In studying this evolution it is useful to note that there is a natural limit on the hierarchical complexity that can be supported by a given amount of experience. Confirmation proceeds more rapidly for higher level rules in a default hierarchy—such rules are sampled more often because they are satisfied more often. A rule with a condition picking out a lower-dimensional hyperplane in the state space will be sampled at a lower frequency. Until a lower-dimensional hyperplane has been sampled, it makes little sense to construct a rule that specifies an action to be taken when that condition is encountered.]

[The heart of the genetic algorithm is an operator that treats rules very like chromosomes under genetic *recombination.* Extensions of Fisher's theorems on evolutionary genetics provide some insights into the working of genetic algorithms. It is important that recombination rather than mutation is the main driving force of the search. (Recombination also plays a major role in providing the repertoire for the immune system, and it seems more and more likely that recombination plays a more important role than mutation in supplying the variants in evolution.) The fundamental theorem about genetic algorithms shows that new rules are generated primarily by the recombination of parts—the building blocks—of rules that have already been useful to the system. These building blocks, called *schemas*, are technically hyperplanes in the space of all possible rules. The most surprising part of the theorem shows that each time M rules are tested and processed by the genetic algorithm, at least M^3 schemas have been implicitly rated and used to bias the construction of new rules (see Holland, 1987). This tremendous "speed-up" goes a long way toward explaining why complex adapted structures can be built up in relatively short times. Classifier systems lend themselves well to simulation on highly parallel computers—currently, simulations involving 8000 rules are being run on the Connection Machine (Hillis, 1986).]

A direct way to apply classifier systems to the study of the global economy is to find an established model from economics that, even in its simplest form, raises some of the central quandries of the global economy (such as the ability of one nation to implicitly tax another by "printing money," or any of the several questions about international debt raised by Mario Simonsen in his lecture in this volume). The model should be easily extendible in several dimensions (number of countries, number of commodities, number of generations, arbitrage procedures, etc.) so that ever more realistic situations can be studied. In my opinion, the two-nation *overlapping generations* model (see the discussion in Sargent, 1981) goes a long way toward meeting all of these criteria. There are several ways of modifying the model that allow comparisons between the standard *equilibrium* solutions, which require perfect one-step *foresight* in the agents, and solutions that would arise if adaptive, rule-based agents were used. In particular, there are counterparts of the *prisoner's dilemma* model studied by Axelrod (in this context, the possibility that both nations print money, each attempting to implicitly tax the other, until the money is worthless). This opens the possibility of employing a genetic algorithm to explore the space of agent strategies, much as Axelrod has done in some of his more recent work on the prisoner's dilemma (see Axelrod, 1987). The object is to locate strategies that are stable against other competing strategies—the *evolutionary stable strategies*—counterparts of the strategy TIT-FOR-TAT in Axelrod's work. Such a study should help us to deepen our understanding of the trajectories (of prices, commodity amounts, etc.) induced by the interaction of a *variety* of limited agents faced with continuing improvements (technological innovations).

REFERENCES

1. Axelrod, R. (1987), "The Evolution of Strategies in the Iterated Prisoner's Dilemma," *Genetic Algorithms and Simulated Annealing*, Ed. L. D. Davis (Los Altos, CA: Morgan Kaufmann).
2. Fahlman, S. E. (1979), *NETL: A System for Representing and Using Real-World Knowledge* (Cambridge: MIT Press).
3. Hillis, W. D. (1985), *The Connection Machine* (Cambridge: MIT Press).
4. Holland, J. H., K. J. Holyoak, R. E. Nisbett, and P. R. Thagard (1986), *Induction: Processes of Inference, Learning, and Discovery* (Cambridge: MIT Press), 385.
5. Holland, J. H. (1986a), "A Mathematical Framework for Studying Learning in Classifier Systems," *Physica* **22D**, 307.317; also in *Evolution, Games and Learning*, Eds. J. D. Farmer, A. Lapedes, N. H. Packard and B. Wendroff (Amsterdam: North Holland).
6. Holland, J. H. (1986b), "Escaping Brittleness: The Possibilities of General Purpose Machine Learning Algorithms Applied to Parallel Rule-Based Systems," *Machine Learning II*, Eds. R. S. Michalski, J. G. Carbonell, and T. M. Mitchell (Los Altos, CA: Morgan Kaufmann), chapter 20.
7. Holland, J. H. (1987), "Genetic Algorithms and Classifier Systems: Foundations and Future Directions," *Genetic Algorithms and Their Applications*, Ed. J. J. Grefenstette (Hillsdale, NJ: Lawrence Erlbaum).
8. Holland, J. H. (1988), "The Dynamics of Searches Directed by Genetic Algorithms," *Evolution, Learning, and Cognition*, Ed. Y. C. Lee (Singapore: World Scientific).
9. Sargent, T. J., and L. P. Hansen (1981), "Linear Rational Expectations Models of Dynamically Interrelated Variables," *Rational Expectations and Econometric Practice* (Minneapolis: U. Minnesota Press).

STUART A. KAUFFMAN
Department of Chemistry, Biophysics and Biochemistry, School of Medicine, University of Pennsylvania, Philadelphia, PA 19174

The Evolution of Economic Webs

I. INTRODUCTION

My aim in this article is to sketch, in admittedly rough form, a web of ideas which I hope bear on the issue of the evolution of economic webs. The opportunity for this effort is the workshop on the global economy, which has encouraged a biologist to turn his mind to a neighboring field.

Even a non-economist, regarding a hunter-gatherer society, or the economy which graced an early agricultural society in Mesopotamia or the American West, and comparing it to the economy of a modern continental nation such as the United States, is struck by the enormous increase in the diversity of means of earning a living which our present economy affords. Mathematical economics, it must be admitted, is a refined way of catching dinner. The coupled web of economic activities, the production and sale of goods and services, has obviously increased in complexity over time. Yet apparently we have no well-developed body of theory for this historical process, even in the presence of the work of economic historians, and the large field of developmental economics. Whether it is fair for an outsider to suppose that no adequate theoretical picture is available is not for me to say. But if it exists, then at least it is not well known to the new economic friends who gathered in Santa Fe.

The "basic image" I wish to discuss is this: an economy is a web of transformations of products and services among economic agents. Over time, "technological evolution" generates new products and services which must mesh together "coherently" to jointly fulfill a set of "needed" tasks. It is this web of needed tasks which affords economic opportunity to agents to sell, hence earn a living, and thus obtain the capacity in money or other form to maintain demand for those very goods and services. Like living systems bootstrapping their own evolution, an economic system bootstraps its own evolution. New technological advances afford new opportunities for yet further tasks and products which may integrate into such an economic web. Old products and tasks, no longer coherent members of the web, no longer useful, hence no longer "profitable," drop away. The web transforms over time, driven by technological advances and economic opportunity. The web at any moment has some structure with respect to its connectivity features: what tasks, goods, and services exist and are sold to which others as intermediate or final products in the economy. Thus we are led to some general questions: (1) What is the web in any given economy? (2) What technological and economic forces govern the transformation of the web over time? (3) May there be "laws" about the statistical features of such webs and their time evolution?

My purpose, then, is to sketch some of the issues which would appear natural components in any adequate theory for these phenomena. This article is organized into the following sections: in the second section, I consider combinatorial optimization on "rugged fitness landscapes" as a crude model of technological evolution. The main result is to show that progressive improvement for the same task typically becomes progressively harder. In the third section, I briefly discuss optimization for "neighboring" uses, an issue which arises in some forms of product differentiation. In the fourth section, I raise briefly the question of what kinds of products can be both complex and "perfectible." This issue is important in its own right in biological evolution, and may harbor hints about economic evolution. In the fifth section, I consider an image borrowed from ecology, that of the coevolution of entities, where the "fitness" of each entity is governed in part by the other entities which form its niche. The hope here is to obtain some general requirements for successful coevolution, and draw parallels to technological coevolution of firms or other economic agents, where the economic usefulness of a good or product depends upon how it fits into the mesh of other goods and products as the latter change. In the sixth section, I broach a central and perhaps unorthodox thesis: the web of economic activities, the linked tasks, can be thought of as harboring internal, endogenous forces which tend "autocatalytically" to drive an increase in the numbers of tasks, goods and services which can "fit" into the growing web. These endogenous forces, I want to suggest, exist in some sense "before" economic considerations come into play. But economic factors certainly come into play in determining which if any among the possible new components of the web are actually added to that web, and which "old" members of the web become economically pointless; hence, they languish and fade. In the seventh section, therefore, I try to point at some of these forces. Here it appears hard to avoid ideas common in biology, that local nonlinear self-reinforcing processes in coevolving clusters of industries, in spatial regions, and so forth, commit an evolving economic system to one of many alternative patterns, the choice of which further constrains subsequent evolution.

I stress with pleasure the fact that many of the ideas presented here have been discussed by Dr. Brian Arthur, and that Arthur has been thinking about the evolution of such webs with care for some time.

II. COMBINATORIAL OPTIMIZATION AS A MODEL OF TECHNOLOGICAL IMPROVEMENT

The first crude bicycle was a rude affair, open to rapid improvement in a number of directions. But with progressive improvement in design, with ramification into different early forms with large front wheels and small back wheels and vice versa, with introduction of gears, improved brakes, the capacity to switch gears, progressive refinement has become harder. The modern bicycle has become specialized to racing bicycle, mountain bicycle, touring bicycle, robust child's bicycle. One has the informal impression of a wide range of early variants, progressive improvement in which less "fit" variants dropped out, and a progressive increase in the difficulty of further improvement. The evolution of the motorcycle evidences the same features. Indeed, Freeman Dyson in his *Disturbing the Universe* recounts the myriad early experiments conducted by riders and manufactures of motorcycles, the gradual accumulation of improvements, and contrasts this with the rigid narrowness in the technological evolution of atomic reactor design and rocket design where little experimentation with radically different designs took place. In part, this reflected the cost of such trials. In an anecdotal way, one has the impression that a first crude design is open to improvements in many directions, but improvement generically becomes progressively harder. At least two issues, therefore, arise. First, is this pattern in fact typical of actual histories of technological evolution? Second, are there theoretical reasons to think such a pattern should be typical? The first question is posed as an empirical issue to be studied. The second can be answered in the affirmative.

The body of theory I describe next was developed with S. Levin,[11] with an eye primarily to biological evolution, but also with a hopeful eye towards technological evolution. In this body of theory we define a concept of "fitness." When later applying fitness to technological evolution of economic webs, fitness will naturally have to be given an economic interpretation in terms of the functional efficiency and goodness of design of a product or service, its cost of production, and market potential due to meshing into the existing economic web. The ideas are first cast in terms of biopolymers, namely proteins.

Proteins are linear strings of 20 kinds of amino acids. Thus, a protein with 100 amino acids might be any one of 20^{100} or 10^{33} kinds. Consider a "protein space" where each protein is a point in the space, and is "next to" all those proteins which differ from it by one amino acid in one of the 100 possible positions. Evidently, any protein has 19 × 100 or 1900 such "neighbors." Now consider measuring how good each of those proteins is at some definite "task," such as catalyzing a specific chemical reaction. Write down "on" each protein how well it does this task, and consider this measure of "goodness" a measure of the "fitness" of that protein with respect to that task. This procedure generates a "fitness landscape."[19] That is, we have assigned a measure of fitness to each protein in the protein space. The fitness can be thought of as a height at each point in the space. Thus, the landscape may be rugged, with many peaks, ridges, and valleys. Or it may be smooth with gentle hills and dales. There may be one massive mountain leading to a single global optimal peak. And at another important extreme, the distribution of fitness values across the landscape may be entirely random. That is, the fitness of one protein may portend nothing whatsoever about the fitness of its immediate neighbors in the space.

Consider now the simplest image of an *adaptive walk* in such a landscape.[5–8,15,17] Start with an arbitrary protein. Assess its 1900 neighbors one at a time. If a fitter neighbor is encountered, move to that neighbor. Iterate the process from that fitter neighbor. Such a procedure is an adaptive walk via immediate neighbors, which will progress until the walk terminates on a protein which is fitter than all its immediate neighbors. Obviously we can consider a similar walk which is allowed to proceed via neighbors which are "2 moves" away in the space, or "j" moves. And we may consider walks which can pass with some probability via less fit neighbors as well.

Natural questions which arise concerning such an adaptive walk include the number of improvement steps before reaching such an optimum, the number of alternative ways to improve at each step on the walk, the number of alternative local optima which can be reached by adaptive walks from any single initial protein, and how "fit" the local optima actually are.

Obviously, we want to know how these properties depend upon how "correlated" the fitness values in the landscape are, that is, on how similar the fitness values of neighboring proteins typically are. Below we develop a general model which allows one to "tune" how correlated a landscape is. For the moment, however, it is useful to concentrate on a completely *uncorrelated* fitness landscape. To be concrete, suppose that the fitness value of each protein is drawn at random from a uniform distribution between 0.0 and 1.0. The resulting landscape is uncorrelated in the precise sense that the fitnesses of immediate neighbors of any protein are a random sample from the same range, 0.0–1.0, as those in any other volume of protein space.

Because I have assumed for simplicity that an adaptive walk only passes to a neighbor which is fitter, no matter how much fitter, it is convenient to replace actual fitness values with rank orders, from worst to best. Thus, the uncorrelated landscape is a random assignment of rank orders to all proteins in the space.

Levin and I[11] have worked out a number of features of adaptive walks on such landscapes:

1. The expected *number* of local optima is $20^{100}/1901$, i.e., the number of proteins in the space divided by 1 plus the number of immediate neighbors of a protein.

2. The *lengths* of walks to adaptive optima are very short, being only $\log_2$ of the number of immediate neighbors; hence, here $\log_2(1900) = 11$.

3. Along an adaptive walk, the expected number of *fitter neighbors* decreases by *half* when each improved variant is found. That is, the *numbers of ways "uphill"* is cut in half at each improvement step.

4. Consequently, if the adaptive process can *branch* to more than one fitter variant, then such adaptive walks have many fitter variants and branch widely at first, but branch less and less frequently as fitness increases, until single "lineages" wend upward to local optima without further branching. Branching is bushy at the base, and dwindles as optima are approached.

5. From any initial protein, only a *small* fraction of all possible local optima are accessible via branching walks. An upper bound on the number of accessible local optima is:

   $$Nop = D^{(\log_2 D - 1)/2},$$

   where D is the number of immediate neighbors. Clearly Nop is a tiny fraction of the expected total number of local optima, given in (1). Thus, even though Nop is an overestimate, it suffices to show that only a small fraction of local optima are accessible from any initial point.

These general results, which I extend below, show that typically, adaptation on rugged fitness landscapes is easy at first, and becomes progressively harder, with fewer ways of improving further as fitness increases. This suggests that the anecdotal image of technological improvement may be generally true, and warrants serious investigation.

UNIVERSAL FEATURES OF ADAPTATION ON CORRELATED BUT RUGGED LANDSCAPES

Real landscapes in combinatorial optimization problems are rugged, but correlated.[10] Fitness values of neighboring points are similar. For example, the famous traveling salesman problem[3,10,14] consists in finding the shortest circuit route through a fixed number of cities on a plain. Any permutation through all cities is a possible solution. Neighboring permutations, obtained by swapping the order of two pairs of cities, typically have similar lengths. Letting length of tour stand for fitness, and locating all tours in a space, with neighbors to each solution being neighboring tours by swapping two pairs of cities defines a fitness landscape. While correlated it is also rugged.

Consider an adaptive process which "jumps" long distances across such a landscape, say, by swapping the orders of half the pairs of cities. This is analogous to a long jump in the Alps. If one steps a meter away, altitude of initial and final points are correlated, with catastrophic exceptions. If one travels 50 Km, the altitude one ends at is uncorrelated with the beginning altitude. Long jump adaptation *samples an uncorrelated landscape.* To understand adaptive walks in such a process we must replace steps to optima with waiting times to find a fitter variant. Levin and I show a simple universal result: after each improvement step, the expected waiting time to find the next improved variant *doubles.* Thus, the expected number of improvement steps S as a function of the number of "tries" G is just

$$S = \log_2(G).$$

This implies that improvement is rapid at first, then slows progressively.

In turn, these observations imply *three* natural time scales for adaptation on correlated but rugged landscapes. In the short time scale, the process begins with an initial entity of low fitness (the first crude bicycle, first guess at a tour through the cities, etc.). On average, half the variants tried are fitter. But consider the difference between nearby variants, obtained by changing only one or two minor features of the entity, and distant long-jump variants. The former are, on average, only slightly fitter or less fit than the initial entity because the landscape is *correlated.* The long-jump variants, by contrast can be very much fitter since they are far enough away to escape the correlation structure. Thus early on, it is easier to find very much fitter variants a long way away. Restated, early on, major improvements via fairly radical changes are readily found. But with each long-jump improvement, the waiting time to find a yet fitter variant doubles. Soon finding such distant major changes which improve matters becomes rare. This ushers in a second phase, when it becomes easier to find fitter variants by taking advantage of the correlation structure of the landscape and looking in the immediate vicinity. The process now refines by a local hill climb. But ultimately, that climb ends on a local optimum, and the process must now await a long jump to a fitter variant on the side of a new hill to yet another local optimum.

It becomes a matter of some interest, therefore, to investigate whether this pattern is often found in technological evolution. If so, we would expect the first crude gun, bicycle, motorcycle, radio, to be open to rapid, dramatic, major improvement in a number of different ways; thereafter, refinements of the major innovations to the basic design will be attained. Ultimately some quite new and dramatic modification will be found, ushering in a new period of refinement.

One economic implication of the above general features of combinatorial optimization might bear on the work of Dr. Brian Arthur,[2] where it tends to argue that saturation in "learning effects" between competing technologies will favor ultimate sharing of markets.

Arthur has considered the pattern of technological improvements in two or more competing technologies to accomplish much the same task, for example, steam and gasoline engine automobiles. The major point Arthur makes is that under conditions where investment in one technology leads to its improvement, it may happen that such investment by chance in technology A lead to a sufficient superiority of A over B, that further investment is drawn into A rather than B. That is, such technological "learning effects" can by chance lead to domination of A over B or B over A. In particular, Arthur argues that if the "learning effect" is roughly linear, such that each unit of investment in technology A (or B) leads to a further unit improvement in that technology, then one expects such a process to lead with probability 1 to eventual total market domination by one or the other competing technology. On the other hand, if the learning effects *saturate,* that is, if further improvements in technology A become progressively harder as investments are made, then the possibility of more rapid improvements in lagging technology B may become sufficiently attractive to warrant investment in technology B. Consequently the competing technologies will come to share the market, but perhaps at unequal ratios reflecting the historical sequence of investment choices. Our view that the generic property of technological improvements for a given function or task shows clear *saturation,* becoming harder with progressive investments, suggests that market sharing may often come to characterize mature competing technologies.

The image of technological evolution presented above is inadequate for a number of reasons. Some are addressed in the next sections.

III. OPTIMIZATION FOR "NEIGHBORING USES"

The previous section considered adaptation on a fixed fitness landscape. In the case of proteins, we measured the capacity of each protein to accomplish the *same* catalytic task. But of course, we might also ask the capacity of all proteins to catalyze each of a set of different reactions. Some reactions are similar to one another, and the concept of a "catalytic task space" can be defined.[13] In this space, neighboring points are neighboring catalytic tasks. We might imagine that, if we knew the fitness landscape across the space of proteins for one catalytic task, that the landscape for quite similar catalytic tasks would be small perturbations away from that first landscape. Modulating across the set of catalytic tasks would progressively deform the fitness landscape across the space of proteins in some as yet unknown, but knowable way.[7,8,13] If that were known, we could define the ways in which requiring organisms to cope with different catalytic tasks would deform the selection pressure on their constituent proteins to meet the newly diverged "needs."

Any adequate theory of technological evolution will likewise ultimately need to envision a sense in which modulation from one to another "design goal," from racing bicycle to mountain bicycle, modifies the previous adaptive landscape. I state this image not because the answers are known. But again they may be both knowable and ultimately explicable in a theoretical framework. We would want here to gain insight into the extent to which altering design goals in some way which would need to be quantifiable, typically allows given fractions of the components to be taken over without modification, or with minor refinements, or which demand major innovations. Part of the importance here may well be that an optimization process which has nearly saturated with respect to one design goal may typically be open to a burst of innovation for a new neighboring task. This simple suggestion follows from the obvious fact that, on average, when the landscape is deformed to that of the new design goal, the old entity which was well designed to the old task will typically be rather poorer at the new task; thus, more ways of improvement lie open.

In so far as this image is likely to be correct in general, it suggests that avoiding learning saturation in industries which depend upon innovation and "staying ahead" of the competing technologies may typically be forced to achieve this strategy by product differentiation for new "tasks," i.e., design goals and new product markets. The strategy can succeed so long as new tasks require sufficient new innovation to stave off saturation of the "learning curve" or improvements with respect to investment.

I suspect that these are not new ideas to economists, of course. It would be interesting in this regard to study the evolution of computer technologies as opposed to canning technologies, where the opportunities for defining new tasks and goals is probably inherently more limited. Indeed, asking whether there is any logical way of understanding the openness to new tasks and goals inherent in different technologies is, I also suspect, an important part of understanding the evolution of economic webs under the drive of technological innovation and economic forces together. I return to this below.

IV. REQUIREMENTS OF COMPLEX ENTITIES ALLOWING PERFECTIBILITY

In Section II, I discussed adaptation on rugged landscapes. But what features of entities, biological or economic, govern the ruggedness of their adaptive landscapes? More concretely, if adaptation occurs on "bad" landscapes, only very poor optima can be found. What features of entities, then, assure that "good" optima can be found? To be Kantian, what features of complex products allows combinatorial optimization fitting parts together in a context of conflicting design requirements, to achieve good solutions? I broach here a simple formal model[12,13] which suggests a simple, sufficient, and perhaps necessary answer: the complex product must be constructed such that each part interacts directly with only a *few* other parts. As we shall see, this may be true of many complex adapting systems, biological, ecological and even economic.

THE NK MODEL

Consider a hypothetical entity with N parts, each of which can be of two "types" 1 and 0. Thus, each entity is a vector of N bits. I shall suppose that the "fitness" contribution of the ith part to the whole entity depends upon itself and K other parts. I may also require that if part i depends upon part j, then j also depends upon i. It turns out that this sensible requirement does not seem to matter as much as one might have thought. I wish to leave all other features of this model random. Thus, I shall assign the K parts which matter to each part, i, at random, perhaps with the reflexivity condition above. Then for each part, i, there are $(K+1)$ parts whose types, 1 or 0 bear on the fitness contribution of that given part, i. Thus, there are $2^{(K+1)}$ combinations of the 1 or 0 types for these parts, and for each of these combinations I shall choose the fitness contribution of the ith part by drawing at random a decimal number between 0.0 and 1.0. Then typically, the ith part, in the context of each of the different combinations of the $K+1$ part types, is accorded a different fitness contribution. Having assigned these values for the ith part, I proceed similarly to assign random values between 0.0 and 1.0 for the $2^{(K+1)}$ conditions for each of the N parts in the system. Having completed that random assignment, then for any specification of the 1 and 0 types for the N parts of the entity, I can find out the fitness of the entire entity by summing the fitness contribution of each part in the context of itself and the K which bear upon it. For simplicity, I shall then normalize fitness to lie between 0.0 and 1.0 by dividing this sum by N. The NK model appears to be closely similar to a class of spin glasses[16] with a sum of monadic, diadic, . . . , k-adic energy contribution terms plus a random external magnetic field.[1]

The question is this: given N and K, what does the fitness landscape look like? This model has been presented elsewhere,[12,13] and I here merely summarize it.

1. For $K = 0$, each part is independent of other parts. By chance the 0 or the 1 type for that part is fitter. Thus, there is a global optimum where each part is in its more favored type. Any other entity is suboptimal, but lies on a connected *pathway* via *fitter variants* to the global optimum by the simple expedient of altering each part to its more favorable type. Further, changing any one part cannot change the fitness of the entity by more than $1/N$; hence, the fitness landscape is obviously highly *correlated.*
2. For $K = N - 1$, each part is connected to all parts. Changing one part from a 1 to a 0 changes the context for all N parts. For each of those, the changed context results in a new random fitness contribution by that part, thus changing any part results in a new fitness contribution by all N parts, hence an entirely random fitness with respect to the original entity. The $K = N - 1$ case corresponds to a completely *uncorrelated* fitness landscape.
3. For $0 < K < N - 1$ the landscape is correlated in some way. Thus tuning the connectivity, K, tunes the correlation structure of the landscape.

TABLE 1 Mean Fitnesses of Local Optima for Various N and k Values (Monte Carlo Results Assuming Random Interactions)

k\N	8	16	24	48	96
0	.65 (.08)	.65 (.06)	.66 (.04)	.66 (.03)	.66 (.02)
2	.68 (.08)	.68 (.05)	.69 (.04)	.69 (.03)	.69 (.02)
4	.69 (.07)	.70 (.05)	.71 (.03)	.72 (.03)	.72 (.02)
8	[.68 (.06)]	.69 (.04)	.70 (.04)	.71 (.03)	.72 (.02)
16	–	[.64 (.04)]	.65 (.03)	.67 (.02)	.69 (.02)
24	–	–	[.63 (.03)]	.65 (.02)	.66 (.02)
48	–	–	–	–	–
96	–	–	–	–	–

Tables 1 and 2 show the results of numerical trials for different values of N and K. Among the results, the following critical feature emerges. For $K = N - 1$, the completely uncorrelated landscape, a *complexity catastrophe* occurs. As N and K increase, the fitness of attainable optima *fall* toward the mean of the space. This feature is absolutely fundamental. In uncorrelated landscapes, as the entities become more complex, here as N increases, the attainable optima become hardly better than chance. It is easy to show that the global optima, by contrast, increases as N increases, thus the divergence between the attainable optima and the global optimum increases monotonically. This result holds even if fitness is not normalized by division by N. Furthermore, even if K is less than N, but is *proportional* to N, the same catastrophe occurs. Thus, if as a system becomes more complex the number of interacting connections increases in proportion to the number of parts, the attainable optima recede toward chance.

However, this complexity catastrophe does *not* occur if K is small and fixed as N grows large. Instead, as shown in the figures, fitness of accessible local optima actually increase with N, but may asymptote. Thus avoiding the complexity catastrophe requires that each part interact with only a few other parts.

I suspect this is a very general result. It obtains in *hierarchical* systems, where the hierarchy assures that parts in each level interact predominantly with parts in their own level, hence with only a fraction of the other parts. It obtains in biological systems, such as proteins where each amino acid interacts dominantly with its few neighbors in the primary chain sequence, and somewhat with distant amino acids due to chain folding. And in technology, real machines are constructed such that each part interacts with a rather small fraction of the other parts of the system. In short, it is a reasonable bet that low connectivity, low K, is a sufficient and perhaps necessary feature in complex systems which must be perfected in the face of conflicting constraints, that allows good designs to be realized by natural selection or by human artifice. In economics, it may apply both to the design of products themselves, and to the design of economic *agents* such as firms, which are themselves complex adapting entities with many parts and processes which must mesh well to meet their external worlds.

I comment that this may imply a "second-order" feature to natural selection, which may achieve the kinds of entities which adapt on "good" landscapes. Similarly, it may imply a kind of selective process in human cultural and economic evolution, in which those kinds of complex systems which are "perfectible" come to exist and persist.

I note also that low connectivity may be a critical feature of *coadaptation* of entities, each adapting on a landscape whose peaks, ridges and valleys depends upon the other entities which comprise its niche. These ideas ought to extend to technological coevolution, as discussed in the next section.

V. COADAPTATION ON COUPLED DANCING LANDSCAPES

The discussion above has pictured adaptive walks on fixed fitness landscapes. But biologists and economists know how inadequate this beginning image is. One organism makes its living in a niche defined by the physical environment and the other members of its own species and all those other species with which it interacts. Such interactions include predation, parasites, symbiotic arrangements, and a host of others. Thus, the fitness of one organism may depend upon the frequency of its own phenotype among its own species, and on the vector of phenotype frequencies of the other species in its niche. As each species makes adaptive "moves," or even

TABLE 2 Mean Walk Lengths to Local Optima for Various N and k Values (Monte Carlo Results Assuming Random Interactions)

k\N	8	16	24	48	96
0	5.5 (1.2)	10.0 (1.9)	13.6 (2.2)	25.3 (3.4)	50.0 (4.6)
2	5.5 (1.8)	9.1 (2.6)	13.7 (3.1)	25.6 (4.5)	49.1 (6.1)
4	4.8 (2.0)	8.6 (2.8)	13.2 (3.2)	24.2 (5.4)	48.0 (6.8)
8	[3.6 (1.5)]	6.6 (2.3)	9.9 (3.0)	19.6 (5.2)	37.6 (6.6)
16	–	[4.2 (1.7)]	5.9 (2.3)	11.7 (4.0)	24.8 (5.6)
24	–	–	[4.5 (1.8)]	9.1 (2.6)	17.6 (4.5)
48	–	–	–	–	–
96	–	–	–	–	–

selectively neutral changes by incorporation of novel mutations into the population, those changes may affect not only the changed species, but literally alter the fitness landscape of all those species having the altered species in their niches. As each species moves on its landscape, its motion deforms the landscapes of its ecological neighbors. The coevolving set of species is jointly walking "up" a set of coupled landscapes which deform under them. But obviously the same problem arises in considering the evolution of competing and cooperating firms under the drive of technological evolution. Consider the coupled industries making semiconductors, chips, computers and software. Alterations in products by one component of the industry alter what it is sensible and profitable for its competitors to do, alters what users of its products or services need and now *can* do, hence generates new opportunities to make and sell new goods and services. The landscape alters for all by the action of each.

Game theory has been one approach to the study of coevolution in biology. Here as in economics, the fundamental image has been of a set of agents with a *fixed* vector of strategies (phenotypes in biology), and a *fixed* payoff matrix for pairwise interactions. The solutions to the game consist in finding fixed ratios of strategies of each agent such than none can improve without decreasing the expected payoff to another player. In biology this use of game theory results in the concept of an evolutionary stable strategy. There, one shows that at such an equilibrium, no gene altering ratios of phenotypes can invade the population. In economics, one speaks of a Nash equilibrium.

But as we all recognize, in real life the vector of strategies is not fixed. Innovations in organisms and technologies assure that the opportunities, challenges and payoffs change continuously. Thus again we might pose another Kantian question: what must complex systems be, and how must they be coupled into an "ecology," such that the set of entities can coadapt and "improve"?

Consider again the NK model. We may take N to represent parts of a product, or as traits in an organism, present or absent. In the biological interpretation we can even take N to represent N genes, each of two types or alleles. Then in biology, K represents "epistatic" interactions, the effect of other genes on the fitness contribution of each gene. Allow me to stay within the biological image for a moment. Then if an organism develops a new trait, that new trait may affect the fitness landscape of the others having that organism in its niche. Consider S species, each with its own adaptive landscape, where that landscape is given by a randomly chosen NK landscape, with N and K fixed. Now define a "niche" matrix, showing which species interact with one another directly. Next define which traits of each species impinge on each of the species that species affects, and via which traits in those affected species. Thus, identify a kind of projection of the traits of each species onto the traits of each species it affects. Next, we may model the effects of each species on those it affects by *augmenting* the NK model in a trivial way. Consider each trait in a species which is affected by traits in other species. Simply augment the table of combinations of conditions, present or absent, to include not only the $K+1$ combinations from within that organism itself, but to include the presence or absence of all combinations of traits from all other species which impinge on that trait. For each combination, choose a random fitness contribution between 0.0 and 1.0. The each specie's fitness landscape has become a function of the location of the other species in its niche on their own landscapes. As each moves, the landscapes of the others deforms.

What does this simple model show us? Suppose that $K = N - 1$ within each species, and the projections between species are rich. Then any move by one species randomizes again the already random fitness landscape of all those it projects upon. As each species moves uphill, all others are *cast back,* on average, *to average fitness.* That is, each landscape deforms so drastically that no species can crawl far uphill before being cast back to average by the move of another player.

Conversely, consider N large and K small, hence strongly correlated landscapes, and the limit where each species' traits project onto few traits of those in its niche. Then typically, a move by one species will only slightly deform the other landscapes, casting the other players back only slightly, on average, towards *average.* The set of coevolving entities can struggle uphill until each is at a height such that, on average, the distance it is cast back by the moves of others can be recovered in the interval before someone else again deforms the landscape.

This is a crude picture of a model which, I think, deserves to be developed in detail. But already it suffices to characterize some main features of coadaptation and provides one answer to the Kantian question. The set of species can coevolve if the landscapes are *weakly coupled.* Thus, within a complex adapting entity with conflicting design constraints, low K—that is, low connectivity—is critical to attain good landscapes. And between coadapting complex entities low connectivity is again good. Weak couplings, in short, as supposed by authors such as Herbert Simon, Ross Ashby, and others, on different grounds again emerges as critical to coadaptation in complex interacting systems. Clearly there is nothing uniquely biological in the above considerations. One can expect them to apply to coadapting firms in an economic web. If each innovation in the semiconductor industry demanded radically new computers for its use, and radically new computers required radically new software, the coevolution of that set of industries would be chaotic indeed.

Granting for the moment the obvious fact that this model of coadaptation is yet crude and remains to be worked out in more detail, does it begin to suggest empirical data which might be gathered in economics and brought to bear? The question posed, of course, is the extent to which an innovation by one firm in a set of competing and cooperating firms in a linked part of an economy, alters the present economic "fitness" of the goods and services provided by the other linked firms. For example, when automatic transmissions were introduced by one of the Detroit Big Three, that act reduced the market for standard transmissions, standard transmission parts, repair and servicing of such parts, and so forth. This reduced the "economic fitness" of that vector of goods and services by suppliers of those goods and services. On the other hand, the introduction of automatic transmissions afforded new opportunities to the same or other suppliers of parts and services for those parts which make up an automatic transmission. Automatic transmissions alter fuel efficiency, hence alter the economic value of petroleum products, etc. In concrete cases it would appear feasible to analyze in specific detail the fraction of goods and services per firm or other economic unit, whose economic usefulness are reduced, and those old goods and services which might happen to have enhanced value. In addition, we can see how many new products and services now naturally enter the web and afford new ways of earning a living by meeting rational "needs" required internally by the logic of the web of goods and services. I shall return below to considering this internal logic.

The aim of such empirical studies would be to attempt to build up a *statistical* picture correlating an "innovation" by one economic agent and the typical mean and variance in the alterations in economic values of old goods and services provided by linked members of the web, and the numbers of new goods and services which now become rational to produce to mesh with the innovation. For it is this loss in value of old goods and services, and the rational requirement for new goods and services which mesh with the innovation which constitutes the major force in driving the evolution of the economic web. In the next section I turn to an attempt to formulate this problem in at least preliminary fashion.

VI. ENDOGENOUS FACTORS DRIVING WEB GROWTH

I note at the outset that the class of issues I want to raise here have been under consideration by Brian Arthur for some time. Further, I have benefited from conversations with him about the material which follows.

The perhaps unorthodox question I wish to pose is whether the actual structure and components of an economic web play an endogenous role in leading to the evolution of that web, even *prior to* economic considerations of whether the new goods and services which may enter the web are economically valuable and warrant investment and effort. I believe the answer is "yes"; that is, I shall suggest that endogenous forces tend to lead to the evolution of economic webs, and that economic issues are among the further factors which determine which *actual* among many *possible* evolutionary pathways are chosen.

I want to consider an economic web as comprised of the union of two slightly different kinds of directed graphs. Nodes in each graph represent distinct goods or services, regardless of the economic agents which produce such goods or services. Arrows in the two graphs shall have slightly different meanings.

1. THE SUBCOMPONENT GRAPH

An automobile is comprised of a number of parts. Just how an automobile might be parsed into its parts is, of course, somewhat arbitrary, but I ignore this problem for the moment. In the Subcomponent Graph, an arrow is directed from a "part" towards a larger component in which it is a member. The directed graph of parts to their "wholes" then immediately represents goods which must be assembled to construct the larger products. Any such arrow clearly represents as well an economic opportunity. Manufacturers of the parts may make a living selling those parts to manufacturers of larger parts, and so on to the wholes. But also, any whole product, automobile or otherwise, affords opportunities for earning a living by servicing and repairing any of its parts, scavenging its parts for resale or junk, etc. Thus, I want to think of the Subcomponent Graph as a directed graph whose arrows represent opportunities to produce and sell, or service portions of other products in the economic web. While stated in terms of parts of "products," the product can instead be a service, with component subservices.

2. THE "NEED" GRAPH

I want to define a *"Need"* Graph. By "need" I do not intend to invoke terms with current economic technical meanings. Consider the example pointed out to me by Brian Arthur. To use a car, roads are needed. To use roads, with traffic conditions, traffic lights are needed. Traffic lights imply rules, and the enforcement of those rules, hence traffic police and even a legal system capable of assessing guilt and innocence, fines and punishment. Automobiles imply a need for gasoline stations at spaced intervals. Travel by automobile implies a need for motels, perhaps restaurants with fast food for the anxious traveler, etc. In turn, gasoline stations need a petroleum industry to find, refine, and deliver gasoline to the motorist at his convenience. Many or most "nodes" in the web require other nodes for the actual "use" or "purpose" of the initial node. The web of needs, when *met,* constitutes a coherent whole: all the joint uses and functions can conjointly be met. The coupled needed tasks are fulfilled.

There is no mystery here. We all recognize this property of economic systems in any concrete case. Just the obvious point is that it is the very *existence of unmet needs which offers new economic opportunities.* An unmet need is a good or service which meshes into the web, a means of earning a living by providing a means to meet the need.

Let me conceive of any good or service as having some number of discriminable *uses.* Thus, a bow and arrows may be used for hunting, or for target shooting in competition or recreation. An airplane can be used to transport people, transport cargo, recreation, stunts and entertainment, as a means to obtain visual information about an enemy disposition in war, etc. Thus, some products or services may have many, other may have only a few uses. Think of each use as radiating *need* arrows, meaning that that use of the good requires the good or service specified at the termination of the arrow. Automobiles radiate arrows to roads and to gasoline service stations. The web is coherent if all the *need* arrows terminate on an existing good or service which fulfills that need. Economic opportunities exist in improving the meeting of the need by a better good or service. And most obviously, when technological innovation yields a *new* product with *new needs,* meeting those new needs with *new* products affords new economic opportunities. Meeting those opportunities transforms the web.

Can we begin to see endogenous processes which may lead to the expansion of the web, even before we consider economic forces? Suppose several *need* arrows end on the *same* product. Then I shall invoke a "jack-of-all-trades,-master-of-none" principle. Typically that product does not meet all the convergent needs very well. Such convergence is obviously a marked stimulus to *product differentiation.* Technological evolution of the original product on more or less neighboring landscapes to meet these diverse needs will ensue. Or at least such differentiation may ensue if economically profitable.

Consider the invention of a new product. If, on average, it has many uses, and each use radiates many needs, then each such invention will require for its many uses many new or modified goods and services. Thus, suppose we were to carry out an empirical study relating the *complexity* of a good or service to the numbers of its *uses,* and the number of *needs* radiated per use. We would find some distribution, with mean and variance for each complexity of good or service. More simply, we would find, lumped over all goods and services, some mean and variance for the number of uses and needs radiated by a good. If the total number of needs radiated is greater than 1.0, then innovation tends endogenously to engender, drive, demand, warrant, yet more innovation in a kind of autocatalytic fashion. That is, I will give a preliminary guess that typically an innovation radiates more than one need arrow, and the entities which fulfill those needs themselves radiate more than one need arrow, hence an endogenous autocatalytic force exists which tends to drive technological change at an accelerating pace merely to meet the logical requirements of coherence in the web. Whether or not any of these innovations are economically feasible is another issue, but to be feasible any such innovation *must* mesh into the evolving web coherently. Coherence is a necessary, but not sufficient conditions for evolution. The point here is that the web's coherent structure endogenously can evolve in such a way that each new innovation engenders the plausibility or need for more than one further innovation. Growth of the *"possible economic web"* is autocatalytic.

The Wright brothers had the wit to join bicycle wheels, airfoils and a gasoline engine to create an airplane. That is, technological advances often *recombine* parts from many different machines or products to form a new product. Consider a terribly naive model of the *ease* of such innovations. Let the economy have N products, with an average of M parts per product. Crudely, any part of any product might be used in combination with any part of any other product to make a new product. Then the number of combinations of pairs of parts is $(NM) \times (NM)$; hence, the ease of making new combinations might scale as something like $k(NM)^2$. Notice the main point: the more complex the economy already is, with more products and services, then the easier it is to invent yet another product or service. In the grossly oversimplified expression above, the ease of new inventions increases as the *square* of the current number of goods and services in the economy. But obviously, it is already terribly interesting if the ease of making further inventions *increases* at all in a systematic way with the increased complexity of the economic web. Any such increasing ease is, again, an autocatalytic force *prior to economic plausibility* considerations, which endogenously tends to drive innovation and expansion of the economic web.

The expression above is almost frivolously naive. But the question is not. There is some way in which the complexity of goods and products and services in an economy abets the invention of further goods and products, not merely by technological evolution *within* one kind of product, but by marriage of parts of more than one product. This obvious point suggests that actual examination of case histories may allow us to see how this abetting actually occurs. In considering such studies, however, care must be taken to distinguish between those innovations which *also* are economically feasible and happen to come to fruition, and the more basic question about how the complexity of the economic system in terms of numbers and kinds of goods and services itself alters the ease of further innovation, whether or not such innovations are also economically profitable. It is worth stressing here that innovation by recombining parts of previous entities, a marriage of "part solutions," is a fundamental feature of biological as well as technological evolution.[6,9]

These three forces ("jack of all trades master of none," greater than 1.0 Needs radiating from innovations, and the combinatorial explosion of possible new products) are surely only a few of the endogenous factors which, upon reflection, we may find tend to underlie the expansion of the goods and services in an economic web.

TOWARDS STATISTICAL THEORIES OF ECONOMIC WEB EVOLUTION

One course for the subsequent development of this kind of theory to take is to consider the development of *statistical theories of web evolution.* Here one would try to gather statistical data on the mean numbers of goods and services at some point in time, and their connectivity, then include equations for the expected jack of all trades force, the needs arrow force, and the probability of new recombination inventions force, and study the time evolution of the subsequent graph of the economic interactions in the absence of economic forces limiting its growth.

A beginning approach might be the following. Represent a product or service abstractly as a *string* of symbols drawn from some alphabet. The *length* of the string represents the complexity of the product or service. The "goodness of design" of the product with respect to its prospective use(s) can be modeled by a generalization of the NK model above. This yields, for each product, a more or less rugged fitness landscape. The Subcomponent Graph can be defined rather naturally as producers of subcomponent strings of larger strings, or repair and service of such substrings. The familiar "parsing" question of the "natural parts" of a string, or automobile arises again. Uses per string, and needs per use are assigned at random from empirically defined distributions. Which strings meet which needs radiated from other strings is decided at random. The sum of Subcomponent and Need graphs defines the present economic web. The statistical connectivity features of the resulting random graph, such as cycles, input-output matrices of intermediate products, graph radius (the shortest distance from an intermediate product to the most distant ultimate product in which it is used), etc., can then be analyzed.

Graph *growth* can then be studied by use of the three forces defined above, "jack of all trades" and the two others, plus additional endogenous forces which may be uncovered. Such graph growth can be thought of as *transformations to nascent possible new economic webs.* The new components and links may then be added and linked in to the existing Subcomponent and Need graphs, to follow web growth.

The emphasis here is on nascent and possible. This picture of web growth in the *absence* of economic forces only *adds* components and links. No economic forces constrain which of the new components and links are actually added, nor lead to loss of old components and links. Those forces must be added to hope to attain a real theory of web evolution.

Before turning to those forces, however, note the main conclusion of this section: endogenous forces exist which tend to amplify the complexity and interconnectivity of economic webs. In light of this, I point out that among the most robust stylized facts of economic evolution during industrialization is an *increased density* of the input output matrices.[4] More intermediate goods and services are formed. The numbers of nodes and transformations increases. This aggregate feature of industrialization suggests that our picture of web dynamics may not be wrong, for it would predict just such an increase.

VII. NONEQUILIBRIUM ECONOMIC FORCES AND CONSTRAINTS ON WEB GROWTH

To a non-economist thinking about economics, perhaps the most astonishing, if obvious realization, is that the web of economic activities is self-generating and self-sustaining. In a global sense, all that exists are people to work, and means of production to obtain or produce some array of goods and services. The production of those goods and services is sustained by adequate demand for the same set of goods and services. In turn, this requires that the people, in one form or another, have the wherewithall via barter or money, to exhibit the requisite demand. Like life itself, bootstrapped by its increasing capacity to harness limited exogenous energy resources, economic systems bootstrap to increased complexity and richness of goods and services. But, equally obviously, since at any time the labor pool and production means are finite, not all possible nascent technologies and new goods and services can be pursued, and not all old technologies can or could be maintained. The actual web is a subset of the possible; and the actual evolution of economic webs must be a path through the ever expanding set of next possible economic webs. Two main components must then be analyzed. Here I only mention them. First, what economic forces constrain, limit and determine which of the nascent technologies that might bud and mesh into the economic web at any point in time actually occur? Or, from the point of view of a statistical theory, how many, and what kinds of new subwebs can grow, mesh into, and transform the current web? Second, what economic forces lead to loss of old goods, services and technologies? And from a statistical point of view, how rich are the subwebs which disappear? Since the actual evolution of the web is caught on the edge of addition and disappearance, under endogenous non-economic and economic forces, an eventual dynamical theory of web evolution must meld all these.

CONSTRAINTS ON NASCENT TECHNOLOGIES

One expects that economic feasibility will constrain the actual set of new web components to a small fraction of the possible. Because I know so little economics, I will dare to comment here only briefly. One point of interest which may be less widely appreciated than a biologist would have expected is that such evolution can lead to multiple possible alternative behaviors. Arthur[2] has noted that when two technologies are competing, early investment in one, say, A, may lead to such improvement that it becomes uneconomical to invest in the lagging technology B, even though both may have been even to start with and B may be ultimately superior. It would appear to be important to ask the same form of question, not simply for two competing technologies, but for the alternative choices available to meet each *need* in the Need Graph, and hence follow the entire linked structure. Thus, it may be the case, since coherence is required, that if technology A wins over B at some point in the web, then the *needs* radiated by A will be called upon to develop, rather than those radiated by B. But further, since any region of the web must be coherent, we can envision somewhat overlapping nascent webs, with alternate choices at many points, and ask whether whole web substructures might not be selected, more or less by chance initial advantage, over other potential web components. It would be interesting to try to use the ideas of Subcomponent and Need graphs to begin to gain insight into how many coupled technologies for using a subweb must "logically" jointly win such a race with other nascent coupled technologies. Simultaneously, some feeling for how extensive such coherent patches of the economic web are, may give a feeling for the requirements to nucleate a self-sustaining fragment of an economy.

LOSS OF OLD COMPONENTS

Old technologies lapse. No one makes Roman siege engines. The pony express, alas, is no more. Why do such goods and services lapse? Partially for economic reasons. Howitzers and air mail are more efficient than their predecessors. The demand lapses for the goods and services, since better means are available. The point is obvious. But the coherent web picture leads to some new questions. Perhaps some "old" products can be individually exchanged for "new" products in a web without destroying the web's coherence. But other products in the same web may be so central, radiating so many need arrows and component arrows to so many other components, that their replacement collapses the entire web. Arthur's example is replacement of the horse by the car. The more general question is a statistical one. Consider a coherent subweb. Replace a fixed fraction of "old" products by "new" ones. As that faction increases, how likely is the entire subweb to lapse into disuse? This is a nucleation of economic elimination question.

TURING'S MORPHOGENETIC MODEL REVISITED: BALANCING POSITIVE AND NEGATIVE FEEDBACK LEADING TO MULTIPLE STATIONARY STATES

A final point. Can old webs persist in performing a web of tasks in old fashioned ways, while a new web or web component of high technology meeting similar underlying wants *coexists* in the same economic system? Biologists are familiar with Turing's model of morphogenesis,[18] in which two chemical species are present in a petri plate. The first, an excitator, catalyses its own formation and that of an inhibitor. The latter inhibits its own formation that of the excitator. Further, the inhibitor diffuses faster than the excitator. Consider an initial condition which is both *spatially homogeneous* and a steady state. Concentrations of each chemical are uniform and unchanging over time. Turing showed that a slight amount extra of the excitator would lead to its autocatalytic growth in concentration at that local point in the petri dish. In turn, the excitator would cause synthesis of the inhibitor, but the latter would diffuse laterally in the dish away from the excitator peak. Consequently, at the peak, the excitator "wins" over the inhibitor and the peak continues to grow. But laterally, the concentration of inhibitor is larger than excitator, hence inhibition wins, and the peak cannot spread. A free-standing peak of excitator and inhibitor stands out above the low plane of steady-state concentrations in the petri dish. That is, autocatalytic growth is locally faster than diffusive dissipation, hence a spatially inhomogeneous pattern emerges and is stable until substrate resources for the excitator and inhibitor are used up.

This now familiar idea in biology seems likely to have wide application in economics, where mixtures of positive feedback and diminishing returns negative feedback are present in the complex system. Such positive feedbacks include the learning effects mentioned above, the growth of an infrastructure in an economy or fragment of an economy, education of skilled labor, assemblage of capital and labor into working wholes, and other points discussed at the conference. But also, such positive feedback is inherent in the idea of a coherent subweb of the economy, whose components are mutually necessary to meet needs. Joint presence is mutually catalytic. This leads to the general question of maintenance of such subwebs against dissipative forces. A particular case follows.

Can a high technology web "live" on a broader low technology web? Brazil was informally described in our meeting as Belgium in India. If so, then there are cases where new technologies and the high technology web will not replace an older web of economic activity. Thus I wonder at the following possibility. Consider a closed economy. Suppose a high technology subweb produces goods and services which are coherent, *and* suppose that the efficiency of production allows a small work force to produce those goods. Sale of the same complex of goods requires adequate demand. Let the high technology workers be paid high wages, while the unskilled tillers of the fields using old technologies receive low pay due to low efficiency, and less value per unit product in the marketplace. Then maintenance of the high technology web requires a sufficient number of high technology workers with enough disposable income to support the aggregate demand for the high technology products. But their efficiency of production is so high that no further workers are needed; hence, the high technology web cannot expand throughout the economy.

Is this reasonable? As a non-economist I surely cannot say. I suppose one evolution of such an economy would envision paying high technology workers less due to the supply of unskilled labor. Then the high technology web would be likely to collapse due to loss of adequate aggregate demand. Perhaps the system evolves towards a mid-technology economy with lower efficiency cottage industries. But alternatively, the high technology workers might form unions, and the high technology web might persist stably in a stratified society. Then old and new technologies would stably coexist. And nucleation of a high technology web might be a poor way to transform a closed economy to a high technology closed economy. Either passage via cottage industries, or an open economy with world markets to maintain demand, as with Korea and the industrialized world, would appear better routes.

I have now stepped so far beyond my competence that I forebare further questions to the economists, save a summary.

SUMMARY

I have tried to suggest that there might be something like a theory of innovation, and technological coevolution. In terms of such an eventual theory, we would hope to understand how and why economic systems tend to become more complex and diversified, due to non-economic and economic forces. The steps towards such a theory sketched here make, as yet, no use of the economist's theories of rational expectations and foresight. But presumably such can also be included, for they must play a role in determining which directions of innovation in coevolving technological webs actually occur. I make no excuse for the crudeness of the ideas presented. But I hope the questions are useful ones which will be better answered by economists in the future.

REFERENCES

1. Anderson, Philip A. (1988), personal communication.
2. Arthur, W. Brian (1988), "Self-Reinforcing Mechanisms in Economics," this volume.
3. Brady, R. M. (1986), "Optimization Strategies Gleaned from Biological Evolution," *Nature* **317**, 804.
4. Chenery, H., S. Robinson, and M. Syrquin, eds. (1986), *Industrialization and Growth: A Comparative Study* (Oxford: Oxford University Press).
5. Eigen, Manfred (1987), "Macromolecular Evolution: Dynamical Ordering in Sequence Space," *Emerging Syntheses in Science. Santa Fe Institute Studies in the Sciences of Complexity*, Ed. David Pines (Reading, MA: Addison-Wesley), vol. 1, 21–42.
6. Ewens, W. (1979), *Mathematical Population Genetics* (New York: Springer-Verlag).
7. Gillespie, J. H. (1983), *Theoretical Population Biology* **23**, 202–215.
8. Gillespie, J. H. (1984), *Evolution* **38(5)**, 1116–1129.
9. Holland, J. (1981), "Genetic Algorithms and Adaptation," *Technical Report #24, University of Michigan Cognitive Sciences Department, Ann Arbor, Michigan.*
10. Johnson, D. S., and C. H. Papadimitriou (1985), "Computational Complexity," *The Traveling Salesman Problem* (Chichester, England: Wiley Interscience, John Wiley and Sons), 37–85.
11. Kauffman, S., and S. Levin (1987), "Towards a General Theory of Adaptive Walks on Rugged Landscapes," *J. Theoret. Biol.* **128**, 11–45.
12. Kauffman, S., E. Weinberger, and A. S. Perelson (1988), "Maturation of the Immune Response via Adaptive Walks on Affinity Landscapes," *Theoretical Immunology,* Santa Fe Institute Studies in the Sciences of Complexity, Ed. A. S. Perelson (Reading, MA: Addison-Wesley), in press.
13. Kauffman, S. (1988), *Origins of Order: Self-Organization and Selection in Evolution,* book submitted for publication.
14. Kirkpatrick, S. and G. Toulouse, "Configuration Space Analysis of the Traveling Salesman Problems," *J. Physique* **46**, 1277.
15. Ninio, J. (1979), "Approches Moleculaires de l'Evolution," *Collection de Biologie Evolutive* (Paris, New York, Masson) **5**, 93.
16. Sherrington, D., and S. Kirkpatrick (1975), "Solvable Model of a Spin Glass," *Phys. Rev. Lett.* **35**, 1792.
17. Smith, J. M. (1970), *Nature* **225**, 563.
18. Turing, A. M. (1952), "The Chemical Basis of Morphogenesis," *Royal Society B.* **237**, 37–72.
19. Wright, S. (1932), "The Roles of Mutation, Inbreeding, Crossbreeding and Selection in Evolution," *Proceedings of the Sixth International Congress of Genetics* **1**, 356–366.

TIMOTHY J. KEHOE
Department of Economics, University of Minnesota, Minneapolis, Minnesota 55455

Computation and Multiplicity of Economic Equilibria

1. INTRODUCTION

The principal model in economic theory is the Walrasian model of general economic equilibrium. In it, consumers choose to demand and supply goods to maximize a utility function subject to the constraint that the value of what they demand must equal the value of what they supply. Producers choose to demand and supply goods to maximize profits subject to the restrictions of a production technology. An equilibrium of this model is a vector of prices, one for each good, which the agents all take as given in solving their maximization problems, such that demand is equal to supply for every good. Economists use this type of model to do comparative statics analysis: they first compute a benchmark equilibrium for the model; they then change a parameter of the model such as a tax rate; finally they compare the new equilibrium with the benchmark. Large-scale empirical models are often used to do policy analysis (see, for example, Kehoe and Serra-Puche, 1983, and Shoven and Whalley, 1984).

In this paper we study several simple, highly stylized versions of the Walrasian equilibrium model. (Debreu, 1959; Arrow and Hahn, 1971; and Mas-Colell, 1985 are good references on the mathematics of general equilbrium theory.) Our emphasis is on the formal properties of these models that make them different from many models in the physical sciences, rather than on the economics of the models. In particular, we discuss the problems involved in proving the existence of equilibria and of computing equilibria. We give special attention to the possibility of multiplicity of equilibria.

2. EXISTENCE OF EQUILIBRIUM AND BROUWER'S FIXED-POINT THEOREM

We begin with the simplest possible model, an *exchange economy* in which the only economic activity is exchange of goods among consumers. We discuss models with production later. There are m consumers and n goods. Consumer i, $i = 1, 2, \ldots, m$, is endowed with nonnegative amounts, $w^i = (w_1^i, w_2^i, \ldots, w_n^i)$, of goods. He also has a *utility function* $u_i : R_+^n \to R$ that is strictly concave and monotonically increasing. When faced by the price vector $p \in R_+^n$, he chooses the consumption plan x^i to maximize $u_i(x)$ subject to the *budget constraint* $p \cdot x \leq p \cdot w^i$ and nonnegativity constraint $x \geq 0$.

The solution to this problem, the consumer's demand function $x^i(p)$, is continuous (at least for strictly positive prices); is homogeneous of degree zero, $x^i(\lambda p) = x^i(p)$ for all $\lambda > 0$ and all p; and satisfies the budget constraint, $p \cdot x^i(p) = p \cdot w^i$ for all p. The *aggregate excess demand function*,

$$f(p) = \sum_{i=1}^{m} (x^i(p) - w^i),$$

therefore is continuous, is homogeneous of degree zero, and satisfies $p \cdot f(p) = 0$ for all p. This final property is known as *Walras' law* and is as close as economics comes to having a law of conservation.

Our price domain is $R_+^n \backslash \{0\}$, the set of all nonnegative price vectors except the origin. There is often a technical problem with continuity of excess demand when some prices approach zero. Some, but not necessarily all, of the corresponding excess demands might then approach infinity. There are simple ways to get around this problem, however (see, for example, Kehoe, 1982). Since nothing of conceptual significance is involved, we shall ignore any problems posed by discontinuity of demand due to zero prices.

An *equilibrium* is a price vector $\hat{p}$ for which $f(\hat{p}) \leq 0$. Notice that Walras' law implies that $f_j(\hat{p}) < 0$ only if $\hat{p}_j = 0$; in other words, we allow supply to exceed demand for a good only if it is free. Walras (1874) himself had two arguments for the existence of equilibrium: first, he counted equations and unknowns in the definition of equilibrium and verified that they are equal. Second, he proposed a dynamic adjustment process that would bring the system into equilibrium from an arbitrary starting price vector. Although each of these approaches provided important insights, neither is correct.

The system $f(\hat{p}) = 0$ involves n equations and n unknowns. Walras recognized that there are two offsetting complications: because of Walras' law, one equation is redundant, but because of homogeneity we can impose a price normalization. In other words, we can either eliminate a variable by choosing a numeraire, a good in whose terms all values are measured, by setting, for example, $\hat{p}_1 = 1$, or we can add an equation, such as $\sum_{i=1}^{n} \hat{p}_i = 1$. Consequently, we are left with a system with the same number of equations and unknowns. Although this clearly does not, as Walras seems to have thought it does, assure us of the existence of a solution, it does, as we shall see, tell us something about local uniqueness of equilibria.

The second approach to existence followed by Walras was a disequilibrium adjustment process that he called *tâtonnement*, or groping. In it, an auctioneer adjusts prices systematically by raising the prices of goods in excess demand and lowering those of goods in excess supply. Samuelson (1941, 1942) formalized this process as the system of differential equations

$$\frac{dp_j}{dt} = f_j\left(p(t)\right).$$

(Walras himself thought of tâtonnement more as a nonlinear Gauss-Seidel method.)

This approach was popular in economics for a time, and many economists searched for conditions under which it leads to convergence (see, for example, Arrow, Block, and Hurwicz, 1959). There are problems in giving the process a real-time interpretation, however, and, in any case, it became less popular after Scarf (1960) constructed an example in which, unless $p(0) = \hat{p}$, the unique equilibrium of his example, the solution converges to a limit cycle.

In a series of papers of increasing generality, Sonnenschein (1973), Mantel (1974), and Debreu (1974) proved that the excess demand function f is arbitrary except for continuity, homogeneity, and Walras' law. Specifically, Debreu proved that for any function f that satisfies these properties, there are n consumers with strictly concave, monotonically increasing utility functions whose individual excess demands sum to f. To see that this implies that the tâtonnement process is arbitrary, let us use homogeneity to normalize prices to lie on the intersection of the unit sphere and the positive orthant, $\sum_{i=1}^{n} p_i^2 = 1$, $p_i \geq 0$. Walras' law implies that $f(p)$ defines a vector field on the sphere:

$$\frac{d}{dt}\left(\sum_{i=1}^{n} p_i(t)^2\right) = 2\sum p_i(t)\,\frac{dp_i}{dt} = 2\sum p_i(t) f_i\left(p(t)\right) = 0.$$

Consequently, continuity, homogeneity, and Walras' law imply no more than that the tâtonnement process defines a continuous vector field on the sphere. The solution for an initial condition $p(0) = p_0$ is fairly arbitrary. With three goods, for example, there can be a stable limit cycle, as in Scarf's (1960) example. See Figure 1. With four or more goods, the tâtonnement process can generate chaotic dynamics.

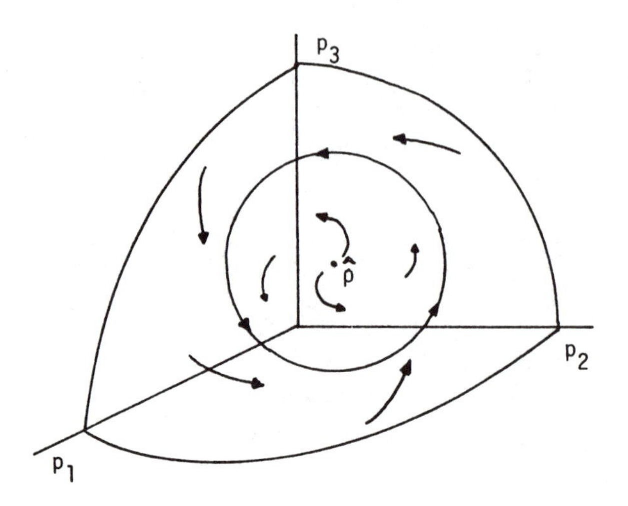

FIGURE 1 Scarf's example.

Although Wald (1936) had proved the existence of equilibria for two special models earlier, Arrow and Debreu (1954) and McKenzie (1954) realized that the existence of equilibria in more general models could be demonstrated using *Brouwer's fixed-point theorem* or some variant. Brouwer's theorem says that a continuous map $g : S \to S$ of some nonempty, compact, convex set S into itself leaves some point fixed, that is, $\hat{x} = g(\hat{x})$ for some $\hat{x} \in S$.

Renormalizing prices, we can use the simplex $S = \{p \in R^n \mid e \cdot p = 1, p \geq 0\}$ as our price domain. (Here $e = (1, \ldots, 1)$.) Walras' adjustment process suggests $g(p) = p + f(p)$ as a map whose fixed points are equilibrium. The problem is that $g(p)$ is not necessarily in S for all $p \in S$. For any $q \in R^n$, let $\pi(q)$ be that point in S that is closest to q in euclidean distance. Define

$$g(p) = \pi\big(p + f(p)\big).$$

Since S is convex, π and, therefore, g are continuous.

We claim that $f(\hat{p}) \leq 0$ if and only if $\hat{p} = g(\hat{p})$. To prove this, notice $g(p)$ solves the problem

$$\begin{aligned} &\min(1/2)\ \|g - p - f(p)\|^2, \\ &\text{s.t. } e \cdot g = 1 \\ &\qquad g \geq 0 \end{aligned}$$

if and only if

$$\begin{aligned} &g - p - f(p) + \lambda e \geq 0 \\ &g \cdot \big(g - p - f(p) + \lambda e\big) = 0 \end{aligned}$$

for some Lagrange multiplier λ. Suppose that $\hat{p} = g(\hat{p})$. Walras' law implies that $0 = \hat{p} \cdot f(\hat{p}) = \hat{\lambda}\hat{p} \cdot e = \hat{\lambda}$. Consequently, $f(\hat{p}) \leq 0$. To prove the converse that

$f(\hat{p}) \leq 0$ implies $g(\hat{p}) = \hat{p}$, merely requires setting $g = p$ and $\lambda = 0$ in conditions that define $g(p)$.

The relationship between the existence theorem and Brouwer's theorem is very close: as we have just argued, the existence of equilibrium follows quickly from Brouwer's theorem. Furthermore, as Uzawa (1962) first noticed, Brouwer's theorem follows quickly from the existence theorem. Suppose that $g: S \rightarrow S$ is continuous. We can construct an excess demand function f that satisfies continuity, homogeneity, and Walras' law and is such that $\hat{p} = g(\hat{p})$ if and only if $f(\hat{p}) \leq 0$. Since we know that aggregate excess demand is arbitrary except for these properties, it follows that, if we know that an equilibrium exists for every economy, then we know Brouwer's theorem. Consequently, except for special cases, Brouwer's theorem or its equivalent is necessary for proving the existence of equilibrium.

An obvious candidate for the excess demand function based on g is $f(p) = g(p) - p$; the problems are that it does not satisfy homogeneity or Walras' law. Homogeneity is trivial, however: if we have a function $f: S \rightarrow R^n$, we can define f on all $R^n_+ \backslash \{0\}$ as $f(\pi(p))$, where, of course, $\pi(p) = (1/e \cdot p)p$. Let us, therefore, without loss of generality, restrict ourselves to the price domain S. To make f obey Walras' law, we define $\lambda(p) = p \cdot g(p)/p \cdot p$ and set

$$f(p) = g(p) - \lambda(p)p.$$

Suppose now that $f(\hat{p}) \leq 0$. This implies that, in fact, $f(\hat{p}) = 0$ since, by Walras' law, $f_j(\hat{p}) < 0$ would imply $\hat{p}_j = 0$, in which case $f_j(\hat{p}) = g_j(\hat{p}) \geq 0$. That $f(\hat{p}) = 0$, however, implies that $g(\hat{p}) = \lambda(\hat{p})\hat{p}$. Since $e \cdot g(\hat{p}) = e \cdot \hat{p} = 1$, this implies that $\lambda(\hat{p}) = 1$ and, therefore, $g(\hat{p}) = \hat{p}$. The converse, that $g(\hat{p}) = \hat{p}$ implies $f(\hat{p}) = 0$, follows from $\hat{p} \cdot g(\hat{p})/\hat{p} \cdot \hat{p} = 1$.

3. ECONOMIC EQUILIBRIUM AND OPTIMIZATION

Unlike many problems in the physical sciences, the economic equilibrium problem cannot usually be solved as an optimization problem: although an equilibrium is the solution to an optimization problem, finding the right optimization problem is, in general, as difficult as finding the equilibrium itself.

Pareto (1909) first realized that the allocation of goods $(\hat{x}^1, \hat{x}^2, \ldots, \hat{x}^m)$ associated with an equilibrium $\hat{p}$ has a property now known as *Pareto efficiency*. There is no alternative allocation $(\bar{x}^1, \bar{x}^2, \ldots, \bar{x}^m)$ that is superior in the sense that is feasible, $\sum_{i=1}^m \bar{x}^i \leq \sum_{i=1}^m w^i$, and $u_i(\bar{x}^i) \geq u_i(\hat{x}^i)$, $i = 1, 2, \ldots, m$, with strict inequality some i. In other words, there is no way to reallocate goods to make some consumer better off that does not make another consumer worse off. The argument, due to Arrow (1951) and Debreu (1954), is simple: suppose to the contrary, that there is an allocation superior to the competitive allocation. Then $\hat{p} \cdot \bar{x}^i \geq \hat{p} \cdot \hat{x}^i = \hat{p} \cdot w^i$ with strict inequality wherever $u_i(\bar{x}^i) > u_i(\hat{x}^i)$; otherwise, $\hat{x}^i$ would not maximize utility subject to the budget constraint. Consequently, $\sum_{i=1}^m \hat{p} \cdot \bar{x}^i > \sum_{i=1}^m \hat{p} \cdot w^i$.

Multiplying the feasibility conditions by $\hat{p}$, however, yields $\hat{p}\cdot\sum_{i=1}^{m}\overline{x}^i \leq \hat{p}\cdot\sum_{i=1}^{m} w^i$, which can be rearranged as $\sum_{i=1}^{m}\hat{p}\cdot\overline{x}^i \leq \sum_{i=1}^{m}\hat{p}\cdot w^i$. This contradiction shows that there can be no allocation that is Pareto superior to the competitive allocation.

Since an equilibrium is Pareto efficient, the associated allocation solves

$$\max \sum_{i=1}^{m} \alpha_i u_i(x^i)$$

$$\text{s.t.} \sum_{i=1}^{m} x^i \leq \sum_{i=1}^{m} w^i$$

$$x^i \geq 0$$

for some nonnegative weights α_i. How do we find the right weights α_i? Associated with the feasibility conditions are nonnegative Lagrange multipliers $p(\alpha) = (p_1(\alpha), p_2(\alpha), \ldots, p_n(\alpha))$. It is easy to show that, for any vector α, the prices $p(\alpha)$ and allocation $(x^1(\alpha), x^2(\alpha), \ldots, x^m(\alpha))$ satisfy all of the conditions for an equilibrium except the individual budget constraints. In the case where u_i is continuously differentiable, for example, this is simply a matter of showing that the necessary and sufficient conditions for a solution to this problem can be rearranged into the conditions for utility maximization. If we give each consumer a net transfer given by the *transfer function*

$$t_i(\alpha) = p(\alpha)\cdot\left(x^i(\alpha) - w^i\right), \qquad i = 1, 2, \ldots, m,$$

then $p(\alpha)$ is an equilibrium of the economy with transfer payments. In this economy budget constraints have the form $p\cdot x^i \leq p\cdot w^i + t_i(\alpha)$.

To find an equilibrium, we must find a vector $\hat{\alpha}$ such that $t(\hat{\alpha}) = 0$. The transfer functions t are continuous, homogeneous of degree one, and sum to zero. The functions $f_i(\alpha) = -t_i(\alpha)/\alpha_i$, in fact, have the same formal properties as excess demand functions. Finding an equilibrium using this approach, which is due to Negishi (1960), involves solving a fixed-point problem in R^m rather than R^n. This is sometimes useful if $m < n$, for example, if m is finite and n is infinite.

4. COMPUTATION OF EQUILIBRIA

Scarf (1967, 1973, 1982) realized that any algorithm that could be guaranteed to compute economic equilibria would have to be able to compute fixed points of arbitrary maps $g: S \rightarrow S$. He developed such an algorithm. Numerous researchers have further improved algorithms of this type, now known as *simplicial algorithms* (see, for example, Eaves, 1972; Todd, 1976; and van der Laan and Talman, 1980).

In R^n, a k-dimensional *simplex* is the convex hull of $k+1$ points, called *vertices*, $v^1, v^2, \ldots, v^{k+1}$, that have the property that the k vectors $v^1 - v^{k+1}, \ldots, v^k - v^{k+1}$ are linearly independent. The price simplex S, for example, has vertices e^i, $i = 1, 2, \ldots, n$, where $e^i_i = 1$, $e^i_j = 0$, $j \neq i$. A *face* of a simplex is a lower-dimensional simplex whose vertices are vertices of the large simplex. In R^3, for example, the point e^1 is a zero-dimensional face of S and the convex hull of e^1 and e^2 is a one-dimensional face. A *subdivision* of S divides S into smaller simplices so that every point in S is an element of some subsimplex and the intersection of any two subsimplices is either empty or a face of both.

Scarf's approach to computation of equilibria is based on a constructive proof of a version of *Sperner's lemma*: assign to every vertex of a simplicial subdivision of S a *label*, an integer from the set $1, 2, \ldots, n$, with the property that a vertex v on the boundary of S receives a label i for which $v_i = 0$. Then there exists a subsimplex whose vertices have all of the labels $1, 2, \ldots, n$.

Scarf's algorithm for finding this completely labeled subsimplex is to start in the corner of S where there is a subsimplex with boundary vertices with all of the labels $2, 3, \ldots, n$. See Figure 2. If the additional vertex of this subsimplex has the label 1, then the algorithm stops. Otherwise, it proceeds to a new subsimplex with all of that labels $2, 3, \ldots, n$: the original subsimplex has two faces that have all of these labels. One of them includes the interior vertex. The algorithm moves to the unique other subsimplex that shares this face. If the additional vertex of this subsimplex has the label 1, the algorithm stops. Otherwise, it proceeds, moving to

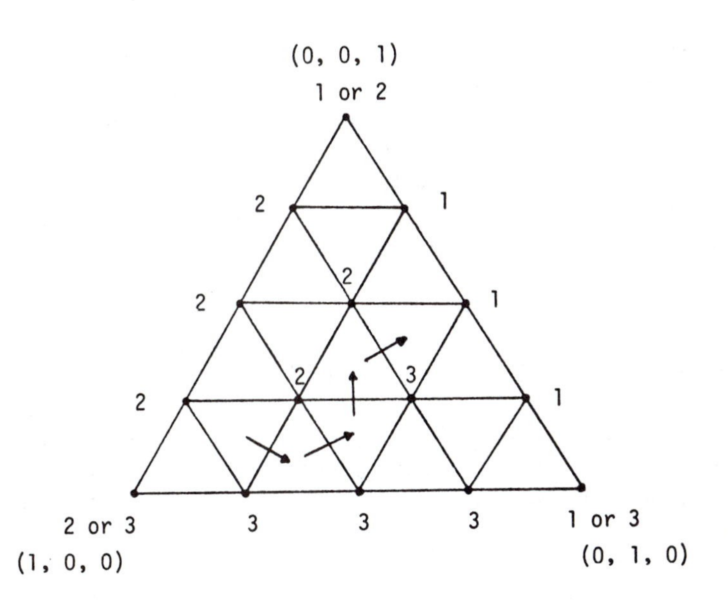

FIGURE 2 Scarf's algorithm for finding completely labeled subsimplex.

the unique subsimplex that shares the new face and has the labels $2, 3, \ldots, n$. The algorithm cannot try to exit through a boundary face. (Think of what labels the vertices of such a face must have.) Nor can it cycle. (To cycle there must be some subsimplex that is the first that the algorithm encounters for the second time, but the algorithm must have previously encountered both of the subsimplices that share the two faces of this subsimplex with the labels $2, 3, \ldots, n$.) Since the subdivision consists of a finite number of subsimplices, the algorithm must terminate with a completely labeled subsimplex.

To see the connection of this algorithm with Brouwer's theorem, we assign a vertex v with a label i for which $g_i(v) \geq v_i$. Since $e \cdot g(v) = e \cdot v = 1$, there must be such an i. Notice that, since $g_i(v) \geq 0$, i can be chosen such that the labeling convention on the boundary is satisfied. A completely labeled subsimplex has vertices $v^1, v^2, \ldots, v^n$ such that $g_i(v^i) \geq v^i_i$. To prove Brouwer's theorem, we consider a sequence of subdivisions whose *mesh*, the maximum distance between vertices in the same subsimplex, approaches zero. Associate each subdivision with a point in a completely labeled subsimplex. Since S is compact, this sequence of points has a convergent subsequence. Call the limit of this subsequence $\hat{p}$. Since g is continuous, we know $g_i(\hat{p}) \geq \hat{p}_i$, $i = 1, 2, \ldots, n$. Since $e \cdot g(\hat{p}) = e \cdot \hat{p} = 1$, $g(\hat{p}) = \hat{p}$.

Scarf did not consider an infinite sequence of subdivisions, which is the nonconstructive aspect of this proof. Instead, he worked with a subdivision with a small mesh. Any point in a completely labeled subsimplex serves as an approximate fixed point in the sense that $\|g(x) - x\| < \varepsilon$ where ε depends on the mesh and the modulus of continuity of g.

An alternative algorithm for computing fixed points was developed by Smale (1976), who called it the *global Newton's method*. It is based on Hirsch's (1963) proof of Brouwer's theorem. Let S now be the disk $\{x \in R^n \mid x \cdot x \leq 1\}$; like the simplex, it is a nonempty, compact, convex set. Smale developed an algorithm for computing fixed points of a continuously differentiable map $g : S \to S$ that has the property that $g(x) = 0$ for every x on the boundary of S, the sphere $\partial S = \{x \in R^n \mid x \cdot x = 1\}$. He also showed how to extend this algorithm to situations where g is an arbitrary continuous map and S is again the simplex.

If S had no fixed points, we could define a map

$$h(x) = \lambda(x)\big(x - g(x)\big)$$

where $\lambda(x) = (x - g(x)) \cdot (x - g(x))^{-1/2}$. This map would be a *retraction* of S into its boundary: it would continuously map S into ∂S and be the identity on ∂S. Hirsh proved that no such map could exist, thereby proving Brouwer's theorem. Smale proposed starting with a regular value of $x - g(x)$, a point $\overline{x} \in \partial S$ such that $I - Dg(\overline{x})$ is nonsingular. *Sard's theorem* says that the set of regular values has full measure. The algorithm then follows the solution to

$$\lambda(x(t))\Big(x(t) - g\big(x(t)\big)\Big) = \overline{x}.$$

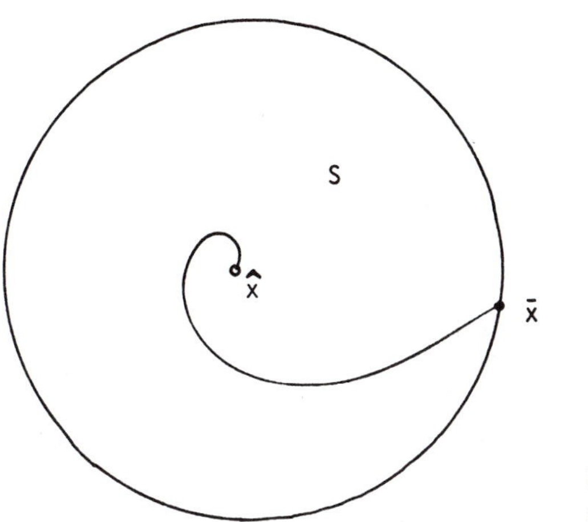

FIGURE 3 Illustration of Smale's algorithm.

Since the path $x(t)$ cannot return to any other boundary point, and since it cannot return to $\overline{x}$ because it is a regular value, it must terminate at a fixed point.

Smale shows that $x(t)$is the solution to the differential equation

$$[I - Dg(x(t))] \frac{dx}{dt} = \mu(x(t)) \Big(x(t) - g(x(t))\Big)$$

where $\mu(x) = \text{sgn}(\det[I - Dg(x)])$. Except for the factor μ, this is a continuous version of Newton's method for solving $x - g(x) = 0$:

$$x_{t+1} = x_t - [I - Dg(x_t)]^{-1}(x_t - g(x_t)).$$

5. MODELS WITH PRODUCTION

We now generalize our model to include a simple production technology. It is specified by a $k \times n$ *activity analysis matrix* A. Columns of A represent feasible production plans: positive entries denote outputs, negative entries inputs. The set of technologically feasible production plans is $Y = \{x \in R^n \mid x = Ay,\ y \geq 0\}$. Here $y = (y_1, y_2, \ldots, y_k)$ is a vector of nonnegative *activity levels.*

This type of linear production technology, initially formalized by Koopmans (1951), but with many antecedents, is a fairly general example of a *constant returns* technology: $x \in Y$ if and only if $\lambda x \in Y$ for all $\lambda \geq 0$. Economists also sometimes work with *decreasing returns* technologies where $x \in Y$ implies $\lambda x \in Y$ for $0 \geq \lambda \geq 1$. In both cases they assume Y is a convex set. Decreasing returns are probably best thought of, both in terms of formal mathematics and of economic intuition, however, as constant returns in an economy with additional goods that are inputs to specific groups of production activities and are in fixed supply. *Increasing returns*, where Y is not convex, are another matter: they are not well handled by general equilibrium theory, since they would cause firms to become very large relative to the size of the economy, in which case the competitive assumption that all agents take prices as given is untenable. We assume that A is such that there is free disposal and that there is no output without any inputs. *Free disposal* means that, if $x \in Y$ and $x' \leq x$, then $x' \in Y$. This can be ensured by including vectors $-e^i$, $i = 1, 2, \ldots, n$ as columns in A. That $-e^i$ is a column of A means that good i can be thrown away without using other inputs. That there is no output is without any inputs means that $Y \cap R^n_+ = \{0\}$. This can be ensured by making sure that $Ay \geq 0$ and $y \geq 0$ imply $Ay = 0$. An *equilibrium* of this model is a price vector $\hat{p}$ such that, for some vector of activity levels $\hat{y} \geq 0$,

$$f(\hat{p}) = A\hat{y},$$
$$\hat{p}A \leq 0,$$
$$e \cdot \hat{p} = 1.$$

Notice that Walras' law and the first equilibrium condition imply that $\hat{p} \cdot f(\hat{p}) = \hat{p} \cdot (A\hat{y}) = 0$. In other words, the *economic profit* made by the production plan $\hat{p} \cdot (A\hat{y})$, revenue minus expenditures, is equal to zero in equilibrium. The second equilibrium condition implies that $\hat{p} \cdot (A\hat{y}) \leq 0$ for any $y \geq 0$. Consequently, the production plan

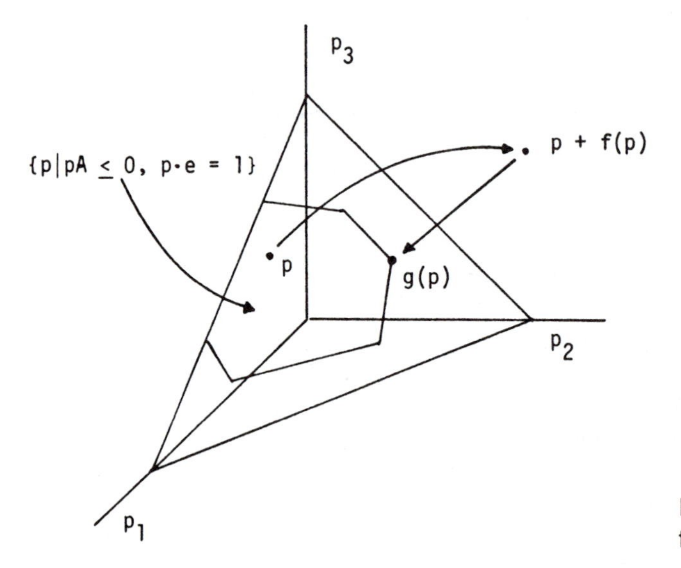

FIGURE 4 Continuous function $g : S \to S$.

$A\hat{y}$ maximizes profits at prices $\hat{p}$. (Activities in decreasing returns technology typically earn positive economic profits; these profits are best thought of as payments to sector specific fixed inputs.)

The analysis of the previous sections extends easily to models with production. To prove the existence of equilibrium, we again construct a continuous function $g : S \to S$. For any $p \in S$, we define $g(p)$ as the solution to

$$\begin{aligned} &\min (1/2)\|g - p - f(p)\|^2 \\ &\text{s.t.} \quad gA \leq 0 \\ &\qquad\quad e \cdot g = 1. \end{aligned}$$

Notice that, because of free disposal, $gA \leq 0$ implies $g \geq 0$. g solves this problem if and only if

$$\begin{aligned} g - p - f(p) + Ay + \lambda e &= 0 \\ g \cdot (Ay) &= 0 \end{aligned}$$

for some scalar λ and vector $y \geq 0$ of Lagrange multipliers. Once again Walras' law implies that $\hat{p}$ is an equilibrium if and only if $\hat{p} = g(\hat{p})$.

Equilibria of the model with production are Pareto efficient: the associated allocation and production plan solves

$$\begin{aligned} &\max \sum \alpha_i u_i(x^i) \\ &\text{s.t.} \sum_{i=1}^{m} x^i = Ay \\ &\qquad x^i \geq 0, \quad y \geq 0. \end{aligned}$$

Once again there are transfer functions $t_i(\alpha)$ such that $t(\hat{\alpha}) = 0$ is an alternative system of equilibrium conditions.

6. MULTIPLICITY OF EQUILIBRIA

Are equilibria unique? If not, are they locally unique? Do they vary continuously with the parameters of the economy? In recent years, economists have used the tools differential topology to investigate these questions.

Debreu (1970) first investigated the questions of local uniqueness and continuity of equilibrium in exchange economies with continuously differentiable excess demand functions. He defined a *regular economy* to be one for which the Jacobian matrix of excess demands $Df(\hat{p})$ with the first row and column deleted, the $(n-1) \times (n-1)$ matrix J, is nonsingular at every equilibrium. The first row is deleted because of Walras' law, the first column because of homogeneity; we are left with a square matrix because, as Walras had pointed out, the number of equations equals the number of unknowns in the equilibrium conditions. The inverse function theorem implies that every equilibrium of a regular economy is locally unique. Since the set S is compact and the equilibrium conditions involve continuous functions, this implies that a regular economy has a finite number of equilibria.

Let us rewrite the equilibrium conditions as $f(p, b) = 0$ where $b \in B$ and B is a topological space of parameters. If f and its partial derivatives with respect to p are continuous in both p and b, then the implicit function theorem implies that equilibria vary continuously at regular economies. Furthermore, in the case where B is the set of possible endowment vectors w^i, Debreu used Sard's theorem to prove that, for every b in an open set of full measure in B, $f(\cdot, b)$ is a regular economy. When B is the function space of excess demand functions with the uniform C^1 topology, an open dense set of B consists of regular economies. Consequently, if we are willing to restrict attention to continuously differentiable excess demand functions, a restriction that Debreu (1972) and Mas-Colell (1974) have shown is fairly innocuous, almost all economies, in a very precise mathematical sense, are regular.

Dierker (1972) noticed that a fixed-point index theorem could be used to count the number of equilibria of a regular economy. Let us define the fixed-point *index* of a regular equilibrium $\hat{p}$ as $\mathrm{sgn}(\det[I - Dg(\hat{p})])$ whenever this expression is nonzero. Dierker showed that the index can also be written as $\mathrm{sgn}(\det[-J])$. The *index theorem* says that $\sum \mathrm{index}(\hat{p}) = +1$ where the sum is over equilibria of a regular economy. This result is depicted in Figure 5 where $n = 2$, $p_1 = 1 - p_2$, and $g_1(p_1, p_2) = 1 - g_2(p_1, p_2)$. Here $\mathrm{index}(\hat{p}) = \mathrm{sgn}(1 - \partial g_2 / \partial p_2)$ and a regular economy is one where the graph of g does not become tangent to the diagonal.

Mas-Colell (1977) showed that any compact subset of S can be the equilibrium set of some economy f. If we restrict ourselves to regular economies and $n \geq 3$,

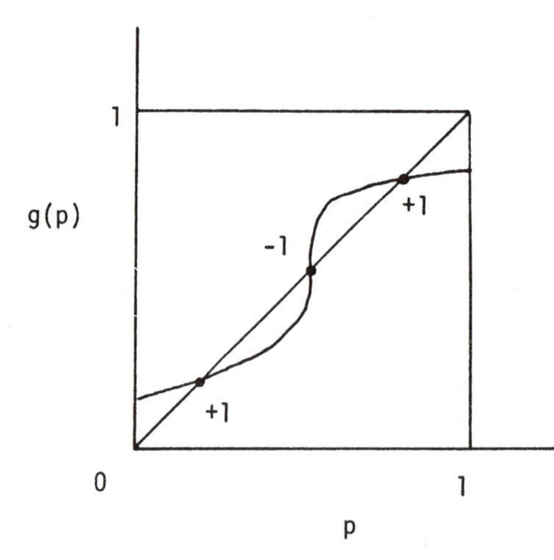

FIGURE 5 Illustration of the index theorem.

then the only restrictions placed on the number of equilibria are those given by the index theorem. (An equilibrium with index -1 must lie between two with index $+1$ if $n = 2$.) This implies that the number of equilibria is odd, and that there is a unique equilibrium if and only if $\text{index}(\hat{p}) = +1$ at every equilibrium.

Mas-Colell (1975, 1985) and Kehoe (1980, 1983) have extended the concepts of regularity and fixed-point index to economies with production. They prove that regular production economies have the same desirable properties as regular exchange economies and that, in a precise sense, almost all economies are regular. Kehoe (1980) further calculates the index of a regular equilibrium as

$$\text{index}(\hat{p}) = \text{sgn}\left(\det\begin{bmatrix} -J & B \\ -B^T & 0 \end{bmatrix}\right).$$

Here B is the submatrix of A formed by deleting the first row from A and any column for which the corresponding activity level is zero. A regular economy, of course, is one for which this expression is nonzero.

Using this formula, Kehoe (1985) has constructed a simple example of a production economy with three equilibria. (Exchange economies with $n = m = 2$ that have multiple equilibria are also easy to construct; see, for example, Shapley and Shubik, 1977.) This example has $n = m = 4$. Consumer i solves

$$\max \sum_{j=1}^{4} \gamma_j^i \log x_j^i$$

$$\text{s.t.} \sum_{j=1}^{4} p_j x_j^i \leq \sum_{j=1}^{4} p_j w_j^i$$

$$x_j^i \geq 0.$$

The utility parameters γ_j^i are given in Table 1.

TABLE 1 Utility Parameters γ_j^i

	Consumer			
Commodity	1	2	3	4
1	0.52	0.86	0.5	0.06
2	0.4	0.1	0.2	0.25
3	0.04	0.02	0.2975	0.0025
4	0.04	0.02	0.0025	0.6875

TABLE 2 Endowment Parameters w_j^i

	Consumer			
Commodity	1	2	3	4
1	50	0	0	0
2	0	50	0	0
3	0	0	400	0
4	0	0	0	400

The endowment parameters w_j^i are given in Table 2.
The aggregate excess demand function f has the form

$$f_j(p_1, p_2, p_3, p_4) = \frac{\sum_{i=1}^{4} \gamma_j^i \sum_{\ell=1}^{4} p_\ell w_\ell^i}{p_j} - \sum_{i=1}^{4} w_j^i, \quad i = 1, 2, 3, 4.$$

As can easily be verified, this function satisfies continuity, homogeneity, and Walras' law. The production side of the economy is given by a 4×6 activity analysis matrix A:

$$A = \begin{bmatrix} -1 & 0 & 0 & 0 & 6 & -1 \\ 0 & -1 & 0 & 0 & -1 & 3 \\ 0 & 0 & -1 & 0 & -4 & -1 \\ 0 & 0 & 0 & -1 & -1 & -1 \end{bmatrix}.$$

TABLE 3 Equilibrium 1

	Consumer			
Commodity	1	2	3	4
1	26.000	43.000	200.000	24.000
2	20.000	5.000	80.000	100.000
3	2.000	1.000	119.000	1.000
4	2.000	1.000	1.000	275.000
u_i	2.948	3.396	4.947	5.204

TABLE 4 Equilibrium 2

	Consumer			
Commodity	1	2	3	4
1	26.000	67.431	48.490	83.089
2	12.754	5.000	12.368	220.771
3	8.249	6.468	119.000	14.280
4	0.578	0.453	0.070	275.000
u_i	2.775	3.804	3.858	5.483

The parameters of this example have been chosen so that $\hat{p}_1 = \hat{p}_2 = \hat{p}_3 = \hat{p}_4 = 1/4$ is an equilibrium with $\text{index}(\hat{p}) = -1$. Consequently, we know that there are multiple equilibria. Two other equilibria, both with index $+1$, have been found using an exhaustive computer search.

EQUILIBRIUM 1

$$p^1 = (0.25000, 0.25000, 0.25000, 0.25000)$$
$$y^1 = (0, 0, 0, 0, 52.000, 69.000)$$
$$\alpha^1 = (0.05556, 0.05556, 0.44444, 0.44444)$$

See Table 3.

EQUILIBRIUM 2

$$p^2 = (0.15942, 0.25000, 0.03865, 0.55193)$$
$$y^2 = (0, 0, 0, 0, 42.701, 81.198)$$
$$\alpha^2 = (0.03105, 0.04869, 0.06023, 0.86003)$$

See Table 4.

TABLE 5 Equilibrium 3

	Consumer			
Commodity	1	2	3	4
1	26.000	39.072	224.362	14.499
2	22.001	5.000	98.768	66.485
3	1.783	0.810	119.000	0.539
4	3.311	1.504	1.857	275.000
u_i	3.002	3.317	5.049	5.070

EQUILIBRIUM 3

$$p^3 = (0.27514, 0.25000, 0.30865, 0.16621)$$
$$y^3 = (0, 0, 0, 0, 53.180, 65.148)$$
$$\alpha^3 = (0.06363, 0.05782, 0.57104, 0.30751)$$

See Table 5.

Each of the equilibria of this example is Pareto efficient and solves a problem of maximizing a weighted sum of utilities subject to feasibility constraints. The weights α_i associated with each of the equilibria are listed above. We emphasize that the multiplicity of equilibria is not due to any sort of nonconvexity in the consumers' or producers' maximization problems. Indeed, for any vector of weights α_i, the social planning problem that produces Pareto efficient allocations always has a unique solution.

7. INTERTEMPORAL MODELS

Let us now consider models in which goods are distinguished by date. Time is discrete, and at each date $t = 1, 2, \ldots$, there are n goods. Although our models have infinite time horizons, our results have strong implications for models with long but finite horizons. For simplicity, we restrict attention to exchange economies: goods cannot be stored between periods because storage is best thought of as a kind of production. Moreover, the models are stationary: their parameters do not depend on the date.

Consider first a model with a finite number of infinitely lived consumers. Consumer i, $i = 1, 2, \ldots, m$, chooses a sequence of consumption vectors $x_1^i, x_2^i, \ldots$ to maximize a utility function, such as $\sum_{t=1}^{\infty} \beta_i^t u_i(x_t)$, where $0 < \beta_i < 1$, subject to $\sum_{t=1}^{\infty} p_t \cdot x_t \leq \sum_{t=1}^{\infty} p_t \cdot w^i$ and $x_t \geq 0$. One way to formulate the equilibrium conditions is to set supply equal to demand; thus results in a system of an infinite number of equations and unknowns. Equilibria of this model are Pareto efficient, however, and we can use the approach using transfer functions to reduce this to a system of $m-1$ equations and $m-1$ unknowns. Using regularity analysis, we could argue that almost all such economies have a finite number of equilibria (see Kehoe and Levine, 1985).

Consider instead a model with an infinite number of consumers, an overlapping generations model of the type first considered by Samuelson (1958). This model has a number of features not shared by the model with a finite number of consumers: it may, for example, have equilibria that are not Pareto efficient. There may also be robust examples with a continuum of equilibria.

Consider an example with a single consumer in each generation born in period t, who lives for three periods. He has utility function $\sum_{\tau=t}^{t+2} \alpha^{\tau-t}(x_\tau^b - 1)/b$, where $b < 1$, and endowment stream (w_1, w_2, w_3). The equilibrium condition in periods $t = 3, 4, \ldots$, has the form

$$f(p_{t-2}, p_{t-1}, p_t, p_{t+1}, p_{t+2}) = 0$$

since it involves demands by consumers born in $t-2$, $t-1$, and t. In addition to these consumers, there is an old consumer, who lives only one period, and a middle-aged consumer, who lives two, alive in the first period. The equilibrium conditions in the first two periods have the form

$$f_1(p_1, p_2, p_3) = 0$$
$$f_2(p_1, p_2, p_3, p_4) = 0.$$

This model, like the one previously described, implicitly assumes *perfect foresight*: consumers know, possible by solving the model themselves, what prices prevail in the future.

The standard approach to proving the existence of, and computing, equilibria involves truncating the time horizon at some date T. (See, for example, Balasko, Cass, and Shell, 1980; Auerbach, Kotlikoff, and Skinner, 1983.) Equilibria then depend on the anticipated values of $p_{T=1}$ and p_{T+2}. To prove existence, we consider a sequence of price sequences $(p_1^k, p_2^k, \ldots)$ that satisfy the equilibrium conditions in the first T_k periods. An equilibrium is the limit of a convergent subsequence in the product topology as $T_k \to \infty$. A computational procedure must stop at a finite T, however. In computing equilibria, a standard assumption is that the equilibrium converges to a *steady state*, a solution to the equilibrium condition in

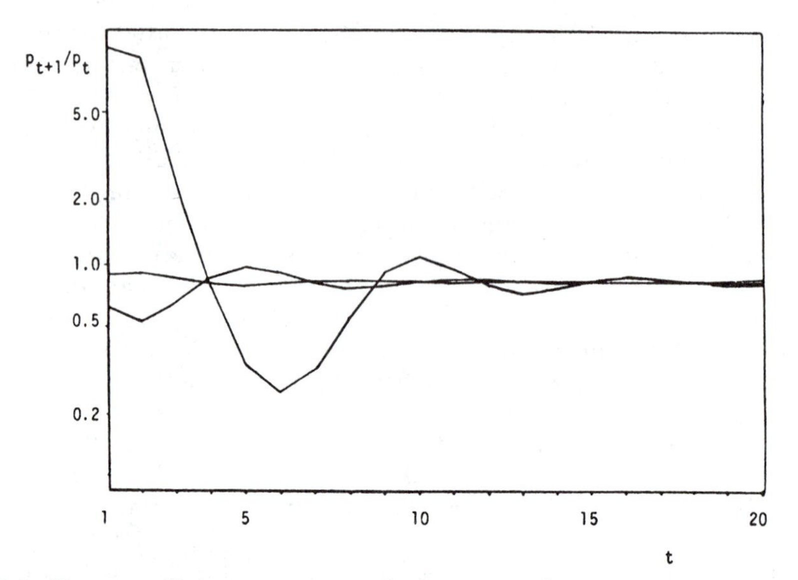

FIGURE 6 Three equilibria converging to the same steady state.

periods $t = 3, 4, \ldots$ of the form $p_t = \beta^{t-1}p_1$. (Physical scientists are often surprised that economists use the word equilibrium to describe something besides a steady state.) In this case, the expectations would be $p_{T+1} = \beta p_T$ and $p_{T+2} = \beta^2 p_T$. Equilibria need not, of course, converge to a steady state. Some may exhibit chaotic trajectories (see, for example, Benhabib and Day, 1982).

Unfortunately, this type of model can have a vast multiplicity of equilibria. Suppose, for example, we choose $a = 0.5$, $b = -3$, and $(w_1, w_2, w_3) = (3, 12, 1)$. Then, choosing the initial old and middle-aged consumers to ensure that the steady state satisfies the equilibrium conditions in periods 1 and 2, we can demonstrate that this economy has a continuum of equilibria that all converge to the steady state where $\beta = 0.7925$ (see Kehoe and Levine, 1987, for details).

Figure 6 depicts three of these equilibria. Notice that, if we slightly perturb the terminal conditions for p_{T+1}, p_{T+2} in, say, $T = 20$, then we produce drastically different equilibria. This illustrates the point that a continuum of equilibria in an infinite horizon model is typically symptomatic of sensitivity to terminal conditions in truncated versions of the model.

ACKNOWLEDGMENTS

I gratefully acknowledge the support of National Science Foundation Grant SES 85-09484.

REFERENCES

1. Arrow, K. J. (1951), "An Extension of the Basic Theorems of Classical Welfare Economics," *Proceedings of the Second Berkeley Symposium on Mathematical Statistics and Probability*, Ed. J. Neyman (Berkeley: University of California Press), 507–532.
2. Arrow, K. J., H. D. Block, and L. Hurwicz (1959), "On the Stability of the Competitive Equilibrium II," *Econometrica* **27**, 82–109.
3. Arrow, K. J., and G. Debreu (1954), "Existence of Equilibrium for a Competitive Economy," *Econometrica* **22**, 265–290.
4. Arrow, K. J. and F. H. Hahn (1971), *General Competitive Analysis* (San Francisco: Holden-Day).
5. Auerbach, A. J., L. J. Kotlikoff, and J. Skinner (1983), "The Efficiency Gains from Dynamic Tax Reform," *International Economic Review* **24**, 81–100.
6. Balasko, Y., D. Cass, and K. Shell (1980), "Existence of Competitive Equilibrium in a General Overlapping Generations Model," *Journal of Economic Theory* **23**, 307–322.
7. Benhabib, J. and R. H. Day (1982), "A Characterization of Erratic Dynamics in the Overlapping Generations Model," *Journal of Economic Dynamics and Control* **4**, 37–55.
8. Debreu, G. (1954), "Valuation Equilibrium and Pareto Optimum," *Proceedings of the National Academy of Sciences* **40**, 588–592.
9. Debreu, G. (1959), *Theory of Value*. (New York: Wiley).
10. Debreu, G. (1970), "Economies with a Finite Set of Equilibria," *Econometrica* **38**, 387–392.
11. Debreu, G. (1972), "Smooth Preferences," *Econometrica* **40**, 603–612.
12. Debreu, G. (1974), "Excess Demand Functions," *Journal of Mathematical Economics* **1**, 15–23.
13. Dierker, E. (1972), "Two Remarks on the Number of Equilibria of an Economy," *Econometrica* **40**, 951–953.
14. Eaves, B. C. (1972), "Homotopies for Computation of Fixed Points," *Mathematical Programming* **3**, 1–22.
15. Hirsch, M. W. (1963), "A Proof of the Nonretractibility of a Cell into Its Boundary," *Proceedings of the American Mathematical Society* **14**, 364–365.
16. Kehoe, T. J. (1980), "An Index Theorem for General Equilibrium Models with Production," *Econometrica* **48**, 1211–1232.
17. Kehoe, T. J. (1982), "Regular Production Economies," *Journal of Mathematical Economics* **10**, 147–176.
18. Kehoe, T. J. (1983), "Regularity and Index Theory for Economies with Smooth Production Technologies," *Econometrica* **51**, 895–917.
19. Kehoe, T. J. (1985), "Multiplicity of Equilibria and Comparative Statics," *Quarterly Journal of Economics* **100**, 119–147.
20. Kehoe, T.J. and D. K. Levine (1985), "Comparative Statics and Perfect Foresight in Infinite Horizon Economies," *Econometrica* **53**, 433–453.

21. Kehoe, T.J. and D. K. Levine (1987), "The Economics of Indeterminacy in Overlapping Generations Models," unpublished.
22. Kehoe, T. J. and J. Serra Puche (1983), "A Computational General Equilibrium Model with Endogenous Unemployment: An Analysis of the 1980 Fiscal Reform in Mexico," *Journal of Public Economics* **22**, 1–26.
23. Koopmans, T. C., ed. (1951), *Activity Analysis of Allocation and Production* (New York: Wiley).
24. van de Laan, G. and A. J. J. Talman (1980), "An Improvement of Fixed Points Algorithms by Using a Good Triangulation," *Mathematical Programming* **18**, 274–285.
25. McKenzie, L. (1954), "On Equilibrium in Graham's Model of World Trade and Other Competitive Systems," *Econometrica* **22**, 147–161.
26. Mantel, R. (1974), "On the Characterization of Aggregate Excess Demand," *Journal of Economic Theory* **12**, 197–201.
27. Mas-Colell, A. (1974), "Continuous and Smooth Consumers: Approximation Theorems," *Journal of Economic Theory* **8**, 305–336.
28. Mas-Colell, A. (1975), "On the Continuity of Equilibrium Prices in Constant-Returns Production Economies," *Journal of Mathematical Economics* **2**, 21–33.
29. Mas-Colell, A. (1977), "On the Equilibrium Price Set of an Exchange Economy," *Journal of Mathematical Economics* **4**, 117–126.
30. Mas-Colell, A. (1985), *The Theory of General Economic Equilibrium: A Differentiable Approach* (Cambridge: Cambridge University Press.)
31. Negishi, T. (1960), "Welfare Economics and Existence of an Equilibrium for a Competitive Economy," *Metroeconomica* **12**, 92–97.
32. Pareto, V. (1909), *Manuel d'Économie Politique*. (Paris: Girard and Briere).
33. Samuelson, P. A. (1941, 1942), "The Stability of Equilibrium," *Econometrica* **9**, 97-120; **10**, 1-25.
34. Samuelson, P. A. (1958), "An Exact Consumption-Loan Model of Interest with or without the Social Contrivance of Money," *Journal of Political Economy* **6**, 467–482.
35. Scarf, H. E. (1960), "Some Examples of Global Instability of Competitive Equilibrium," *International Economic Review* **1**, 157–172.
36. Scarf, H. E. (1967), "On the Approximation of Fixed Points of a Continuous Mapping," *SIAM Journal of Applied Mathematics* **15**, 1328–1343.
37. Scarf, H. E. (with the collaboration of T. Hansen) (1973), *The Computation of Economic Equilibria* (New Haven: Yale University Press).
38. Scarf, H. E. (1982), "The Computation of Equilibrium Prices," *Handbook of Mathematical Economics*, Eds. K. J. Arrow and M. D. Intriligator (New York: North- Holland), Vol. II, 1006–1061.
39. Shapley, L. S. and M. Shubik (1977), "An Example of a Trading Economy with Three Competitive Equilibria," *Journal of Political Economy* **85**, 873–875.

40. Shoven, J. B. and J. Whalley (1984), "Applied General Equilibrium Models of Taxation and International Trade," *Journal of Economic Literature* **22**, 1007–1051.
41. Smale, S. (1976), "A Convergent Process of Price Adjustment and Global Newton's Methods," *Journal of Mathematical Economics* **3**, 1–14.
42. Sonnenschein, H. (1973), "Do Walras' Identity and Continuity Characterize the Class of Community Excess Demand Functions?" *Journal of Economic Theory* **6**, 345–354.
43. Todd, M. J. (1976), *The Computation of Fixed Points and Applications* (New York: Springer-Verlag).
44. Uzawa, H. (1962), "Walras' Existence Theorem and Brouwer's Fixed Point Theorem," *Economic Studies Quarterly* **13**.
45. Wald, A. (1936), "Über einige Gleichungssysteme der Mathematischen Ökonomie," *Zeitschrift für Nationalökonomie* **7**, 637–670; translated as Wald, A. (1951), "On Some Systems of Equations in Mathematical Economics," *Econometrica* **19**, 368–403.
46. Walras, L. (1874), *Élements d'Économie Politique Pure*. (Lausanne: Corbaz); translated by W. Jaffe (1954), *Elements of Pure Economics* (London: Allen and Unwin).

NORMAN H. PACKARD
Center for Complex Systems Research, and the Physics Department, University of Illinois, 508 South Sixth Street, Champaign, IL 61820

Dynamics of Development: A Simple Model for Dynamics away from Attractors

ABSTRACT

An argument is made for the necessity of using "open" dynamical models in certain economic contexts, especially when modeling developing countries. These models are open in the sense that the system is initialized to have just a few dimensions of its state space active (non-zero), and then is allowed to explore new dimensions without restriction. A simple model is proposed to study this type of expansion dynamics.

A CRITIQUE OF DYNAMICAL SYSTEMS MODELS FOR ECONOMIC SYSTEMS

Modeling economic systems is difficult at best, and all the more so when the size of the system being modeled is large. Modeling on the largest scale, i.e., on the scale of interacting countries, has particular difficulties associated with the fact that the set of relevant dynamical variables is constantly changing with time. The problem is particularly vivid for the case of developing countries, where the formation of new industries is a crucial part of the economic dynamics. Modeling the large-scale dynamics of developed countries has similar problems, though, because new industries come, old ones go and others merge.

Conventionally, dynamical systems modeling begins with the identification of the relevant dynamical variables, which form the state space of the system. The time evolution of the system is then represented as an orbit in the state space. The equations of motion provide an explicit rule for generating an orbit from any initial point. If the set of relevant variables changes with time, then the state space is itself changing with time, which is not commensurate with a conventional dynamical systems model.

There are two ways of coping with this problem. One way is to consider the state space to be an infinite-dimensional (or very high-dimensional) space with not only the currently relevant variables, but also all possible other relevant variables. The problem with this approach is that the equations of motion must then take into account all the coupling between all possible potential variables. If the set of potential variables is very large, chances are small that equations of motion can be written down from first principles. The other approach is to add an element to the modeling process, a "meta-dynamic" that specifies a rule for adding new components to the system. In this paper, the first approach will be used, and the only reason that the equations of motion can be specified is that some drastically simplifying assumptions are used in their construction.

The reason this is a problem for dynamical systems theory is that quite a lot of results pertain to situations where an orbit is moving on or near an invariant set, and on that set the orbit has stationary statistical properties. If the system is coupled to new variables as time passes, then the system is unlikely to remain on a fixed invariant set, and it is possible for the statistics to change as well.

It may seem too severe to throw out all conventional dynamical systems models for economic systems, and indeed, there are many cases where the relevant variables remain fixed, and the conventional approach works fine. This may happen for at least two different reasons. One possibility is that the time scale over which the system is observed is short compared to the time scale of new variables being created within the system. An example might be that the fluctuations of stock prices are statistically stable over time scales less than that of takeovers and mergers.

The other possibility is that the state space is in fact changing constantly, but that there remains a set of global modes that executes motion near an invariant set, with the new variables playing the role of adding small fluctuations. This possibility is analogous to the phenomenon of coherent structures in fluid flow, where it is possible to have large-scale turbulence, with eddies constantly being created and destroyed, but with consistent motion of a few global modes. Jupiter's red spot may provide an example. In the case of an economic model, the business cycle might be a reflection of an analog of a coherent structure in the dynamics of the national economy. Unfortunately, there is very little in the way of either theory or models for such phenomena.

In this paper, we concentrate specifically on modeling dynamics of a system that does not remain within a fixed subspace, but rather one that constantly explores new dimensions in a very large space of possible dimensions.

A SIMPLE MODEL

The model we consider is a dynamic network, with random couplings. Each node of the network contains a "strength" variable. The crude analogy to be made with large-scale economic systems is that the strength variable represents the level of presence of a particular industry in the economy. In order to study the dynamics of development we consider all but a few of the nodes to start with zero strength. Then as time proceeds, expansion of "industry" can be observed by an increasing number of nodes becoming active.

The idea behind the present model is to create the simplest possible dynamics for the node strengths that allows the network to to display the dynamics of expansion. A discussion of how more realism might be included in the dynamics will be in the following section.

The strength variables are assumed to lie between 0 and 1. Each node is coupled to n neighbors, thus n may be treated as a parameter that varies the connectivity of the network. The current model uses diffusive coupling between nodes, and a simple "reaction" dynamic at each node:

$$x_i^{t+1} = F(x_i^t + k \sum_{j=1}^{n} x_j^t),$$

where $F(\alpha)$ is a map that has two fixed-point attractors at $\alpha = 0$ and $\alpha = 1$ and an unstable fixed point at $\alpha = .5$. The exact form of the map used here is

$$F(\alpha) = \alpha - A \sin(2\pi\alpha),$$

where A is a parameter that measures the strength of the attractors at 0 and 1. We take $F(\alpha)$ to be 0 for $\alpha < 0$ and 1 for $\alpha > 1$. This form for F implies that there will be an excitation threshold for the strength of any node, and once the excitation drives the strength of a node above the threshold, the node becomes active.

We have made a preliminary investigation of the dynamics of this model as a function of k, the strength of coupling to neighbors, n, the number of neighbors each node is coupled to, A, the nonlinearity parameter, and R, the size of the set of nodes that is initially stimulated. The results are shown in figures 1–4.

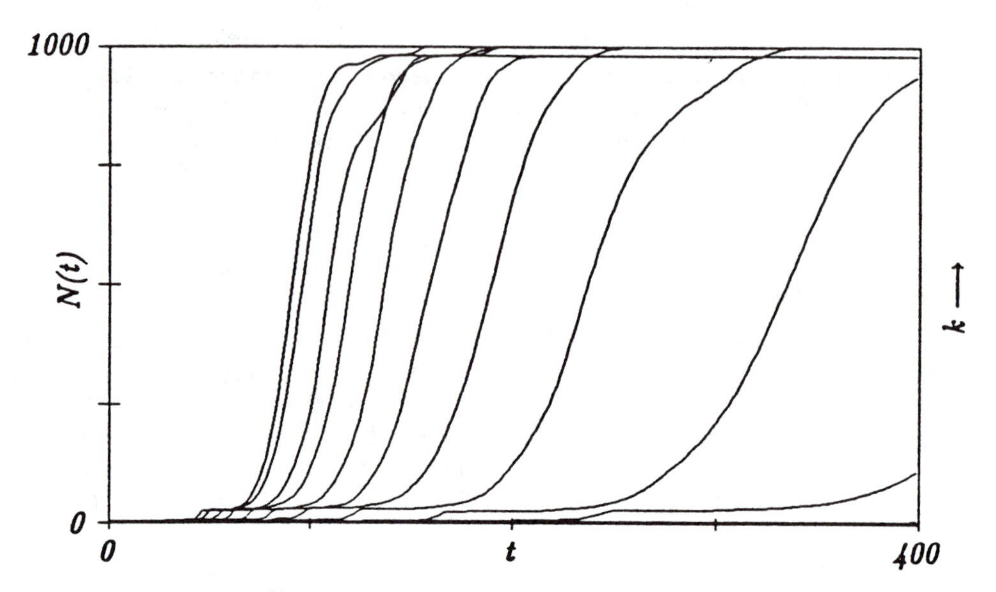

FIGURE 1 Expansion dynamics for the network with the nonlinearity parameter turned to $A = 0$. Other parameter values were $R = 1$, $n = 5$, and k was varied from .001 to .01. $N(t)$ represents the number of active nodes.

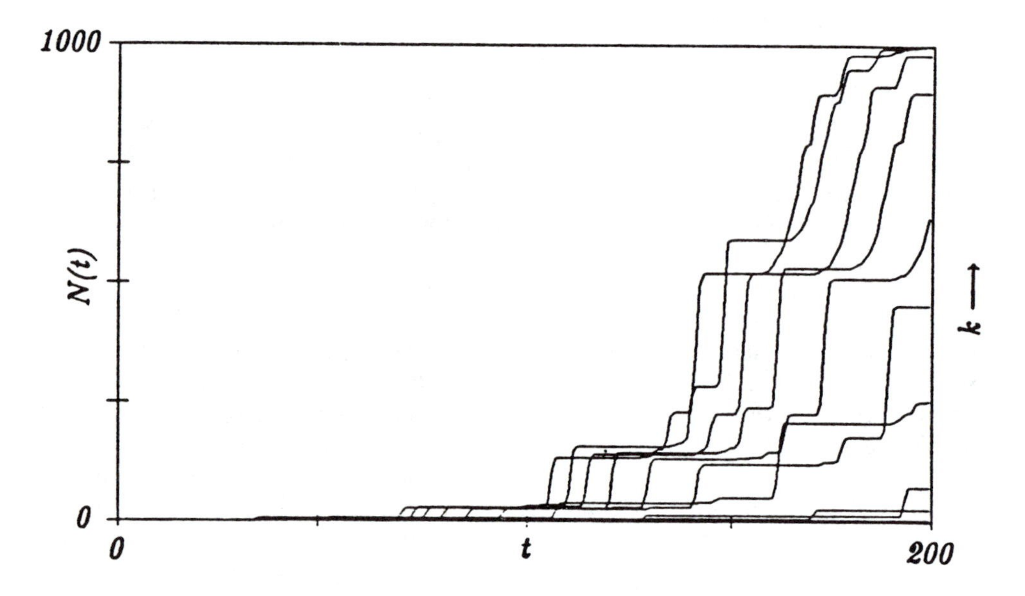

FIGURE 2 Expansion dynamics as in figure 1, but with $A = .1$. Other parameters are the same, and k was varied from $.0001$ to $.001$.

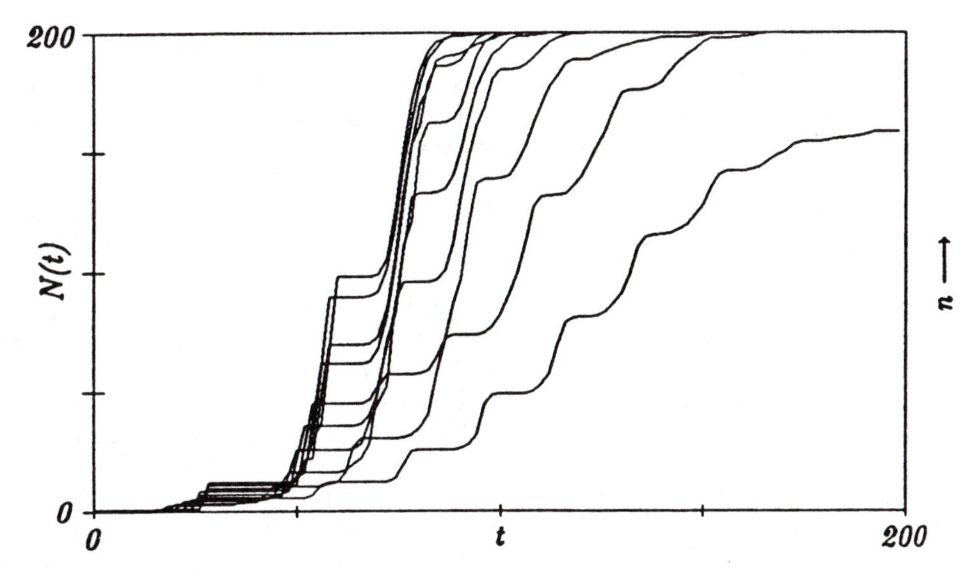

FIGURE 3 Expansion dynamics as a function of the graph connectivity; n varies from 2 to 10. Other parameters are $A = .1$, $k = .0005$, and $R = 1$.

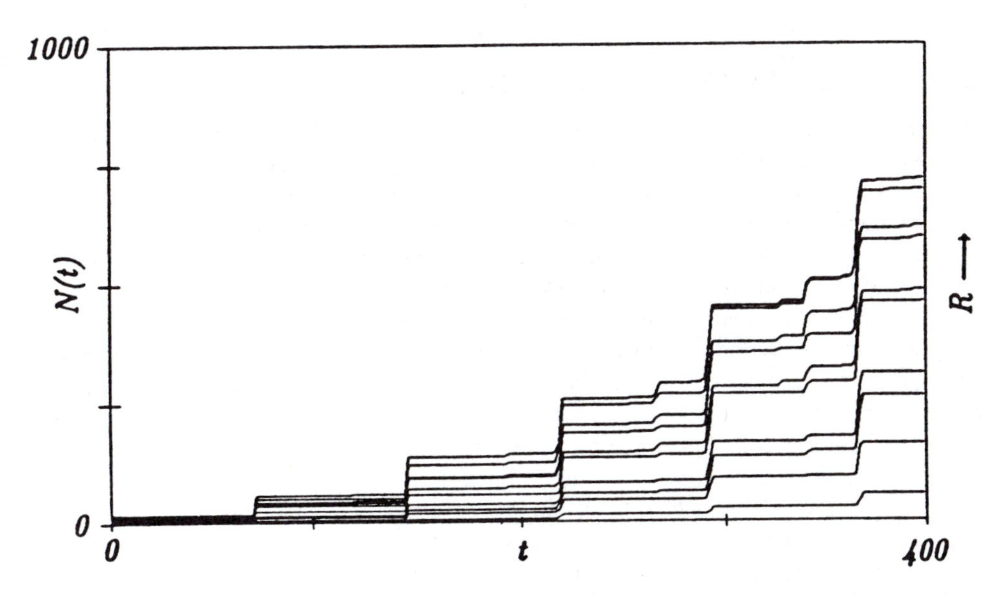

FIGURE 4 Expansion dynamics as a function of the size of the initially active set of nodes; R varies from 1 to 10. Other parameters are $A = .1$, $k = .0005$, and $n = 5$.

Figure 1 illustrates the growth pattern when nonlinearity is not present ($A = 0$). This implies that the node strengths are not attracted to either 0 or 1. The number of active nodes $N(t)$ is seen to increase exponentially until the network saturates. The family of curves has been obtained by varying the coupling strength k.

In Figure 2 and all succeeding figures, the nonlinearity has been raised to $A = .1$. Figure 2 shows a family of curves as strength is varied. We see that the threshold dynamics causes plateaus in $N(t)$. The typical time scale of the plateaus is proportional to A.

Figures 3 and 4 illustrate the effect of varying the network connectivity n, and the size of the initially activated set of nodes R. In each case, we see that as n, R, and k is increased, the rate of growth for $N(t)$ also increases. The increase is smooth, in the sense that there appears to be no phase transition (i.e., a singularity or abrupt increase in the growth rate of $N(t)$). This is reasonable because there is no inhibitory terms in the dynamics; the diffusion of strength continuously increases to activate new nodes, however slowly. The addition of inhibitory terms in the equations of motion, as discussed below, may very well cause a phase transition to appear. One point that is not conclusively decided from the present numerical experiments is that as R is varied, $N(t)$ seems to have average linear growth, and as n and k are varied, $N(t)$ appears to have exponential growth until the graph saturates.

COMPARISON WITH OTHER MODELS

Models with "open" dynamics similar to those displayed by the present network are just beginning to be used. The closest model to the present one is a model of chemicals interacting with each other through a network of catalyzed reactions.[1–3] The chemicals are represented by strings of characters (analogous to amino acids), the strength variables represent the concentration of any given species, and the autocatalytic couplings in the interaction graph are chosen at random, with a certain probability of catalysis P. The parameter P determines the average connectivity of the graph, and is analogous to the parameter n in the present model. Another model, perhaps even more similar in spirit to this one, has been proposed by Rössler,[4,5] whose idea was to abstract the dynamics of autocatalytic reactions into a switching network of automata, where each automaton would represent an autocatalytic loop.

The autocatalytic network also has a parameter that determines a concentration threshold above which a chemical can participate as a catalyst. This is roughly analogous to the switch in the present model from an inactive node (a node staying near the attractor at $x_i = 0$) and an active node (a node shifting to the attractor at $x_i = 1$), and the relevant parameter that determines the strength of these attractors is A, the nonlinearity parameter.

The time evolution of the network begins with an initial condition consisting of a few simple chemicals (short strings), and through a series of reactions, the set of chemical grows, analogous to the expansion of activity in the present network model. The time evolution may be thought of as having two components. One is the growth of the "logical graph", the set of potential couplings between chemicals without considering concentration dynamics. The other type of growth uses the chemical kinetics to evolve the chemical concentrations, and only a finite piece of the logical graph is explored.

In the autocatalytic network model there is a phase transition in the logical graph, a transition between graphs that are asymptotically finite to graphs that expand forever. When the kinetics are added, the transition appears to be softened. In the present model, the logical graph is specified from the beginning by choosing the random couplings between nodes. The graph appears to be connected even for small values of n, and so there is no sharp transition. Rather, there seems to be a fairly continuous tradeoff between R, n, and k.

ENHANCEMENTS AND CONCLUSIONS

The primary enhancement that could be made in this model is the inclusion of more realistic dynamics between the nodes. The simplest addition to the dynamics would be to change the coupling between nodes from diffusive to a stimulation-inhibition dynamic, possibly similar to a Lotka-Volterra system with quadratic nonlinearity in the strength variables. Even more realism could be added by making the node dynamics a continuous flow rather than a map. More economic intuition could be built in by adding more dimensions to the internal state of each node, corresponding to presence of different types of commodities, and the dynamics of the strength variables established implicitly by rationality and self-consistency assumptions.[1]

The point of the present simulations is to isolate those aspects of the dynamics that are due purely to the system's ability to explore increasingly large regions of a high-dimensional state space. A complete picture of the dependence of the dynamics with more realistic equations of motion will be reported in the future.

The model discussed here is manifestly distant from real economic systems, but there are some aspects of the dynamics that may be relevant. We have seen that there is a more or less continuous tradeoff between the three parameters varied in the model: the connectivity n, the interaction strength k, and the size of the initially excited nodes R.

In the context of a developing country, there may be a natural interpretation for these parameters in spite of their simplicity. The connectivity is analogous to the level of communication and interdependence of different industrial sectors, the interaction strength is analogous to the actual level of flow of money between industries, and the size of the initial condition is the set of industries that is initially stimulated, e.g., by foreign aid. There is clearly a tradeoff between the rate of

[1]The last approach is being carried out by the author with M. Boldrin and J. Scheinkman.

growth of new industrialization, and these parameters. If it is possible to make even a crude identification between the model parameters and the parameters that characterize a real system, estimates could possibly be made for setting values of these parameters through the design of aid packages, to insure a desired growth rate for a developing country.

ACKNOWLEDGMENTS

This work was stimulated by the Santa Fe Institute workshop on complex dynamics of the global economy. I am grateful to conversations with all the participants there. In addition, I have appreciated conversations with Otto Rössler, and Rob Shaw, and Kunihiko Kaneko. This work has also been supported by the National Center for Supercomputing Applications at the University of Illinois, and the National Science Foundation, grant number PHY86-58062 PYI.

REFERENCES

1. Farmer, J. D., S. A. Kauffman, and N. H. Packard (1986), "Autocatalytic Replication of Polymers", *Physica* **22D**, 50.
2. Farmer, J. D., S. A. Kauffman, N. H. Packard, and A. S. Perelson (1987), "Adaptive Dynamic Networks as Models for the Immune System and Autocatalytic Sets", *Ann. N.Y.A.S.* **204**, 120.
3. Kauffman, S. A. (1969), *J. Theo. Bio.* **22**, 437.
4. Rössler, O. (1974), "Chemical Automata in Homogeneous and Reaction-Diffusion Kinetics," *Notes in Biomath.* **B4**, 399–418.
5. Rössler, O. (1983), "Deductive Prebiology," *Molecular Evolution and the Prebiological Paradigm*, Ed. K. L. Rolfing (New York: Plenum).

RICHARD PALMER
Department of Physics, Duke University, Durham NC 27706

Statistical Mechanics Approaches to Complex Optimization Problems

1. INTRODUCTION

In recent years a number of methods developed in statistical mechanics have been applied to complex optimization problems. The results tend to be more in the form of general theorems and bounds than specific results for particular problems. They are nonetheless interesting, especially for the new viewpoints they bring to old problems. These new viewpoints, and possibly specific methods, may also prove of use in economic problems.

The origin of the current applications of statistical mechanics to diverse "complex systems" lies mainly in the development of *spin glass* theory. In the mid-1970's, physicists became very interested in the magnetic alloys known as spin glasses, principally because they displayed a rather sharp phase transition *without* developing a spatial magnetic ordering pattern. The individual "spins" (atoms that act like microscopic bar magnets) settle down into directions that are more or less fixed individually but not part of an overall spatial pattern. This phenomena was rather surprising in the light of previous experience, and new methods had to be developed to cope with it.

Two essential ingredients are *randomness* and *frustration*. Randomness arises because the magnetic atoms (spins) are distributed randomly in the material among a larger number of other spinless atoms. Frustration refers to the conflict or competition among the interactions between pairs of spins. Although the interaction between a given pair tends to produce a very definite alignment, with three or more spins interacting the individual pairwise alignments often cannot be simultaneously achieved. These ingredients lead not only to the lack of spatial order in the frozen state, but also to the existence of *many possible frozen states.* Frustration and randomness prevent there being a single optimum state, or a few equivalent ones related by symmetry, as is found in most non-random systems. Instead, there are many possible optima that are about equally good, but unrelated by symmetry.

Combinatorial optimization problems studied in computer science and operations research share many of the features of spin glasses. As an example, consider the well-known *traveling salesman problem* (TSP). We are given a set of N cities $n = 1, 2, \ldots, N$ and all inter-city distances d_{mn}, and are asked to find a closed tour n_j of those cities (visiting each once) that minimizes the total distance

$$D = \sum_{j=1}^{N} d_{n_j n_{j+1}} \tag{1}$$

(where $n_{N+1} = n_1$ by definition). The system is random, or at least irregular, in that the cities do not in general lie on any regular grid. It is also frustrated in the sense that good local sub-tours may not join together well globally and may in effect exclude one another. Finally, although in principle there may be a single best tour, in practice there are found to be a large number (for large N) of tours that are almost as good, roughly equivalent, and not related by symmetry.

With these similarities in mind, it was natural to explore applications to optimization problems of the *concepts* and *methods* developed in statistical mechanics to deal with spin glasses. These concepts and methods have in fact proved useful in a number of other fields. Application to complex optimization problems has occurred in parallel with applications to neural networks, "real" (structural) glasses, dynamics and folding of biomolecules, and prebiotic evolution. Anderson's paper[1] in an earlier volume in this series describes some of these.

It is not yet clear whether any of the ideas common to spin glasses and complex optimization problems will have detailed applications in economics. Much economic modeling involves optimization (e.g., of utility functions), but the *complex* aspect is not obviously present. Of course, economic problems are *complicated*, but "complex" means more than that in this context. While there is no generally agreed definition, the existence of many possible states or optima is normally regarded as essential, with frustration and randomness coming second and third in importance. This is more than a matter of empty semantics; the "complex systems" currently accepted as such exhibit many common features and even a degree of universality.

Frustration does, of course, occur in various economic contexts. It is obvious to some extent in the interactions between individuals, between firms, and between nations. There is also randomness, in the sense of irregularity, in the distribution of resources and goods, in present values of economic variables, in exogenous shocks, and elsewhere. While these ingredients might reasonably be expected to lead to many possible economic states or equilibria, it is not clear whether or not they do so in practice. The situation is somewhat obscured by the tendency in economics to look only for unique solutions, and to reject or modify models that do not provide them.

In the remainder of this paper I do not address any economic issues directly. I limit myself to an overview of the application of statistical mechanics methods to combinatorial optimization problems. Though important, I do not treat neural network approaches to optimization.[2] For the most part, I assume *no* prior knowledge of statistical mechanics or of combinatorial optimization. After discussing in general terms combinatorial optimization problems (section 2) and their cost surfaces (section 3), I present a short summary of the relevant parts of statistical mechanics (section 4). I then describe two types of statistical mechanics approaches that can and have been used in combinatorial problems: simulated annealing (section 5), and various analytical approaches (section 6). In the conclusion (section 7), I point to the more generalizable ideas.

2. COMBINATORIAL OPTIMIZATION PROBLEMS

We have already defined the Traveling Salesman Problem (TSP). Although of some practical relevance (e.g., in the trucking and shipping industries), the main interest in such problems is that they are *hard*. If one wants to find the very best tour with certainty, there is no known way that does not take computer time t that increases at least exponentially with the number of cities N. Exhaustively searching all the $(N-1)!/2$ possible tours gives approximately $t \propto \exp(N \ln N)$ for example. This makes such a computation impractical for large N. It is generally believed (but not yet proven) that a fast method, one that is only polynomial in N, will never be found, because the TSP belongs to the class of related problems called *NP-complete*.[3] There is no polynomial algorithm known for any NP-complete problem, and if one of them had a polynomial solution, then they all would have. Note, however, that this applies only to finding the absolute best solution in a worst-case instance. There are often polynomial-time algorithms for good (but not guaranteed best) solutions, or for typical (but not worst) cases.

Because NP-complete problems are hard, one is interested in approximate solutions, bounds on solutions, and in general anything that can help to refine our understanding of the problem or classify cases. It is in these areas that the methods described below become useful. The best practical methods for solving particular problems still rely on clever algorithms invented for the specific problem.

It is worth mentioning two more examples of combinatorial optimization problems. In the *Graph Bipartitioning* problem, we are given a graph with vertices V_n, $n = 1, 2, \ldots, N$ (N even), and a set of edges E_{mn}. The problem is to find a partition of the vertices into two subsets of equal size $N/2$ such that the number of edges connecting the two subsets is minimized. This problem is also NP-complete. Most of the recent work on applying statistical mechanics to complex optimization problems has used this example.[4–10] Closely related is the problem of finding the *Ground State of a Spin Glass.* In a simplified model, the spins S_n $(1 \leq n \leq N)$ are allowed to point in only one of two directions $(S_n = \pm 1)$ and the pair interactions $J_{mn} = J_{nm}$ between spins m and n are three-valued, $J_{mn} \in \{0, +1, -1\}$, representing respectively no interaction, $S_m = S_n$ for minimum energy, and $S_m = -S_n$ for minimum energy. Then the ground state problem consists of minimizing

$$E\{S\} = -\sum_{(mn)} J_{mn} S_m S_n \tag{2}$$

with respect to the spin directions $\{S\}$ for fixed $\{J\}$. Three spins l, m, and n are mutually frustrated if $J_{lm} J_{mn} J_{nl} = -1$. This problem is also NP-complete unless special conditions are imposed on the J_{mn} interactions.

Combinatorial optimization problems in general involve a problem *size* N, a set of possible *instances* $i \in I$, a set of possible *configurations* $x \in X$, and a *cost function* $C_i(x)$. For the TSP, an instance is a particular set of city locations, specified by the distances d_{mn}, a configuration is a particular tour n_j of those cities, and the cost function is given by Eq. (1). For the spin glass, an instance is a set of interactions $\{J\}$, a configuration is a set of spin orientations $\{S\}$, and the cost function is given by Eq. (2).

Given this framework, there are various problems that we might attempt to solve. In roughly decreasing order of utility, and increasing ease of solution, are:

a. Find the configuration x that minimizes the cost function $C_i(x)$ for a particular instance i. This is the practical problem.

b. Find an x that *approximately* minimizes $C_i(x)$ for given i. For example, a configuration that brings $C_i(x)$ within 5% of its minimum might be good enough. Many heuristic algorithms do this, though often without a known percentage accuracy.

c. Find $\min_x C_i(x)$ for given i. This specifies how good a result can be achieved, without providing a particular configuration.

d. Find $\overline{\min_x C_i(x)}$, where the overbar stands for averaging over an ensemble of possible instances i. Clearly a measure $\rho(i)$ over instances is needed. This specifies how good a result can be achieved on average, or for a typical case *if* the average is dominated by such cases.

e. Find $\lim_{N\to\infty} \overline{\min_x C_i(x)}/N^\alpha$, where α is chosen to capture the leading behavior.

Analytic methods in statistical mechanics typically answer questions of type e. or perhaps d. There may however be other benefits of such analysis, perhaps ultimately more important:

f. Find *how many* different solutions x can be expected with $C_i(x) < c$ for given c, either for given i or for the average.

g. Classify the type and difficulty of the problem. This is discussed further in the next section. It may provide suggestions for useful heuristic approaches to solve specific problems.

3. COST SURFACES

It is often useful to consider the cost function $C_i(x)$ as a landscape on the configuration space X. The problem is then to find the deepest valley. It is clear that knowledge of the geometry of the surface will aid in locating the global minimum, as well as answering questions about the number of minima and their distances from one another. In particular, knowing about the height and distribution of the barriers between minima is of crucial importance in designing algorithms to explore the landscape, allowing for example a choice between local improvement and large jumps.

Adopting a landscape viewpoint, or more generally considering the topology of the configuration space, was a crucial step in understanding spin glasses. It led to, or was part and parcel of, most of the modern ideas about multiple states, broken ergodicity,[11] and ultrametricity.[12] Of course, such a viewpoint does not solve problems in itself, but it does make it easier to ask the right questions.[4,13]

Strictly speaking, we cannot consider $C_i(x)$ as a surface, even in many dimensions, unless x is a continuous variable. This is rarely the case in traditional combinatorial optimization problems, which normally involve a discrete set of possibilities such as the $(N-1)!/2$ possible tours on N cities in the TSP. Nevertheless, we can loosely think in terms of a landscape if a suitable metric d on X can be found, so that $C_i(y)$ is close to $C_i(x)$ whenever $d(x, y)$ is small. The choice of metric is not necessarily unique (besides scale factors), however, and in some cases quite different topologies can be placed on the same problem. This *is* a matter of significance, because most practical algorithms work in terms of gradual improvement by making steps that are short in the chosen metric. In the TSP for example, we might first think of a metric based on the number of transpositions to convert one tour (permutation n_j) to another. But in fact this is a terrible metric, since a single interchange of cities can produce a large change in $C_i(x)$. A much better metric can be based on interchange (crossing) of two *links*, involving four cities (e.g., $nm, pq \to np, mq$).[14]

While on the subject of metrics, it is worth mentioning the notion of *ultrametricity*. A space is said to be ultrametric when, given any three points x, y, and z, two of the three distances $d(x, y)$, $d(y, z)$, and $d(z, x)$ are equal while the third is less than or equal to the other two. At first sight, this property is surprising, but it is actually a natural consequence of a *hierarchical* organization, with the metric monotonic in the *generation number* of the last common ancestor. It is always possible to define a metric with the ultrametric property on a hierarchical tree, and to construct such a tree from a given ultrametric metric. The significance of ultrametric spaces is that the different minima in the spin glass energy function, Eq. (2), are known to be organized ultrametrically in the "infinite range" model of Sherrington and Kirkpatrick.[15] There is speculation that many minima separated ultrametrically should be the normal situation in NP-complete problems, though it is certainly possible to construct exceptions.[16] In any case the degree of ultrametricity in a problem is a natural classification tool with possible deeper significance.

Returning to the cost surface, we consider some examples of possible geometries in Figure 1. The horizontal and vertical axes represent configuration x and cost function $C_i(x)$ respectively. The sketches are grossly simplified, because the horizontal axis should really have very many dimensions, typically of order $\exp(aN)$. For the spin glass model of N spins with $S_n = \pm 1$, we should consider 2^N points on an N-dimensional hypercube. But simple two-dimensional sketches may be better than nothing.

If our cost surface resembles Figure 1a, the optimization problem is easy and uninteresting. We can simply slide down the surface to the global minimum using gradient descent (or "hill climbing" with the opposite sign convention). On the other hand Figure 1b, which looks somewhat similar, can represent an extremely hard problem (including an NP-complete one[16]); if there is one hole in an otherwise flat "golf course," then there is no better general method than exhaustive search of the whole configuration space X.

Figure 1c represents a case common in physics and elsewhere, in which there are two or more *equivalent* minima related by symmetry. Again the optimization

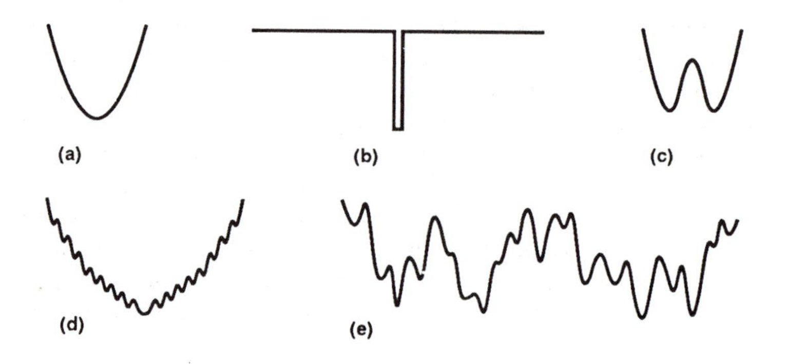

FIGURE 1 Cost Surface Landscapes.

problem is easy, since sliding down to the nearest minimum from any starting point will provide a solution as good as any other. This type of situation is said to exhibit symmetry breaking because the solution does not possess the full symmetry of the problem. It also represents a *lock-in* phenomenon, in which a system becomes stuck in one of several states depending on starting conditions.

A less straightforward problem is shown in Figure 1d. Here, although there is only one global optimum, there are many local optima in which one can become stuck. Continuous gradient descent will obviously fail, but it could succeed if used with discrete steps large enough to get past the fine structure. Even so, there are tricky questions about the optimum step size, and about how to design an algorithm with a self-adjusting step size. A better basis would be an algorithm that is able to accept some jumps that increase $C_i(x)$ somewhat, while preferring jumps that give a decrease. This is the essence of the simulated annealing method discussed in section 5.

Finally, Figure 1e represents a *rough*[17] cost surface, as is often suggested for a spin glass. It may even be self-similar, so that a magnified image of a part would be much like the whole surface, with structure on every scale. There are typically—at least in the cases of particular interest—a large number of roughly equivalent minima *not* related by symmetry. An algorithm to find one of the deep minima must be able to take long jumps or climb considerably hills, at least occasionally, and must not be satisfied with *any* local minimum until considerably more exploration has taken place around it. There are many qualitative and quantitative properties of such landscapes that are relevant to algorithm design for rough cost surfaces.[4,13] Just a few are:

a. What is the distribution of barrier heights? Or more specifically, what is the minimum height below which all minima are connected? And how does this scale with N?

b. What is the distribution of the depths of the minima? Knowing this is crucial if we want to be reasonably sure of being, say, within 5% of the global optimum.

c. What correlation is there between the depth of a given minimum and its width (hypervolume) at some higher $C_i(x)$? Finding deep valleys is much easier if they tend to be broad; at the other extreme lies the golf course (Figure 1b).

d. Are the locations of deep minima correlated? Are their separations approximately ultrametric? If two deep minima are known to be close to one another, is it better to look for a deeper one in that neighborhood or a long way away?

Most of these questions remain unanswered even in the best studied cases. Asking them is nevertheless important, and in the case of spin glasses has been rather productive.

4. STATISTICAL MECHANICS

The application of statistical mechanics to optimization problems relies on the following correspondences:

Statistical Mechanics	**Optimization**
System size N	Problem size N
Microstate x	Configuration x
Sample $\{J\}$	Instance i
Energy $E(x)$	Cost $C_i(x)$
Temperature T	Pseudo-temperature T

Microstates of physical systems correspond to particular positions, orientations, and internal states of the atoms (or other parts) of which the system is composed. There are normally of order $\exp(aN)$ microstates. Each microstate x has an associated energy $E(x)$. The analogue of *instances* in optimization problems only arises in statistical mechanics for *quenched random systems*, in which disorder is presumed frozen into the system and not treated dynamically. As discussed later, one then often averages one's results over possible samples with randomly chosen disorder.

The most obvious difference between optimization and statistical mechanics problems is the presence of *absolute temperature* T as a natural variable in statistical mechanics. There is generally no natural analogue in optimization problems, so an abstract temperature (or *pseudo*-temperature) variable is normally introduced to make the analogy complete. This is, perhaps, the most fundamental step made in the merging of the two fields.

Given a set of states $x \in X$, an energy function $E(x)$, and a temperature T, statistical mechanics predicts the occupation probability $p(x)$ of a state x as

$$p(x) = e^{-E(x)/kT}/Z. \tag{3}$$

Here k is Boltzmann's constant, which converts temperature to energy units, and Z is the *partition function*,

$$Z = \sum_x e^{-E(x)/kT}, \tag{4}$$

which correctly normalizes the probability. Eq. (3) may be derived in several ways, each with its own interpretation of the probability $p(x)$. The most standard one comes from imagining the system to be in equilibrium with a large *heat-bath* with which it can exchange energy, and assuming that the system plus heat bath, visits over time every possible joint state with equal probability. A joint state means a combination of a state x of the system and a state y of the heat bath. $p(x)$ then means the frequency of finding the system in a particular state x (irrespective of y) over a long time trajectory.

The set of possible energies $\{ E(x) \mid x \in X \}$ normally has a unique lowest value E_0 corresponding to the *ground state* x_0. Above E_0 the energy levels become closer and closer together, so the *density of states* $g(E)$ rises rapidly with E. Here $g(E)$ is defined so that the number of states x in the range $[E, E + \delta E]$ is approximately $g(E)\delta E$. The probability of finding the energy in the same range is given by

$$P(E)\delta E = Z^{-1} e^{-E/kT} g(E)\delta E \tag{5}$$

and turns out to be peaked over a very narrow range of E, because $e^{-E/kT}$ is rapidly falling while $g(E)$ is rapidly rising. The energy is then well defined on a macroscopic scale, normally with average and variance both proportional to N; the relative fluctuation is only of order $1/\sqrt{N}$.

The partition function Z is also related to thermodynamic quantities through the relationship

$$F = U - TS = -kT \ln(Z). \tag{6}$$

Here $F(T)$ is the *free energy*, $U(T)$ is the *internal energy*, and $S(T)$ is the *entropy*. U is also the mean energy,

$$U = \int EP(E)\, dE. \tag{7}$$

F, U, and S are all normally of order N (or *extensive*). To this order, the entropy S is also be related to the number W of distinct states x under the peak of P(E) according to

$$S = k \ln W. \tag{8}$$

As the absolute zero of temperature ($T = 0$) is approached, the free energy F and the internal energy U both approach the ground state energy E_0. We can thus find the ground state energy via the partition function:

$$E_0 = - \lim_{T \to 0} kT \ln Z = - \lim_{T \to 0} \left[kT \ln \left(\sum_x e^{-E(x)/kT} \right) \right]. \tag{9}$$

Eq. (9) can be viewed as an identity for $E_0 = \min_x E(x)$, and as such can be applied to a cost function $C_i(x)$ in an optimization problem:

$$\min_x C_i(x) = - \lim_{T \to 0} \left[T \ln \left(\sum_x e^{-C_i(x)/T} \right) \right]. \tag{10}$$

Here the pseudo-temperature T is simply a parameter with no direct physical meaning, and we choose the units of T so that no k is needed.

Eq. (10) apparently gives us a new way to find the minimum cost (though not the actual best configuration) in an optimization problem, but is it useful in practice? For a *particular* instance i, the answer is probably *no*, because there is no way to evaluate the right-hand side of Eq. (10) without already knowing the low-lying configurations. But if we are satisfied with an answer averaged over instances, there are many techniques of statistical mechanics that can allow practical use of Eq. (10). We return to these in section 6. Before that, in section 5, we discuss a technique of computer simulation that allows approximate computation of $U(T)$ via Eq. (7).

It is worth asking whether there are other useful representations for a minimum besides Eq. (10). One could obviously replace $\exp(z)$ and $\ln(z)$ by another suitable function and its inverse. Very different integral representations might also be possible. The principal reason for concentrating on Eq. (10) is that physicists and others have a vast amount of experience with such expressions, and have developed many useful tricks and tools for their evaluation.

5. SIMULATED ANNEALING

There are various ways to perform computer simulations for statistical mechanics problems. The most obvious way is to follow the dynamical equations that the atoms (or other parts) are believed to obey; this is called *molecular dynamics.* Less obvious is *Monte Carlo simulation*, in which an artificial trajectory $x(t)$ through the possible states in X is constructed so as to visit each state x with frequency given by $p(x)$ in Eq. (3) for a chosen temperature T. Here t is time, or iteration number in the computer program. The internal energy can then be found by averaging over the trajectory (*cf.* Eq. (7)),

$$U(T) = \frac{1}{t_{\max}} \int_0^{t_{\max}} E(x(t))\, dt, \tag{11}$$

and similar averages give other quantities. The key to the generation of the trajectory $x(t)$ is a set of transition rates $R(x,y)$ between pairs of states x and y; $R(x,y)$ gives the probability per unit time of passage to any other state y if the system is currently in state x. If the transition rates are chosen to satisfy the *detailed balance condition*

$$p(x)R(x,y) = p(y)R(y,x) \tag{12}$$

(with $p(x)$ given by Eq. (3)), then under rather general conditions[18] the resulting trajectory will have the desired property of visiting x with frequency $p(x)$.

Eq. (12) still leaves a lot of freedom in the choice of $R(x,y)$. For each x one usually selects a small subset of possible y's to which transitions will be allowed, and sets $R(x, y) = 0$ for the rest. In most spin glass work, for example, only *single* spin-flip transitions ($S_n \to -S_n$ for *one* n) are allowed. Then the remaining $R(x, y)$'s are usually set according to the Metropolis[19] rule

$$R(x, y) = \begin{cases} 1, & \text{if } E(y) < E(x); \\ e^{-\{E(y)-E(x)\}/kT}, & \text{if } E(y) > E(x). \end{cases} \tag{13}$$

This leads to a fast simple algorithm. Note that Eq. (13) allows transitions both upward and downward in energy, but prefers the latter more and more strongly as temperature is lowered.

The difficulty with this otherwise splendid approach is that the computer time $t_{\max}$ required to get accurate averages may be prohibitively long. If the algorithm is not run for a long enough time, one obtains averages over only part of the configuration space X, often corresponding to a metastable state of the system. At small T the problem becomes severe, and the algorithm can become stuck for extremely long times in a local minimum of the energy surface $E(x)$. Not surprisingly this situation is worst when there are many local minima, as with rough energy surfaces (Figure 1e).

If one specifically wants to find the ground state x_0, it is little use running the Metropolis Monte Carlo algorithm at $T = 0$, where it only takes downhill steps. A better approach is *simulated annealing*,[20,21] in which one starts at high T (where it is easy to run for an adequate $t_{\max}$) and then gradually lowers T towards zero according to an *annealing schedule* $T(t)$. Information from the current averages can be used to adjust the annealing schedule so that more time is spent in certain temperature regimes where, for example, many degrees of freedom are freezing out. A perfect answer is only obtained with infinitely slow annealing, but a practical schedule can give reasonable results, within a few percent of the global minimum.

Simulated annealing can obviously be applied to optimization problems as well as to statistical mechanics problems. One introduces a pseudo-temperature T and gradually lowers it while running a Monte Carlo simulation. Consider the traveling salesman problem for example. The set of moves with non-zero $R(x, y)$ can be taken as two-link interchanges ($nm, pq \to np, mq$). Starting with a random tour, at high temperature the simulation wanders randomly through very long tangled tours, because there are far more of these than the near-optimum tours; entropy (number of tours) is more important than energy (length) at high T. As T is lowered the length of the tour becomes more important, and untangling begins. Eventually only reasonable tours are generated in the trajectory, with finer discrimination appearing as T approaches zero.

Simulated annealing works reasonably well if the deep minima occur in the widest valleys, which seems to be the case for at least some of the examples that have been studied. It does particularly well on graph bipartitioning, but is not so good for the TSP. The performance of the algorithm is rarely as good as the most cleverly designed heuristic algorithms for the same problem (e.g., the Lin-Kernighan[22] algorithm for the TSP), but it is often easier to implement because a detailed analysis is not needed. It *is* generally much better than naive algorithms that just do local optimization. Some industrial applications have been found useful, particularly in circuit design and placement.[23]

The simulated annealing approach can often be enhanced by extending the configuration space X. By relaxing certain constraints on the physical situation, the algorithm can explore more widely and become stuck in a bad situation less easily. By adding new dimensions to the cost surface $C_i(x)$, one connects regions or valleys that were previously only accessible via high passes. As an example, in graph bipartitioning one might allow configurations with K and $N-K$ vertices in the two subsets, without insisting that $K = N/2$. Or when trying to place chips on a circuit board with minimum wire length, one might allow chip locations to overlap.[20] These unacceptable states can be penalized by adding a new term to the cost function, such as $\lambda(2K - N)^2$ for the graph bipartitioning example. The strength λ of the penalty can be adjusted as a function of T to exclude the unacceptable states progressively as annealing proceeds. Such penalty functions must be treated very carefully however, and can give rise to false minima.[4,10]

6. ANALYTIC APPROACHES

We return now to the traditional tools of statistical mechanics. As already mentioned, there is little hope of evaluating the partition function analytically for a *particular* instance i. There are of course numerical methods, including simulated annealing and perhaps certain *renormalization* procedures, but we focus here on analytic work. For a disordered system, or an optimization problem with many irregular parameters (e.g., city locations in the TSP), it is essential to average over possible samples or instances to make the problem tractable. Results are then in the form of bounds, theorems, and average predictions. Some of these can be checked by numerical methods.[6]

The averaging must be a *quenched* average over physical quantities such as the cost function (or energy), *not* an *annealed* average over the partition function itself. These are not equivalent because $\overline{\ln(Z)} \neq \ln(\overline{Z})$. Averaging Z itself would correspond, besides normalization, to extending the sum over states in Eq. (4) from $\sum_x$ to $\sum_{x,i}$. This would treat instances on the same basis as configurations, as dynamical variables, which is not at all what is wanted. At low temperatures in the TSP, for instance, the dominant terms would put the cities all very close together. Unfortunately, a quenched average $\overline{\ln(Z)}$ is usually much harder to perform than an annealed average $\ln(\overline{Z})$.

A probability measure on instances is obviously needed to define an average. The choice is not always obvious, and is actually often chosen more for mathematical interest or tractability than for physical relevance. For example, the following have been considered for the TSP:

a. Cities distributed uniformly over some planar region, usually square.

b. D-dimensional analogues of (a): cities located randomly in a hypercube.

c. The *random link* model:[24] inter-city distances d_{mn} chosen independently from a distribution $\rho(d) = d^r e^{-d}/r!$. This does not satisfy the triangle inequality, but does agree with the $\rho(d)$ implied by (b) as $d \rightarrow 0$ if $r = D - 1$.

d. Other $\rho(d)$ functions, also non-metric. Both uniform and Gaussian distributions have been tried; negative d_{mn} links do not cause any harm.

A wide range of tricks and tools from statistical mechanics have been applied to computing averages such as $\overline{\min_x C_i(x)}$. Space precludes discussing many in detail, so it seems more appropriate to survey a number of techniques briefly:

1. Study scaling with respect to problem size N, dropping everything but leading order as $N \rightarrow \infty$.[24,4] The partition function, a sum of exponentials, often becomes something like $\int \exp\{Nf(\psi)\}\, d\psi$ for some auxiliary variable(s) ψ, and saddle-point methods can be used for large N.

2. Derive the properties of a system of size $N+1$ from those of one of size N, and then make the results self-consistent.[25,5]

3. Use *mean field theory*.[26,27,25,5] This has many variants, and many representations. Except for special situation, it is an approximate theory. It can be interpreted as replacing a locally varying quantity by its average. The same can sometimes be done for the variations (or *fluctuations*) around the average, giving higher order approximations. The key step is deciding which quantity to average. One often picks an *order parameter*, whose average is formally zero by symmetry (given Eq. (3)), but because of broken symmetry or broken ergodicity[11] should really have a non-zero value (see the discussion of Figure 1c).

4. Expand in powers of $\beta = 1/kT$.[28] The terms in the partition function are of the form $\exp(-\beta C)$, so an expansion around $T = \infty$ is possible, whereas one around $T = 0$ is not. These expansions can be carried to high order with diagrammatic techniques and symbolic computer codes. The resulting series can then sometimes be used to make predictions about the $T = 0$ region with continuation techniques such as Padé approximants and Borel transforms.

5. Introduce more freedom, expanding the configuration space, and add appropriate penalty functions.[4] See the preceding section.

6. Introduce *auxiliary fields*, usually for an integral representation.[25] For example, a quadratic term in an exponential may be linearized using

$$e^{z^2} = \frac{1}{\sqrt{\pi}} \int_{-\infty}^{\infty} e^{-s^2+2sz}\, ds. \tag{14}$$

7. Alter the problem to make it tractable or to relate it to something known.[4,7] For example, the TSP has been considered not only with the various instance measures mentioned above, but also on hierarchical spaces[29] and fractals.[30]

Relations with Potts models[8,13] and electrons in disordered media[27] have also been explored.

8. Use field theory methods. Methods that have been used include the $m \to 0$ trick from polymer physics,[31] where local variables are m-component vectors and the limit $m \to 0$ is taken after a calculation; this implements a self-avoiding condition.[25–28] Grassman algebras, with anti-commuting variables, have also been employed.[26,27]

9. Use *zero-temperature scaling*. Here one looks at energy (or cost) scale as a function of length scale using methods derived from the theory of phase transitions. Scaling theories can be surprisingly powerful; Moore[32] was able to derive an error estimate for the TSP cost when treated by k-opt moves (reorganizing k links at a time) in D dimensions:

$$C_i(x)^{\text{estimate}} - \min C_i(x) \sim k^{-(1-1/D)}. \tag{15}$$

10. Estimate the high- and low-temperature limits of $U(T)$ and $S(T)$ separately. This can be done with simple heuristic arguments, emphasizing the N dependence. In contrast to physical systems in statistical mechanics, optimization problems do not always produce proper extensive quantities ($U \propto N$, $S \propto N$). Combining the low- and high-temperature results can give information about different behavior regimes and their crossover temperatures.[33,24,27]

11. Use an annealed average.[33,24] This is definitely not correct, but does give a lower bound for the quenched average. At high temperature, the difference may be small (or even sub-dominant in N), and the annealed average has been used as a starting point for extrapolation from high to low temperature.

12. Use the *replica method*.[34,26,4,27,28] In many ways this is the most important of the new methods developed to deal with disordered and complex systems. It is only been left until last to permit a more detailed discussion.

The replica method consists essentially of replacing a quenched average, involving $\overline{\ln Z}$, by a limit of annealed averages. The central identity is

$$\overline{\ln Z} = \lim_{n \to 0} \left(\overline{Z^n} - 1 \right) / n. \tag{16}$$

For *integer* n the quantity $\overline{Z^n}$ is the annealed average for a system consisting of n independent identical copies of the original system. This is relatively easy to evaluate, performing the average over instances before completing the sum over states ($\sum_x$, which becomes $\prod_{\alpha=1}^{n} \sum_{x^{(\alpha)}}$). The essential problem is then the extrapolation from integer n to the neighborhood of $n = 0$, which turns out not to be unique in the interesting cases.[35] A very large amount of work has been done on this problem, and reasonably reliable recipes now exist for producing answers, although a firm mathematical foundation has yet to be derived. An essential idea is that of *replica symmetry breaking*; although all replicas start out as equivalent, as $n \to 0$ it is often necessary to treat them inequivalently, and even organize them in a hierarchical manner. This is thought to reflect the ergodicity breaking in the real system, which gets stuck in one of many possible inequivalent states. It may also be related to NP-completeness.[36,26]

7. CONCLUSION

This brief survey cannot possibly do more than paint a general picture of the burgeoning connections between statistical mechanics and combinatorial optimization. The main intent has been to introduce all the major concepts and terms, and to give some feeling the types of tools available. There *are* a lot of interesting ideas offered by the statistical mechanics approach, but they are generally *not* very useful in solving particular instances of optimization problems. Some of the more important lessons learned (not necessarily new) are:

a. Study the cost surface. This is one of the best ways of understanding a problem, and perhaps ultimately of generating better heuristic procedures. Some interesting properties were listed at the end of section 3.

b. Step outside the problem. Add new degrees of freedom and parameters (especially T), relax constraints, and abandon consistency. These all allow exploration of a larger space in which cost minima tend to be more easily found. It may also be possible to relate new problems to old ones. Eventually the constraints can be put back in, generally in a gradual fashion.

c. Average over instances. If the particular problem is insoluble, try to find an average or a typical answer, ignoring unlikely special cases. This may drive particular salesmen away empty-handed, but will attract plenty of physical scientists in their place.

ACKNOWLEDGMENTS

I thank P.W. Anderson and the Santa Fe Institute for inviting me to present this material, and Y. Fu and D. Stein for helpful suggestions and references.

REFERENCES

1. P. W. Anderson(1986), "Spin Glass Hamiltonians: A Bridge Between Biology, Statistical Mechanics and Computer Science," *Emerging Syntheses in Science*, Ed. D. Pines (Reading, MA: Addison-Wesley).
2. J. J. Hopfield and D. Tank (1985), "Collective Computation with Continuous Variables," *Biol. Cybernet.* **52**, 141.
3. M. R. Garey and D. S. Johnson (1979), *Computers and Intractability: A Guide to NP-Completeness* (New York: Freeman).
4. Y. Fu and P. W. Anderson (1986), "Application of Statistical Mechanics to NP-complete Problems in Combinatorial Optimization," *J. Phys. A* **19**, 1605.
5. M. Mézard and G. Parisi (1987), "Mean-Field Theory of Randomly Frustrated Systems with Finite Connectivity," *Europhys. Lett.* **3**, 1067.
6. J. R. Banavar, D. Sherrington, and N. Sourlas (1987), "Graph Bipartitioning and Statistical Mechanics," *J. Phys. A* **20**, L1.
7. D. Sherrington and K. Y. M. Wong (1987), "Graph Bipartitioning and the Bethe Spin Glass," *J. Phys. A* **20**, L785.
8. I. Kanter and H. Sompolinsky (1987), "Graph Optimization Problems and the Potts Glass," *J. Phys. A* **20**, L673.
9. W. Liao (1987), "Towards the Solution of the Graph Bipartitioning Problem," *J. Phys. A* **20**, L695.
10. W. Liao (1988), "Replica Symmetric Solution of the Graph Bipartitioning Problem with Fixed, Finite Valence," *J. Phys. A* **21**, 427.
11. R. G. Palmer (1982), "Broken Ergodicity," *Adv. Phys.* **31**, 669.
12. R. Rammal, G. Toulouse, and M. A. Virasoro (1986), "Ultrametricity for Physicists," *Rev. Mod. Phys.* **58**, 765.
13. J. P. Bouchard and P. Le Doussal (1986), "Ultrametricity Transition in the Graph Colouring Problem," *Europhys. Lett.* **1**, 91.
14. S. Lin (1965), "Computer Solutions of the Travelling Salesman Problem," *Bell Syst. Tech. J.* **44**, 2245.
15. D. Sherrington and S. Kirkpatrick (1975), "Solvable Model of a Spin Glass," *Phys. Rev. Lett.* **32**, 1792.
16. E. Baum (1986), "Intractable Computations without Local Minima," *Phys. Rev. Lett.* **57**, 2764.
17. J. L. van Hemmen (1986), "What is a True Mean-field Spin Glass?," *J. Phys. C* **19**, L379.
18. K. Binder (1978), ed., *Monte Carlo Methods in Statistical Physics* (Berlin: Springer).
19. N. Metropolis, A. Rosenbluth, A. Teller, and E. Teller (1953), "Equation of State Calculations by Fast Computing Machines," *J. Chem. Phys.* **21**, 1087.
20. S. Kirkpatrick, C. D. Gelatt Jr., and M. P. Vecchi (1983), "Optimization by Simulated Annealing," *Science* **220**, 671.
21. S. Kirkpatrick (1984), "Optimization by Simulated Annealing: Quantitative Studies," *J. Stat. Phys.* **34**, 975.

22. S. Lin and B. W. Kernighan (1973), "An Effective Heuristic Algorithm for the Travelling Salesman Problem," *Oper. Res.* **21**, 498.
23. P. Siarry and M. Dreyfus (1984), "An Application of Physical Methods to the Computer Aided Design of Electronic Circuits," *J. Phys. (Paris) Lett.* **45**, L39.
24. J. Vannimenus and M. Mézard (1984), "On the Statistical Mechanics of Optimization Problems of the Travelling Salesman Type," *J. Phys. (Paris) Lett.* **45**, L1145.
25. M. Mézard and G. Parisi (1986), "Mean-field Equations for the Matching and the Travelling Salesman Problems," *Europhys. Lett.* **2**, 913.
26. H. Orland (1985), "Mean-field Theory for Optimization Problems," *J. Phys. (Paris) Lett.* **46**, L763.
27. G. Baskaran, Y. Fu, and P. W. Anderson (1987), "On the Statistical Mechanics of the Travelling Salesman Problem," *J. Stat. Phys.* **45**, 1.
28. M. Mézard and G. Parisi (1986), "A Replica Analysis of the Travelling Salesman Problem," *J. Phys. (Paris)* **47**, 1285.
29. D. Stein and G. Baskaran (1986), "The Travelling Salesman Problem on an Ultrametric Space," (unpublished).
30. R. M. Bradley (1986), "Statistical Mechanics of the Travelling Salesman Problem on the Sierpinski Gasket," *J. Phys. (Paris)* **47**, 9.
31. P. G. de Gennes (1972), "Exponents for the Excluded Volume Problem as Derived by the Wilson Method," *Phys. Lett.* **A38**, 339.
32. M. A. Moore (1987), "Zero-temperature Scaling and Combinatorial Optimization," *Phys. Rev. Lett.* **58**, 1703.
33. S. Kirkpatrick and G. Toulouse (1985), "Configuration Space Analysis of Travelling Salesman Problems," *J. Phys. (Paris)* **46**, 1277.
34. M. Mézard and G. Parisi (1985), "Replicas and Optimization," *J. Phys. (Paris) Lett.* **46**, L771.
35. J. L. van Hemmen and R. G. Palmer (1979), "The Replica Method and a Solvable Spin Glass Model," *J. Phys. A* **12**, 563.
36. D. L. Stein (1986), *Proceedings of the 1986 Les Houches Summer School "Chance and Matter," Statics and Dynamics of Complex Systems*, Eds. J. Souletie, J. Vannimenus and R. Stora (Elsivier Sc. Pub. BV).

DAVID RUELLE
I.H.E.S., 91440 Bures-sur-Yvette, France; written while visiting the California Institute of Technology, Pasadena, California

Can Nonlinear Dynamics Help Economists?

ABSTRACT

This paper reviews the ideas of "chaotic" dynamics, and their possible applications to economics.

1. INTRODUCTION: APERIODIC TIME EVOLUTIONS

Many natural phenomena show a complicated, aperiodic dependence on time. Examples in the field of economics and finance are easy to find. The problem then poses itself of explaining how complicated time evolutions arise and, better still, of extracting useful information from aperiodic, "chaotic" time series.

Now, it is quite clear that aperiodic time evolutions may arise and do arise in several rather different manners. To discuss this point, we must remember first that a natural system is never isolated from the rest of the universe; that is painfully obvious in particular in economics. Therefore, what we call a natural system is really an idealization, where a certain number of parameters x_i are chosen to describe the system, and the rest of the universe appears as a perturbing noise ω. We have thus a time evolution of the form

$$\frac{d}{dt}\,x(t) = F\big(x(t),\omega(t)\big)$$

or, discretizing the time,

$$x(t+1) = f\big(x(t),\omega(t)\big)\,.$$

In these equations, $x(t)$ is a vector with a smaller or larger number of components x_i. The external noise ω is called "shocks" by economists. Keep in mind that there is some arbitrariness as to what variables are put in $x(t)$, and what are put in $\omega(t)$. For instance, one can eliminate ω by studying the evolution of the entire universe, but the resulting equation $dx/dt = F(x)$ is obviously quite intractable. Let me also mention that economists try to make time evolutions more manageable by "detrending" (i.e., removing the effect of long-term economic growth); I do not want to discuss this point further here.

Clearly, the presence of noise—or shocks—produces an aperiodic time evolution. On the other hand, in the absence of noise, i.e., for an autonomous time evolution

$$\frac{d}{dt}\,x = F(x) \tag{1}$$

aperiodic behavior can also be obtained easily if Eq. (1) is in a high-dimensional space. It suffices to represent by Eq. (1) a collection of a large number of oscillators with independent frequencies; the incoherent oscillations produce a complicated, "turbulent" time evolution. It would seem that we have thus more than enough ways of explaining complicated aperiodic behavior, and it may come as a surprise that we have to take into account something else. This something else goes by a variety of names, like strange attractors, deterministic noise, or low-dimensional chaos.

Here is what it is: an autonomous time evolution (defined by an equation of type (1)) in low dimension (3 or more) may and often does exhibit aperiodic behavior with *sensitive dependence on the initial equation.* In other words, a small change $\delta x(0)$ of initial condition produces a change $\delta x(t)$ which grows exponentially as t increase: $\delta x(t) \sim \delta x(0)e^{\lambda t}$, where $\lambda > 0$ as long as $\delta x(t)$ remains small). The trajectory $x(t)$ lies asymptotically for large t on a set called a *strange attractor,* and the type of behavior just described is called *chaos.* It turns out that at the turn of the century, such people as J. Hadamard, P. Duhem, and H. Poincaré knew about what is now called chaos, and about its philosophical implications. This knowledge however was lost somehow by physicists until relatively recently. Chaos reappeared in connection with meteorology (Lorenz[1]) and hydrodynamic turbulence (Ruelle and Takens[2]). At this time, because of the advent of electronic computers and modern experimental techniques, chaos intruded in physics, not just as a philosophically important idea, but as something which could be tested experimentally.

Indeed, detailed experiments have shown that chaos is present in hydrodynamic turbulence.

Now, what is the relevance of all this to economics? Compared to physics where one can do careful and precise experiments, the situation in economics is poor; one lacks long time series of good quality, and shocks are certainly present. Furthermore, the existence of well-defined, nonlinear evolution equations is unproved—to say the least. Yet, some of the things which we have learned about chaos are rather robust facts, which should be significant for economics, at least at the qualitative level. And in the best cases (i.e., for some financial data), we shall see that a quantitative estimate of certain ergodic parameters characterizing chaos (information dimension, characteristic exponents) appears to be meaningful. In Section 2, I present a very brief survey of nonlinear dynamics; in Section 3, possible applications to economics. Section 4 discusses how to extract ergodic parameters from time series. Finally, Section 5 is on the delicate problem of predictions.

2. DIFFERENTIABLE (NONLINEAR) DYNAMICS

Differentiable dynamics is the study of deterministic time evolutions of the form

$$\frac{dx(t)}{dt} = F\big(x(t)\big) \tag{2}$$

where the time t is continuous ($t \in \mathbf{R}$, possibly restricted to $t \geq 0$), or of the form

$$x(t+1) = f\big(x(t)\big) \tag{3}$$

where the time t is discrete ($t \in \mathbf{Z}$, possibly restricted to $t \geq 0$). The point $x = x(t)$ runs over a "phase space" which may be $\mathbf{R}^n$, or a manifold (possibly of infinite dimension). The functions F and f are assumed to be differentiable.

There is a large body of high-grade mathematical literature on differentiable dynamics. The currently fashionable topic of chaos concerns applications of differentiable dynamics (under the name of nonlinear dynamics) to natural phenomena or computer experiments. The literature on chaos is huge, but mostly low grade. My own views on the subject have been expressed in Ruelle[3] and Eckmann and Ruelle[4]; an easy introduction to the whole field is provided by the book of Bergé, Pomeau and Vidal.[5] Anybody seriously interested in chaos is strongly advised to read the original papers conveniently collected in Cvitanović[6] and in Hao Bai-Lin[7] (there is not much overlap between the two). Here, only a brief review can be given.

The large time behavior of solutions of Eq. (2) or (3) falls into several categories, which we describe successively.

1. POINT ATTRACTORS. As $t \to \infty$, we have $x_t \to \overline{x}$, where $\overline{x}$ does not depend on time (and is thus a fixed point for time evolution). In mechanics and physics one would often say that $\overline{x}$ is an *equilibrium* (in fact, a stable equilibrium), but note that the economists use the word "equilibrium" to denote a state which is not necessarily time independent. There is a natural psychological tendency to assume that a system which is not submitted to time-dependent forces will asymptotically reach some equilibrium (the ecologists have the word *climax* for such equilibria), but more complicated asymptotic time behavior is possible and, indeed, common.

2. PERIODIC ATTRACTOR. As $t \to \infty$, the distance of x_t to $\overline{x}_t$ tends to zero, where $\overline{x}_t$ is periodic, i.e., $\overline{x}_{t+T} = \overline{x}_t$ where T is the period. Periodic attractors occur both for continuous-time and discrete-time dynamical systems.

3. MORE COMPLICATED ATTRACTOR. When $t \to \infty$, the orbit (x_t) may lie asymptotically on a more complicated set (attractor) than a point or a loop (periodic attractor). We shall assume that the average time spent in various regions of phase space is well defined; it is represented by a probability measure $\rho(dx)$ invariant under time evolution. In a moment we shall classify the more complicated attractors as strange or nonstrange.

If we make a small change $\delta x(0)$ of the initial condition, the change $\delta x(t)$ of the solution of Eq. (2) or (3) satisfies

$$\frac{d}{dt}\,\delta x(t) = \big(D_{x(t)}F\big)\,\delta x(t)$$

or

$$\delta x(t+1) = \big(D_{x(t)}F\big)\,\delta x(t)\,.$$

We are interested in the rate of exponential growth of $||\delta x(t)||$ or, better, of the length $||x(t)||$ of a *tangent vector* $u(t)$ satisfying the linearized equation

$$\frac{d}{dt}\,u(t) = \big(D_{x(t)}F\big)\,u(t)$$

or

$$u(t+1) = \big(D_{x(t)}f\big)\,u(t)\,.$$

The rate of exponential growth λ of $||u(t)||$ is called a *characteristic exponent.* If $\lambda > 0$, one says that the time evolution is *chaotic*, or has *sensitive dependence on the initial condition,* or that the corresponding attractor is a *strange attractor.*

Point attractors and periodic attractors are not strange, and there are other nonstrange attractors. For instance, an attractor which represents a collection of uncoupled periodic oscillators is a *quasi-periodic attractor* and does not have sensitive dependence on the initial condition. Introducing couplings between the oscillators may however produce chaotic behavior, and this is observed in experiments and in computer simulations.

It is now known that chaotic behavior is quite frequent among dynamical systems provided that the dimension n is high enough. One needs

$n \geq 3$	continuous time
$n \geq 2$	discrete time (invertible map f)
$n \geq 1$	discrete time (noninvertible map f)

For instance, to obtain chaotic behavior by the coupling of n oscillators, one needs $n \geq 3$.

3. POSSIBLE APPLICATIONS TO ECONOMICS

The fact that by introducing couplings between subsystems, one tends to produce complicated temporal behavior and chaos, is very important. This is a very general and robust fact, and survives if one introduces shocks, or relaxes differentiability.

In the light of this, let us look at the standard wisdom that free trade between different countries (or other entities) produces a more advantageous equilibrium than when there are economic barriers. Free trade introduces a coupling between the different national economics which will tend to introduce nontrivial time dependence, possibly with sensitive dependence on the initial condition. Having time dependence, in particular living on a strange attractor with a future which is practically unpredictable, may be disadvantageous, and this disadvantage may offset the advantages of a "better equilibrium."

The issue is slightly confused by the fact that "equilibrium" as understood by economists does allow for time dependence in the form of anticipation of the future. Here, however, we question some standard economic assumptions, and specifically the "perfect foresight" of economic agents in the presence of sensitive dependence on the initial condition.

To digress a bit to biology, let me mention that there also the interreaction between different species easily produces time-dependent situations, including chaos. This is true on a short time scale where species are constant: the standard situation of ecology (see May[8]). It should also be true on longer time scales where species evolve (coevolution).

Consider now a dynamical system depending on a parameter μ. As μ is varied, the qualitative behavior of the system may change, and such changes are called *bifurcations*. Sequences of bifurcations leading from stationary states ("equilibria") via periodic oscillations to chaos have been much studied. Various scenarios are known: the quasi-periodic scenario of Ruelle-Takens, the period-doubling sequence of Feigenbaum, and the "intermittent" scenario of Pomeau-Manneville are some, but not the only possibilities. For a review, see Eckmann.[9] Remember that if several frequencies appear in a system through successive bifurcations, then chaos may occur through coupling of three or more frequencies.

A possible application of these ideas to economics is as follows (see Ruelle[3]). Let μ describe the level of technological development in a certain economy. One natural scenario could be that by increasing μ, one goes from a stationary state to a periodic state, possibly to a quasi-periodic state (several frequencies), and then to a chaotic situation. This rough succession of states is rather general, and robust to shocks and non-differentiability. It also does not appear incompatible with what is known of the history of business cycles.

4. DIAGNOSIS OF TIME SERIES

In the previous section, we have seen how nonlinear dynamics can provide ideas of qualitative interest to economics. Quantitative help appears also possible in the analysis of time series which are sufficiently long and accurate. At least, the situation is sufficiently encouraging that one wants to look more closely into the matter.

Let me start with the problem of reconstructing dynamics. Imagine that we have a time evolution $t \to x(t) \in$ attractor $A \subset \mathbf{R}^n$ and that we know only $u(t) = \Psi(x(t))$ where Ψ is some scalar "observable." Very often, the time evolution

$$t \to \tilde{x}(t) = \Big(u(t), u(t+\tau), \ldots, u\big(t+(m-1)\tau\big)\Big) \in \mathbf{R}^m$$

gives a faithful image of the original time evolution if τ is well chosen and the *embedding dimension* m is large enough. This is the *time-delay method* to reconstruct the dynamics. [Historically, people first tried to reconstruct a vector $\tilde{\tilde{x}}(t) = (u(t), \dot{u}(t), \ddot{u}(t), \ldots)$ using time derivatives. My proposal to use the form $\tilde{x}$ rather than $\tilde{\tilde{x}}$ remained unpublished but is referred to—as their footnote 8—in the paper of the Santa Cruz collective[10] which first discussed the method in writing.]

The reconstruction of an attractor from a time series $(u(t))$ leads to the possibility of determining experimentally various dynamical parameters: dimensions, characteristic exponents, etc. This has been done for a variety of data from physics, chemistry, and computer experiments. One financial time series has also been analyzed, consisting of weekly returns on the value-weighted portfolio of the Center for Research in Security Prices at the University of Chicago (CRISP).

A very natural idea is to try to estimate the dimension of the reconstructed attractor: for sufficiently large embedding dimension, this should become equal to the dimension of the true attractor. Actually, several different dimensions can be defined, and the *correlation* dimension obtained from the Grassberger-Procaccia algorithm is particularly well known. Scheinkman and Le Baron[11] have obtained a value of about 6 for this dimension, using the CRISP data. This is encouragingly low.

One can also estimate the rates at which nearby trajectories diverge, i.e., the *characteristic exponents.* Doing this for the CRISP data, one finds apparently two positive characteristic exponents with values of the order of 0.15 and 0.10 per week. (See Eckmann, Oliffson-Kamphorst, Ruelle and Scheinkman,[12] these proceedings.)

Let us stress that the analysis of the CRISP data from the point of view of nonlinear dynamics has to be judged very cautiously (even though, by economics standards, this series is of particularly good quality, and fairly long). Shocks are certainly important, and we still have no proof of the existence of an underlying deterministic dynamics.

If there is an underlying deterministic dynamics, which changes slowly over time (drift), then this drift would affect dimension estimates more than it would affect the larger characteristic exponents. In fact, drift would essentially amount to introducing a new characteristic exponent close to zero. In view of this, the determination of characteristic exponents may be more useful than the determination of dimensions for the time series of economics.

5. PREDICTIONS

The most desirable thing to extract from the study of time series in economics and finance is a reliable prediction of the future. This is also the least likely thing that we shall obtain from such studies. Clearly there cannot be such accurate forecasts of the stock market that they would allow everyone to become very rich through speculation! Apart from our basic ignorance as to possible equations governing the time evolution in economics, there are shocks, probably sensitive dependence on the initial condition, and the necessity to retain many dynamical variables—not just a few.

Nevertheless, the mathematical problem of prediction is quite interesting, and we shall discuss it briefly, assuming a deterministic low-dimensional time evolution. Basically, there are two ways of proceedings: global nonlinear or local linear.

The following easy result gives a justification of the global nonlinear approach.

THEOREM If the time delay reconstruction in $\mathbf{R}^m$ yields a faithful (i.e., one-to-one) representation A of a compact attractor, then there is a continuous function $\phi : A \to \mathbf{R}$ such that

$$u(t + m\tau) = \phi\Big(u(t), u(t+\tau), \ldots, u\big(t + (m-1)\tau\big)\Big) \tag{4}$$

This result has been presented in Ruelle.[13] Note that ϕ can be uniformly approximated by polynomials (Weierstrass), and one can thus approximate the dynamics by polynomial dynamics. (But one has to make sure that one stays on or close to A, which has no reason to be an attractor for the polynomial dynamics.) Crutchfield and McNamara[14] have used polynomial approximations to predict time series, and have obtained encouraging results. But since the method requires the determination of many polynomial coefficients, its use is limited to fairly long time series of fairly high quality.

The local linear approach to prediction is very simple and natural (see Grassberger[15]; Farmer and Sidorovitch[16]; see also Lapedes and Farber[17]). If $x(1), x(2), \ldots, x(N) \in \mathbf{R}^n$ are known, $x(N+1)$ is predicted by linear interpolation: write

$$x(N) = \sum_i \alpha_i \, x(n_i)$$

where $\sum \alpha_i = 1$ and the $x(n_i)$ are close to x_N, then

$$x(N+1) = \sum_i \alpha_i \, x(n_i + 1) \, .$$

This process can be repeated, but the prediction will soon become uncertain due to sensitive dependence on the initial condition.

A natural problem related to prediction is that of its economic cost when there is sensitive dependence on the initial condition. [This question was raised by Kenneth Arrow at the Santa Fe workshop on "the economy as an evolving complex system," September 1987; the following remarks are based on discussions at Santa Fe, essentially with Kenneth Arrow and José Scheinkman.] Suppose that we have a dynamical system depending on a control parameter μ, such that at time t

$$x(t) = f_\mu^t(x)$$

where x is the initial condition. We want to maximize the following discounted utility function

$$U(x, \mu) = \int_0^\infty e^{-\delta t} \, u(f_\mu^t(x)) \, dt \, .$$

The basic remark is that U is not necessarily differentiable. In fact, trying to differentiate with respect to x, we get

$$D_x \, U = \int_0^\infty e^{-\delta t} \, D_{x(t)} u \cdot D_x f_\mu^t$$

where $\|D_x f_\mu^t\|$ grows like $\exp \lambda_1 t$, if λ_1 is the largest characteristic exponent. Therefore, we do not expect differentiability if $\delta < \lambda_1$ (or existence of a second derivative if $\delta < 2\lambda_1$). Actually, since characteristic exponents depend on the choice of an ergodic measure, the role of this measure should be made explicit, but we refrain from doing that here. One can prove that, even if U is not differentiable, it is Hölder continuous with respect to x and μ.

Suppose now that there is an imprecision ϵ on the knowledge of the initial condition x, i.e., we believe that the initial condition is $\tilde{x}$, when it is really x (and $| \tilde{x} - x | \approx \epsilon$). Choosing $\mu = \tilde{\mu}$ to make $U(\tilde{x}, \mu)$ maximum, we make a loss

$$\Delta U = \max_\mu \, U(x, \mu) - U(x, \tilde{\mu}) \, .$$

Normally, we would expect that $\Delta U \sim A\epsilon^2$, where A is the amount invested, but if U is not *twice* differentiable, we shall get only $\Delta U \sim A\epsilon^\alpha$, with $\alpha < 2$. Suppose now that the cost of knowing x with precision ϵ is $\sim \epsilon^{-\beta}$ (where $\beta = 2$ would be a natural choice). The economic equilibrium is of the form

$$A\epsilon^\alpha = \text{const } \epsilon^{-\beta}$$

which leads to

$$\text{cost} = \text{const } A^{\frac{\beta}{\alpha+\beta}} .$$

Sensitivity to the initial condition manifests itself in this formula by $\alpha < 2$.

REFERENCES

1. Lorenz, E. N. (1963), "Deterministic Nonperiodic Flow," *J. Atmos. Sci.* **20**, 130–141.
2. Ruelle, D., and F. Takens (1971), "On the Nature of Turbulence," *Commun. Math. Phys.* **20**, 167–192; **21**, 21–64.
3. Ruelle, D. (1981), "Strange Attractors," *Math. Intelligencer* **2(3)**, 126–137.
4. Eckmann, J.-P., and D. Ruelle (1985), "Ergodic Theory of Chaos and Strange Attractors," *Rev. Mod. Phys.* **57**, 617–656.
5. Bergé, P., Y. Pomeau and Chr. Vidal (1986), *Order in Chaos* (New York: J. Wiley).
6. Cvitanović, P. (1984), *Universality in Chaos* (Bristol: Adam Hilger).
7. Bai-Lin, Hao (1984), *Chaos* (Singapore: World Scientific).
8. May, R. (1976), "Simple Mathematical Models with Very Complicated Dynamics," *Nature* **261**, 459–467.
9. Eckmann, J.-P. (1981), "Roads to Turbulence in Dissipative Dynamical Systems," *Rev. Mod. Phys.* **53**, 643–654.
10. Packard, N. H., J. P. Crutchfield, J. D. Farmer and R. S. Shaw (1980), "Geometry from a Time Series," *Phys. Rev. Letters* **45**, 712–716.
11. Scheinkman, J. A., and B. Le Baron, "Nonlinear Dynamics and Stock Returns," *J. Business,* to appear.
12. Eckmann, J. P., S. Oliffson Kamphorst, D. Ruelle and J. Scheinkman (1988), "Lyapunov Exponents for Stock Returns," these proceedings.
13. Ruelle, D. (1987), "Theory and Experiment in the Ergodic Study of Chaos and Strange Attractors," *8th Intl. Congress on Mathematical Physics,* Ed. M. Mebkhout and R. Sénéor (Singapore: World Scientific), 273–282.
14. Crutchfield, J. P., and B. S. McNamara (1987), "Equations of Motion from a Data Series," *Complex Systems* **1**, 417–452.
15. Grassberger, P, preprint.
16. Farmer, D., and S. Sidorovitch, "Predicting Chaotic Time Series," *Phys. Rev. Lett.*, to appear.
17. Lapedes, A., and R. Farber, "Nonlinear Signal Processing using Neural Networks: Prediction and System Modeling," *Technical Report LA-UR-87-2662, Los Alamos National Laboratory.*

MARIO HENRIQUE SIMONSEN
Brazil Institute of Economics, Fundacao Getulio Vargas, Prara de Botafago, 18 190, Rio de Janiero, Brazil

Rational Expectations, Game Theory and Inflationary Inertia

1. INTRODUCTION

What is rational behavior under strategical interdependence, that is, when each individual's payoff depends not only on his actions, but also on other people's decisions? This is the central question that Game Theory proposes to answer. Since the pioneer work of von Neumann and Morgenstern, much light has been thrown on how to approach a number of problems involving conflict and interdependence. Yet, except for a limited class of games, a convincing concept of rationality has never been established. Game theorists are not to be blamed for that. One must simple recognize that their program was too ambitious.

Let us concentrate on non-cooperative games, which describe how a market economy works in the absence of a Walrasian auctioneer. The key equilibrium concept has been established by Nash, being an extension of Cournot's solution to the oligopoly problem. A Nash equilibrium is defined as a set of strategies, one for each player, such that no player can improve his expected utility by unilaterally changing his strategy. At first glance, it sounds like a satisfactory description of rational behavior in non-cooperative games. In fact, as will be shown in Section 2, the underlying assumption of the rational-expectations hypothesis is that rational participants in a non-cooperative game immediately locate a Nash equilibrium.

Some reflection shows that rational behavior, pragmatically understood as how intelligent people do behave and not how game theorists would like them to behave, is a much more intriguing issue. In fact, Nash equilibria are defined for games in normal form, which are defined as games of imperfect information: each player must choose his strategy before knowing other participants' choices. Once this point is made explicit, the true meaning of a Nash equilibrium comes on stage: except as far as the states of nature are concerned, it is nothing but *ex post* wisdom, in the sense of non-repentance.

To what extent a Nash equilibrium also corresponds to what really matters, namely, *ex ante* rationality, is a much more intricate question. In most games with incomplete information, where each player knows his own payoff but ignores the payoffs of other participants, players simply have no means of locating immediately a Nash equilibrium. And even in games with complete information and one unique Nash equilibrium, choosing a Nash strategy may seriously harm some player should the others fail to choose Nash strategies. Since decisions must be taken before knowing other people's decisions, prudence may be a serious obstacle to the prompt location of a Nash equilibrium.

The possibility of conflict between prudent playing and Nash behavior is illustrated by the game of half of the average: "one-hundred students are called to write on a slip of paper a real number in the closed interval [0,1], each ignoring other students' choices; the purpose of the game is to guess half of the average of the individual choices. Accordingly, whoever has written a number equal to half of the average of the one-hundred individual indications, gets as a premium a 100-dollar bill. No premium or penalty is imposed on a student that has written a number exceeding half of the average, but those whose choice is below half of the average must pay a 100-dollar fine." In the unique Nash equilibrium, all students should write zero, and each one would collect a 100-dollar bill. The problem is that being a Nash strategist is too risky, since one single outlier can transform the 100-dollar premium into a 100-dollar fine. How students actually play this game is an empirical issue that can be easily tested in a classroom.

The foregoing discussion leads to the classification of non-cooperative games into two groups, A and B games, as described in Section 3. A games are those in which most prudent playing by all participants yields a Nash equilibrium and, conversely, where every Nash equilibrium is a combination of max-min strategies. These can be viewed as easy games, in the sense that playing defensively is the route to *ex post* wisdom. Zero-sum, two-person games with a saddle point, as well as games where a dominant strategy is available for each player (such as the prisoner's dilemma) belong to this class, where Nash equilibria can be fully accepted as describing *ex ante* rational behavior.

The challenging question is what should be understood as rational playing in a B game, where prudence must be left aside to hit a Nash equilibrium. Let us start with games of complete information and one unique Nash equilibrium. To act as a Nash strategist, every player must bet that all others will also behave as Nash strategists. Whether taking such a bet should be classified as rational behavior or as irresponsible decision-making depends on two issues: (a) to what extent each player can trust the others as Nash strategists; (b) how much each player can lose if he chooses a Nash strategy and some others do not. The question is intricate enough to bear no definite answer, but common sense suggests that B-games should be subdivided into two groups: (i) B_1 games, which are B games involving a small number of experts, where complete information is available to all, and where there exists a unique Nash equilibrium; (ii) B_2 games, where either the number of participants is too large to make them all trust the others as Nash strategists, or where information on payoffs is incomplete. Hitting a Nash equilibrium in one single move is a strong possibility in a B_1 game, but not in B_2 games, where players are expected to act much more cautiously. It should be noted that games of complete information involving a few experts, but with multiple Nash equilibria, must often be treated as B_2 games, as will be shown in Section 3.

How to describe the cautious approach to a B_2 game is, once again, an intricate question, since no definite answer is available. As a paradigm, we shall assume that all players start defensively by choosing a max-min strategy, that is, a strategy that maximizes each individual's payoff in the most unfavorable combination of other participants' strategies.

The next step is to describe the dynamics of a repeated B_2 game. We shall concentrate on games with a large number of small participants, leaving no space for signalling, threats or attempts to achieve Stackelberg dominance. Unilateral changes in each move strategies are here a true possibility, and will certainly occur until a Nash equilibrium has been located. A simple trial-and-error model assumes that in the nth move, each player tries to maximize his payoff by assuming that other players will repeat the strategies of the preceding move. The outcome is a sequence of strategies that may converge or not. If it converges (B_{21} games), the limit is a Nash equilibrium. The non-convergence case (B_{22} games) is an open field for further research.

Section 4 presents a simple general equilibrium price-setting model under monopolistic competition. As most economic games, it is a B game. From the purely formal point of view, it may be treated either as a B_1 or as a B_2 game, but common sense suggests that it should be interpreted as a B_2 game, that under iteration can be classified as a B_{21} game. The discussion brings on stage the central weakness of rational expectations macroeconomics: the micro units are aggregated as if they were playing an A or a B_1 game, while they are actually playing a B_2 game.

Section 5 revisits the problem of inflationary inertia, using the price-setting model of Section 4. In the later sixties and early seventies, inflationary inertia was explained by combining the natural unemployment rate hypothesis with Cagan's adaptive expectation formula. Although empirically convincing, the explanation was soon eclipsed by the rational expectations revolution that dismissed backward-looking inflationary expectations as an "*ad hoc*" assumption. Painless inflation cure appeared as a strong possibility, provided a credible monetary rule was announced, and provided some gradualism was accepted to bridge the temporary inertia caused by staggered wage and price setting. Empirical evidence never supported this optimism, but it appeared as the outcome of sound economic theory. Once wage- and price-setting decisions are viewed as a play of a B_2 game, the rational-expectations optimism goes to pieces. Even if the Government commits itself to a credible program of nominal output stabilization, price-setters have no serious reason to stop price increases until they are assured that price-setters will behave as Nash strategists. "Expectations" now appear as an imprecise concept that attempts to summarize how rational economic agents face strategic interdependence problems. In fact, what is involved is how to play a B_2 game. From this point of view, the old-fashioned adaptive-expectations hypothesis deserves some rehabilitation, since it describes a trial-and-error approach to a Nash equilibrium in a B_{21} game.

Section 6 discusses the role of income policies in anti-inflationary programs. Their central role is to orchestrate the immediate location of a Nash equilibrium in a B_2 price-setting game. They may be dangerous insofar as they constrain individual decision-making, but they may be extremely useful in signalling to each agent how the others will behave.

Of course, wage-price controls without aggregate demand discipline can only lead to an inflationary disaster, as has been known since the times of the Roman Emperor Diocletian. Yet the converse is also true. Trying to fight a big inflation from the demand side only may lead to such stagflation that policy-makers may conclude that life with inflation is preferable to life with a monetarist anti-inflationary program. Worse yet, they often reach that conclusion only after a prolonged recession.

Staggered wage setting (which even rational-expectations macroeconomics recognizes as a temporary source of inflationary inertia, since the pioneer works of Stanley Fischer and John Taylor) is revisited in Section 7, now under the perspective of a B_2 game. Section 8 finally describes the potential for monetary reforms coupled with income policies and fiscal austerity in a package to fight a big inflation. From this point of view, both the Argentine Austral Plan and the Brazilian Cruzado Plan proved themselves to be highly innovative. The fact that neither of them achieved true success (the Cruzado Plan turning into a dramatic failure) should not be used as evidence against income policies, but simply as an alert against the use of income policies without aggregate demand discipline.

2. RATIONAL EXPECTATIONS AND NASH EQUILIBRIUM

Rational-expectations macroeconomics is based on a particular game's theoretical framework developed by Lucas and Sargent in the following way. They let h denote a collection of decision rules of private agents. Each element of h is itself a function that maps some private agent's information about his state at a particular point in time into his decision at that point in time. Consumption, investment, and demand functions for money are all examples of elements of h. Lucas and Sargent let f denote a collection of elements that forms the "environment" facing private agents. Some elements of f represent rules of the game or decision rules selected by the government, which map the government's information at some date into its decisions at that date. For example, included among f might be decision rules for fiscal and monetary policy variables. The principle of strategic interdependence establishes that h is a function of f:

$$h = T(f)\,. \tag{2.1}$$

The mapping T represents "cross-equation restrictions" since each element of h and of f is itself a decision rule or equation determining the choice of some variable under some agent's control.

Under Lucas and Sargent's principle of strategic interdependence, economics can be viewed as a two-person game, where government and private agents interact. Active economic policies mean that government acts as a dominant Stackelberg player: to maximize a social utility function $U(h, f)$, government strategies are chosen so as to maximize $U(T(f), f)$. The possibility of Stackelberg warfare between government and private sector is ruled out, since private agents are assumed to be numerous and dispersed. (Sargent argues, nevertheless, that once government is disaggregated between monetary authorities and fiscal authorities, Stackelberg warfare becomes a true possibility. In fact, this is how he views the policy mix of the Reagan administration, combining tight monetary with loose fiscal policies.)

That treating the private sector as one single player involves an aggregation process needs no explanation. The assumption that, as opposed to government, it cannot assume the role of a Stackelberg dominant player because private agents are numerous and dispersed, clearly reveals that the principle of strategic interdependence summarized by Eq. (2.1) transforms what should be described as an $(n+1)$-person game into a two-person game. Which are the underlying aggregation assumptions is a question worth discussion.

Let us assume that there are n private agents and let us denote by h_i the optimum set of decision rules by the ith agent. Aggregate private sector decisions are obviously a function of each individual agent's choices:

$$h = G(h_1, h_2, \ldots, h_n)\,. \tag{2.2}$$

Optimum decision-making by each private agent depends on his perception f_i of the "environment" as well as on the strategies chosen by the other private agents:

$$\begin{aligned} h_1 &= H_1(h_2, \ldots, h_n, f_1) \\ h_2 &= H_2(h_1, h_3, \ldots, h_n, f_2) \\ &\vdots \\ h_n &= H_n(h_1, \ldots, h_{n-1}, f_n)\,. \end{aligned} \tag{2.3}$$

Eqs. (2.2) and (2.3) provide the microeconomic foundations of the principle of strategic interdependence. To derive the Lucas and Sargent's Eq. (2.1), three assumptions are needed:

1. The "environment" vector is common knowledge, namely, $f_1 = f_2 = \ldots = f_n = f$. This means that, besides sharing the same perception of how the government will act, all private agents acknowledge that their expectations on government policies are shared by all other private agents.
2. With $f_1 = f_2 = \ldots = f_n = f$, there exists one unique solution $(h_1, h_2, \ldots, h_n)$ to Eqs. (2.3).
3. Private sector agents are able to solve Eqs. (2.3) under the assumption $f_1 = f_2 = \ldots = f_n = f$ and choose their decision vectors by making $h_i = \hat{h}_i \cdot (i = 1, 2, \ldots, n)$.

The correspondence between rational-expectations macroeconomics and Nash equilibria now becomes evident: for each environment vector f, private agents solve their internal strategic interdependence problem by immediately locating a Nash equilibrium. Under this assumption, the original $(n+1)$-person game is then transformed into a two-person game.

The assumption that the environment vector f is to be treated as common knowledge may be challenged on the grounds that information is not a free good, and traditional criticism of rational-expectations macroeconomics has focused on this point. Yet this is nothing but a convenient hypothesis for simple modeling, one which can be relaxed in more complex rational-expectations exercises. The central weakness of the rational-expectations macroeconomic hypothesis is to be found in a much more sophisticated point. It implicitly assumes that rational participants in a non-cooperative game with millions of players promptly move to a Nash equilibrium. As stressed in Section 1, in B games this is nothing but a confusion between *ex ante* and *ex post* rationality, and as previously noted, most economic problems are to be regarded as B games.

3. A AND B GAMES

As suggested in Section 1, non-cooperative games may be divided into two groups: (i) A games, namely, those in which any Nash equilibrium is a combination of max-min strategies; (ii) B games, namely, all non-A games. The former are to be viewed as easy games, in the sense that most prudent playing yields to nonrepentance. B games are far more complicated in the sense that to reach a Nash equilibrium, prudence must be left aside. Let us discuss a few examples.

Example I (The prisoner's dilemma). This can be described as the bimatrix game:

$$\begin{array}{cc} & \begin{array}{cc} Y_I & Y_{II} \end{array} \\ \begin{array}{c} X_I \\ X_{II} \end{array} & \begin{pmatrix} (-6;-6) & (-2;-10) \\ (-10;-2) & (-3;-3) \end{pmatrix} \end{array}$$

X_I is the dominant strategy for player X, Y_I the dominant strategy for player Y. A dominant strategy is also a max-min strategy and a combination of dominant strategies yields Nash equilibrium. Hence, the prisoner's dilemma is an A game. In the unique Nash equilibrium, both players get their max-min -6.

Example II. Consider the bimatrix game:

$$\begin{array}{cc} & \begin{array}{cc} Y_I & Y_{II} \end{array} \\ \begin{array}{c} X_I \\ X_{II} \end{array} & \begin{pmatrix} (10;8) & (2;5) \\ (3;5) & (0;0) \end{pmatrix} \end{array}$$

Once again X_I and Y_I are dominant strategies. Players' max-mins are $v_X = 2; v_Y = 5$. In the unique Nash equilibrium $(X_I; Y_I)$, each player's payoff exceeds the max-min, indicating this possibility in A games.

Example III. Constant-sum, two-person games with a saddle point:

$$\begin{array}{cc} & \begin{array}{ccc} Y_I & Y_{II} & Y_{III} \end{array} \\ \begin{array}{c} X_I \\ X_{II} \\ X_{III} \end{array} & \begin{pmatrix} (7;3) & (0;10) & (2;8) \\ (6;4) & (5;5) & (4;6) \\ (7;3) & (5;5) & (3;7) \end{pmatrix} \end{array}$$

The combination $(X_{II}; Y_{III})$ of max-min strategies yields the unique Nash equilibrium, where players' payoffs coincide with their max-min. This is a general property of constant-sum, two-person games, which are a particular group of A games. To locate a Nash equilibrium, each player is only required to be prudent or to assume his opponent to be prudent. A player can only benefit if he chooses a max-min strategy and the opponent does not.

Example IV. Consider the bimatrix game:

$$\begin{array}{cc} & \begin{array}{ccc} Y_I & Y_{II} & Y_{III} \end{array} \\ \begin{array}{c} X_I \\ X_{II} \\ X_{III} \end{array} & \begin{pmatrix} (3;7) & (0;4) & (3;2) \\ (2;3) & (3;5) & (4;1) \\ (5;7) & (2;9) & (3;5) \end{pmatrix} \end{array}$$

Of course, if the game is repeated, the max-min combination cannot be looked upon as a stable equilibrium, since each player will perceive that he may be better off by unilaterally changing his strategy. If each player chooses his nth move's strategy assuming that the other will repeat his preceding move, the game will proceed as follows:

First move:	$(X_{III}; Y_{III})$
Second move:	$(X_{III}; Y_I)$
Third move:	$(X_{II}; Y_I)$
Fourth move on:	$(X_{II}; Y_{II})$

Since the Nash equilibrium is located in the fourth move, the game is to be classified in the B_{21} group.

Example V (The game of half of the average). In a classroom with n students, each is asked to choose a real number in the closed interval [0;1]. Indicating by x_i the number chosen by the ith student, his payoff will be:

$$\begin{cases} 0, & \text{if } x_i > s \\ 100, & \text{if } x_i = s \\ -100, & \text{if } x_i < s \end{cases}$$

where $s = (1/2n)\sum_{i=1}^{n} x_i$.

If each student were able to guess the average choice y_i of the remaining students, namely:

$$y_i = \sum_{j \neq i} x_j/(n-1)$$

he would be able to collect the 100-dollar bill by choosing

$$x_i = \frac{n-1}{2n-1}\, y_i \,. \tag{3.1}$$

The problem, of course, is that each student must choose x_i before knowing y_i. In the unique Nash equilibrium, all students would write $x_i = 0$ and each of them would receive a 100-dollar bill. What makes the game appear as a B_2 game is that choosing $x_i = 0$ is an imprudent strategy, at least in a large classroom. In fact, if somebody else makes a different choice, the Nash strategist, instead of collecting the 100-dollar premium, will have to pay a 100-dollar fine.

Since, in Eq. (3.1), $0 \leq y_i \leq 1$, any choice

$$x_i \geq \frac{n-1}{2n-1}$$

is a max-min play. The best max-min strategy is to turn the inequality into an equality, since it guarantees that the student will suffer no loss and still allows the possibility of a 100-dollar premium should all other students choose $x_j = 1$.

Let us now assume that the game is repeated, and that every student starts the first move with the best max-min strategy:

$$x_{i0} = \frac{n-1}{2n-1} .$$

Assuming that students choose their tth-move strategy so as to maximize their payoffs, given other students' choices in the preceding move:

$$x_{it} = \frac{n-1}{2n-1} y_{i,t-1} .$$

Since $y_{it} = x_{it}$,

$$x_{it} = \left\{ \frac{n-1}{2n-1} \right\}^t .$$

This is to say that the game of half of the average belongs to the B_{21} group. In fact, x_{it} converges to zero, that is, to the Nash strategy.

Example VI. Consider the bimatrix game:

$$\begin{array}{c} \\ X_I \\ X_{II} \end{array} \begin{array}{cc} Y_I & Y_{II} \\ \left(\begin{array}{c} (-20;-20) \\ (-15;15) \end{array} \right. & \left. \begin{array}{c} (15;-15) \\ (12;12) \end{array} \right) \end{array}$$

This is a highly intriguing B game. There are two Nash equilibria $(X_I;Y_{II})$ and $(X_{II};Y_I)$, the first very favorable to player X but highly unfavorable to player Y, the second highly rewarding to Y but adverse for X. Moreover, if each player tries to force his preferred Nash strategy, the outcome will be the worse of the worlds (X_I, Y_I) where both will maximize their losses. Common sense suggests that players should agree on choosing the max-min combination $(X_{II};Y_{II})$ that yields a payoff equal to 12 for both of them. Yet the common sense suggestion has nothing to do with the concept of Nash equilibrium, unless the game is repeated indefinitely. Moreover, if the game is repeated with the starting point $(X_{II};Y_{II})$ and if each player chooses his nth move strategy assuming that the other will repeat the strategy of the precedent move, the outcome will be $(X_{II};Y_{II})$ in odd periods and $(X_I;Y_I)$ in even ones. This means that we are facing a B_{22} game.

The formerly described approach to a Nash equilibrium in a repeated B_{21} game cannot escape an obvious criticism; each player chooses his nth strategy move on the basis of the false assumption that other players will repeat the strategies of the preceding move. The vein is the same as the classic critique of Cournot's oligopoly model as well as the attack of the new classic economics on the adaptive expectations hypothesis. Yet, if pushed to an extreme point, the critique strikes the very concept of a Nash equilibrium, at least in a game with few persons. In fact, rational players might try to achieve Stackelberg dominance; the result would be nothing but Stackelberg warfare, an inconsistent combination of individual decisions.

There are, of course, many other ways of describing what is a cautious approached to a Nash equilibrium in a repeated B game, and the preceding B_{21} dynamics is nothing but a conventional rule for distinguishing B_{21} from B_{22} games. (The distinction might change if a different rule were adopted.) The central problem, however, is that in B games, one cannot escape a dilemma: assumptions about other people's behavior are either false or imprudent. This is to say that false assumptions may be convenient, at least as long as they are more prudent than false.

4. A PRICE-SETTING GAME

Let us assume an economy with a continuum of non-storable goods, each one produced by an individual price-setter, and where the nominal output R is controlled by government. In line with monetary theory, one may assume that the Central Bank controls some monetary aggregate that determines R. The nominal output if pre-announced by a credible administration, but prices must be set simultaneously, each agent ignoring how others will decide, except for the fact that no good is expected to be priced above $P_{\max}$. This means that both R and $P_{\max}$ are common knowledge.

Production starts after prices have been set, according to consumers' orders. Since goods cannot be stored, this rules out the possibility of excess supplies. Supply shortages may indeed occur, but are not anticipated by consumers.

All individuals have the same utility function:

$$U_x = L_x^b \left\{ \int_0^1 q_{xy}^a dy \right\} 1/a \qquad (0 < a \leq 1/2;\ b > 0) \tag{4.1}$$

where U_x stands for utility, L_x for leisure time, and q_{xy} for the consumption of good y by individual x $(0 \leq x \leq 1)$.

The supply of good x equals the number of daily working hours by individual x:

$$S_x = 24 - L_x\,. \tag{4.2}$$

Hence, indicating by P_x the price of good x and by R_x the nominal income of individual x:

$$R_x = P_x S_x\,. \tag{4.3}$$

Individual x's budget constraint is expressed by:

$$\int_0^1 P_y q_{xy} dy = R_x . \tag{4.4}$$

Since nominal output equals the sum of individual incomes:

$$\int_0^1 R_x dx = R . \tag{4.5}$$

Let us first determine demand of good y by individual x. Since utility function (4.1) leaves no room for corner equilibria, marginal utilities must be proportional to prices:

$$a q_{xy}^{a-1} = \lambda_x P_y .$$

Taking into account budget constraint Eq. (4.4):

$$q_{xy} = \frac{R_x P^m}{P_y^{m+1}} , \tag{4.6}$$

where

$$m = \frac{a}{1-a} \tag{4.7}$$

and where the consumer price index P is determined by:

$$P^{-m} = \int_0^1 P_y^{-m} dy . \tag{4.8}$$

It should be noted that, since $0 < a \leq 1/2$,

$$0 < m \leq 1 . \tag{4.9}$$

Taking into account Eq. (4.5), the total demand for good y will be given by:

$$Q_y = \frac{R P^m}{P_y^{m+1}} . \tag{4.10}$$

Combining Eqs. (4.1), (4.6) and (4.7) and leaving aside supply shortages, individual x's utility will be expressed by:

$$U_x = L_x^b \frac{R_x}{P} \tag{4.11}$$

or, introducing Eqs. (4.2) and (4.3),

$$U_x = \frac{P_x}{P} S_x(24 - S_x)^b \,. \tag{4.12}$$

Let us now determine the supply S_x of good x. According to the hypotheses of the model, the individual x first sets P_x and then receives consumers' orders. The latter may be fully met or not, according to individual X's preferences. This is to say that S_x is chosen so as to maximize the right side of Eq. (4.12), taking P_x and P as given, and under the constraint $S_x \leq Q_x$. Easy calculations, combined with Eq. (4.10), yield

$$S_x = \min\left\{\frac{RP^m}{P_x^{m+1}}; \frac{24}{1+b}\right\} . \tag{4.13}$$

Market clearing requires

$$\frac{RP^m}{P_x^{m+1}} \leq \frac{24}{1+b} \tag{4.14}$$

for all $0 \leq x \leq 1$. One cannot guarantee *a priori* that this inequality will hold for all markets, ruling out the possibility of supply shortages. The fact, however, is that no individual considers the hypothesis of having his possibility consumptions being restricted by such shortages. Hence, Eq. (4.12) stands as individual x's notional utility, namely, the function he will try to maximize in the price-setting game. Introducing Eq. (4.13),

$$U_x = \begin{cases} \frac{RP^{m-1}}{P_x^m}\left[24 - \frac{RP^m}{P_x^{m+1}}\right]^b, & \text{if } \frac{RP^m}{P_x^{m+1}} \leq \frac{24}{1+b} \\ \frac{P_x}{P} b^b \left[\frac{24}{1+b}\right]^{1+b}, & \text{if } \frac{RP^m}{P_x^{m+1}} > \frac{24}{1+b} \end{cases} . \tag{4.15}$$

Individual x is assumed to know these expressions, and as easy calculations show, for a given P, the optimum price-setting rule makes:

$$P_x^{m+1} = \frac{1+b'}{24} RP^m \,, \tag{4.16}$$

where

$$b' = \frac{m+1}{m} b \,. \tag{4.17}$$

In a Nash equilibrium, all price-setters should follow Eq. (4.16) with perfect foresight on P. Taking into account Eq. (4.8), one immediately concludes that there exists a unique Nash equilibrium, where

$$P_x = P = \frac{1+b'}{24} R \,. \tag{4.18}$$

The problem of locating such an equilibrium is that price-setters play a B_2 game where the only *a priori* information on P is that $0 \leq P \leq P_{\max}$. According to the previously discussed paradigm, the game will start with max-min strategies. Calculations are made easy by our assumption $0 < m \leq 1$: according to Eqs. (4.15), for each P_x, U_x is a decreasing function of P. Hence, the max-min strategy is to use Eq. (4.16) taking $P = P_{\max}$. The use of simple, instead of mixed, strategies should not be a source of concern, since in our specific game:

$$\min_{P} \max_{P_x} U(P_x, P) = \max_{P_x} \min_{P} U(P_x, P)$$

which is to say that two-person, zero-sum translation of the price-setting game, as viewed by each individual, has a saddle-point equilibrium.

To sum up, in the initial move all producers set:

$$p_{x0}^{m+1} = \frac{1+b'}{24} R P_{\max}^{m}$$

and the observed general price level will be:

$$P_0^{m+1} = \frac{1+b'}{24} R P_{\max}^{m}.$$

Let us assume that the game is repeated and that price-setters use Eq. (4.16) assuming, in the tth move, $P = P_{t-1}$, where P_{t-1} stands for the general price level observed after the preceding move. Price dynamics will be described by

$$P_t^{m+1} = \frac{1+b'}{24} R P_{t-1}^{m}$$

implying that P_t will converge to the Nash equilibrium level described by Eq. (4.18). This is to say that the price-setting game under discussion is to be viewed as a B_{21} game.

Consistency of assumptions requires

$$P_0 \leq P_{\max}$$

or, equivalently,

$$P_{\max} \geq \frac{1+b'}{24} R$$

which is to say that general price level's upper bound cannot be less than the Nash equilibrium value. Accordingly, P_t will be a non-increasing sequence: prices will be cut down approaching the Nash equilibrium. Market clearing conditions will hold throughout the game, since

$$\frac{RP_t^m}{P_{xt}^{m+1}} \le \frac{RP_{t-1}^m}{p_t^{m+1}} = \frac{24}{1+b'} < \frac{24}{1+b}\,.$$

Let us now finally interpret the market-clearing equation, Eq. (4.14), since it will throw some light on our forthcoming discussion on income policies.

One should first note that

$$Q'_x = \frac{24}{1+b}$$

is the competitive supply of good x, namely, how much of that good would be supplied if individual x behaved as a price-taker. Competitive markets would equal supply and demand by making, according to Eq. (4.10),

$$\frac{RP^m}{P_x^{\prime m+1}} = \frac{24}{1+b} \qquad (4.19)$$

P'_x indicating the competitive price of good x. What Eq. (4.14) means is that excess demand for good x can only occur if individual x, underestimating the general price level P, sets P_x below what he would receive as a price-taker.

As a further step, let us calculate competitive equilibrium prices, namely, those that would clear all markets if all producers decided to behave as price-takers. Combining Eqs. (4.8) and (4.19):

$$P'_x = P' = \frac{1+b}{24}\,R\,. \qquad (4.20)$$

Competitive prices would be lower than Nash equilibrium prices determined by Eq. (4.18). This should not come as a surprise, since the price-setting model discussed above assumes a regime of monopolistic competition: individual x is the only producer of good x. As such, price controls can squeeze monopoly profit margins without creating supply shortages, as long as prices do not fall below competitive equilibrium levels. In short, monopolistic competition prices equal competitive prices plus a certain spread. How income policies may deal with such a spread is an essential point that will be explored in Section 6.

5. INERTIAL INFLATION REVISITED

The price-setting model discussed in the preceding section is based on heroic assumptions on utilities and production functions. Yet the central conclusions hold under much less restrictive hypotheses. All one needs to assume is the following:

1. There is a continuum of goods, each of them produced by a single individual price-setter.
2. The nominal output R is controlled by the government.
3. Prices must be set simultaneously, each agent ignoring how others will decide, except for the fact that no good is expected to be priced above a certain maximum level, nor below a certain minimum figure.
4. The general price level P is expressed by

$$P = g\left\{\int_0^1 z(x, P_x)dx\right\} \tag{5.1}$$

where $g(z(x, P_x))$ is an increasing function of P_x and where, for any $\lambda \geq 0$:

$$g\left\{\int_0^1 z(x, \lambda P_x)dx\right\} = \lambda g \int_0^1 z(x, P_x)dx \ .$$

5. Individual x's notional utility is a function of P_x, P and R. For each pair (P, R), there is an unique P_x that maximizes individual x's notional utility:

$$P_x = f_x(P, R)\,, \tag{5.2}$$

where $f_x(P, R)$ is a continuous real function, homogeneous of degree one, and increasing on both its variables. Moreover, for any positive P,

$$f_x(P, 0) = 0; \quad f_x(P, \infty) = \infty \ .$$

Let us indicate that

$$h(P, R) = g\left\{\int_0^1 z(x, f_x(P, R))dx\right\} \ . \tag{5.3}$$

It follows from the preceding assumptions that $h(P, R)$ is a continuous function, homogeneous of degree one, increasing in both its variables, and that for any positive P,

$$h(P, 0) = 0; \quad h(P, \infty) = \infty.$$

As a result, for any positive P there exists one unique positive R such that

$$P = h(P, R)\,. \tag{5.4}$$

Since $h(P, R)$ is homogeneous of degree one, this equation is solved by

$$R = \frac{P}{c}$$

where c stands for a positive constant.

The existence and uniqueness of a Nash equilibrium can now be proved immediately. In a Nash equilibrium, all price-setters must correctly locate P. According to the foregoing discussion, a Nash equilibrium is hit if and only if all price-setters behave according to Eq. (5.2) taking P as the solution of Eq. (5.4). This solution exists and is unique:

$$P = cR\,. \tag{5.5}$$

Moreover, if the nominal output is fixed and if price dynamics are described by

$$P_t = h(P_{t-1}, R), \tag{5.6}$$

the general price level will converge to its Nash equilibrium. In fact, since $h(P, R)$ is increasing in both its variables, and since $cR = h(cR, R)$, it follows that if $P_{t-1} > cR$, then

$$P_{t-1} = h\left(P_{t-1}, \frac{P_{t-1}}{c}\right) > h\left(P_{t-1}, R\right) > h(cR, R) = cR\,,$$

which is to say that $P_{t-1} > P_t > cR$. Prices will follow a decreasing bounded sequence, and hence converge to cR. A similar argument proves the convergence to the Nash equilibrium if $P_{t-1} < cR$. The convergence characterizes the price-setting game as a B_{21} game.

Let us now revisit the problem of inertial inflation. The starting point is a chronic inflation at a constant rate per period. For $t \leq 0$, the government has been expanding the nominal output at a constant rate r, namely

$$R_t = R_0(1 + r)^t \qquad \text{for } t \leq 0\,.$$

Price-setters have already adjusted by chronic inflation and its moving Nash equilibrium, by using Eq. (5.2) with $P_t = cR_t$ for $t \leq 0$. At the end of period $t = 0$, a new administration, whose credibility is beyond any doubt, announces nominal output stabilization; namely, it will make $R_t = R_0$ for $t \geq 1$.

In rational-expectations models, inflation would immediately stop, since new Nash equilibria would make $P = cR_0 = P_0$ for $t \geq 1$. The problem is that, even if all price-setters are convinced that the government will actually stabilize the nominal output at R_0, they may suspect that other prices will continue to increase. In fact, the only available information is that P_1 will fall in some point of the closed interval $P_0 \leq P_1 \leq P_0(1 + r)$.

If this was an A game, max-min strategies would set $P_{x1} = f_x(P_0, R_0)$ and inflation would stop immediately. Yet price-setting games are usually B_2 games, prudence requiring $P_{x1} = f_x(P_x, R_0)$ where $P_x > P_0$. This is enough to make $P_1 > P_0$. Inflation will continue in period one, although at a rate less than r. Assuming that the government keeps the nominal output unchanged at R_0, prices will eventually recede to the Nash equilibrium figure $P_0 = cR_0$. How quick or how slow will be the adjustment process is an open question, but the possibility of strong inertia leading to dismal recession cannot be ruled out.

As an example, in the price-setting model discussed in the preceding section, the starting point is

$$P_0 = \frac{1+b'}{24} R_0$$

Max-min strategies for period one require price-setters to use Eq. (4.16) assuming $P = P_0(l + r)$. This would yield

$$p_1^{m+1} = \frac{1+b'}{24} R_0 \big(P_0(1+r)\big)^m$$

or equivalently

$$P_1 = P_0(1+r)\frac{m}{m+1} \, .$$

Prices would then gradually fall according to the difference equation:

$$P_t^{m+1} = \frac{1+b'}{24} R_0 P_{t-1}^m = P_0 P_{t-1}^m$$

converging to the Nash equilibrium P_0.

The preceding discussion provides a compromise between adaptive and rational expectations. Formally, the above-described price dynamics are very similar to that of an old-fashioned adaptive-expectations model. Yet the economic hypotheses that lead to the gradual approach to the Nash equilibrium are completely different from those of a traditional model of backward-looking expectations, insensitive to changes in policy rules. First, the whole discussion skips the very idea of "expectations," a somewhat vague concept often used to escape a much more complex issue, namely, how strategic interdependence problems should be dealt with. Second, inflation falls below the historical rate r in period one, not because of recession, but simply because price-setters actually believe that the government will stabilize nominal output at R_0. The reason why this is not enough to promote immediate price stabilization is that price-setters have no guarantee that other prices will stop increasing.

One may argue that the ensuing price dynamics, where price-setters assume that the general price level in period t will equal that of the previous period, is nothing but an "*ad hoc*" assumption that implicitly brings back on stage a very old-fashioned style of adaptive expectations. Moreover, it commits economic agents to systematic errors, since throughout the adaptation period the general price level is always overestimated. Why this criticism is not to be taken too seriously has been explained before. Decision-making in B_2 games cannot escape a dilemma: it must be based either on false or on imprudent assumptions. Which is to say that false assumptions may be a convenient reference point, as long as they lead to prudent errors.

In short, adaptive expectations should be reconsidered as a useful tool to describe a cautious approach to the Nash equilibrium of a B_{21} game. As such, for each specific problem they should be properly modeled, taking into account individual payoffs, which obviously depend on economic policy rules. Moreover, they should not only describe how players converge to the Nash equilibrium, but how this convergence can be the outcome of a cautious approach to uncertainty under strategic interdependence.

"*Ad hoc*" assumptions cannot be escaped in adaptive-expectations models. Uncomfortable as they might appear, they simply describe a psychological framework in a B_2 repeated game. They should be accepted with some fair play and humility, since Game Theory has proved unable to solve the question: what is "*ex ante*" rationality in a B_2 game?

6. THE ROLE OF INCOME POLICIES

The foregoing discussion provides the rationale for income policies: governments should play the role of the Walrasian auctioneer, speeding up the location of Nash equilibria, that is, using the visible hand to achieve what rational-expectations models assume to be performed by the invisible hand. In the price-setting model of the preceding section, if the government, besides stabilizing the nominal output at R_0, decided to freeze all prices at their levels in period $t = 0$, inflation would stop immediately, with no recession and no shortages, simply because price-setters would be assured that neither the government would expand nominal output nor would other agents continue to increase prices. In short, price-setters would behave like Nash strategists since they would be assured that other players would act in the same way.

Except for the game's theoretical framework, there is nothing new in this interpretation of wage-price inertia and how it can be overcome by income policies. Keynes' explanation (in Chapters 2 and 19 of the *General Theory of Employment*) of why workers resist a nominal wage cut is very much in the same vein. Keynes also perceived that a nominal wage cut could be much less painfully achieved by a decree of a totalitarian government than by the working of free market forces. Wage inertia, as described by Keynes, is pushed to the extreme point where all players implicitly insist on max-min strategies in a repeated game, which turns out to be wise behavior in a game with no Nash equilibrium. The Keynesian inertia may appear as too rigid and too simplistic fifty years after the *General Theory* was published. Yet it reveals a formidable insight in a field that was only to be developed by von Neumann and Morgenstern almost ten years later, Game Theory.

Keynes' perception was right, in the sense that the central function of income policies is not to constrain individual decision-making, but to clear up uncertainties in a B_2 game, telling each actor how the others will play. This, incidentally, dismisses a traditional argument against income policies, i.e., that governments are not better equipped than free markets to identify individual Nash strategies. In fact, the central problem in a B_2 game is not to discover such strategies, but rather to coordinate their simultaneous playing.

This also explains why, in a second stage of a stabilization program, wage-price controls should be removed gradually, in successive sectorial steps, and not in a one-shot manner. In fact, even if income policies are successful enough to bring the economy to a Nash equilibrium, there is no way of conveying such information to the participants in the game. Each player finds himself in equilibrium but does not know if the same applies to other players. If controls are lifted sector by sector, players will realize that no participant in the game will increase prices even when allowed to do so. On the other hand, the one-shot approach would simply bring back the uncertainty of individual players as to what other players will do, perhaps triggering large, defensive, wage-price increases.

Of course, the chances of hitting a Nash equilibrium through income policies are extremely remote. Staggered wage-setting and price-setting may be formidable obstacles to a wage-price freeze, since they imply that there is no calendar date when relative prices are in equilibrium. This is to say that, before being frozen, wages and prices must be realigned, an issue that will be discussed in the next two sections. Yet, even if the synchronization problem is solved, the fact that wage-price controls yield some supply shortages should not come as a surprise. In fact, policy-makers can easily detect when relative prices are visibly out of equilibrium, but can never perfectly identify when the equilibrium has been reached. Even if they could, and then decree the wage-price freeze, the following week they would be wrong, since equilibrium prices move up and down with the time shifts in demand and supply.

The central question in a program intended to fight a big inflation is to choose what is preferable in terms of welfare costs: a few product shortages, that eventually may be overcome by imports, or massive unemployment, which is nothing but a shortage of jobs. From this point of view, objections to income policies should not be taken too seriously, at least when the problem is to fight a big inflation with strong inertial roots. This is all the more so because income policies can be managed with appropriate flexibility, substituting price administration for price freezes.

A more fundamental contention is that the temporary success of income policies may lead policy-makers to forget that price stability can only be sustained with aggregate demand discipline. The temptation is to misread price stability and produce a boom. The misleading signals are a true risk, as is known from uncountable examples in history.

The model presented in Section 4 indicates where these misleading signals may come from. Let us assume that, in period $t = 0$, the government freezes prices at P_0 without preventing further expansion of nominal output. Initially, the economy might experience a euphoric expansion in consumption, employment and output, as long as the increase in nominal income is matched by a cut in monopolistic competition profit margins. Then, in line with a long list of income policy failures, the program will collapse because of generalized shortages. In the model in Section 4, generalized shortages would occur if the nominal output were expanded beyond

$$\frac{R}{R_0} = \frac{1+b'}{1+b}$$

the right side in the above equation corresponding to the ratio between monopolistic competition and perfectly competitive equilibrium prices.

The fact that after a price freeze, some nominal output expansion is possible without yielding supply shortages helps to explain why price controls are usually so successful in the very short run. Price-setters are converted into price-takers, thus being forced to accept some squeeze in their profit margins. As long as the latter do not fall below perfect-competition margins, nominal output expansion leads to both output growth and increased welfare. Yet this euphoric start is nothing but an income policy trap. In fact, once price controls are lifted, producers will restore their previous margins by increasing prices and reducing quantities. Moreover, one ominous possibility is that policy-makers, misinterpreting price signals, might become convinced that once prices have been frozen, price stability can be reconciled with sustained economic growth through relentless aggregate demand expansion. The rise and fall of the Cruzado Plan should serve as a case on how naive policy-makers backed by highly sophisticated economic advisors may fall into the income policies trap.

Assuming that policy-makers are wise enough to understand that income policies should be the counterpart of a fiscal austerity plan that keeps aggregate demand under control, there is still a challenging issue to be solved, namely, how income policies should be designed. While a low-calorie menu seems enough to break inertia after a true hyperinflation, when prices expand by factors of millions or billions a year, much more sophisticated cooking is required when the problem is to stop a three- or four-digit annual inflation.

In fact, in a true hyperinflation (where, according to Cagan's definition, prices continuously increases above 50% a month), money not only loses its function as a store of value, but also ceases to be serviceable as a unit of account, wages and prices being usually set in some foreign currency. In this case, exchange-rate stabilization serves as a universal incomes-policy tool. The classic success story is the German stabilization plan of November 1923, where inflation was exorcized by a successful combination of supply-side and demand-side policies. The latter was the creation of an independent Central Bank that could not print money to finance the public sector deficit. The former was stabilizing the exchange rate.

Complications arise when inflation is high enough to trigger widespread backward-looking indexation arrangements (since instantaneous price-index links, fashionable as they might have been in economic literature, never proved to be feasible), but not to the point of quoting wages and prices in foreign currencies. The unit of account, in such cases, is neither the domestic currency, nor the exchange rate, but some lagged price index. Here, imbalanced income policies may lead to a disaster, even when matched by fiscal austerity, as happened with Chile in 1979. The exchange rate was pegged at 39 pesos per dollar, the fiscal deficit was turned into a surplus, but wages continued to be indexed for past inflation. As a result, the only force to dampen inflationary inertia was real exchange-rate overvaluation. Annual inflation rates actually fell from 40% in 1978 to 9% in 1981. Yet exchange rate overvaluation led to dismal recession and to a wasteful increase in foreign debt. In short, income policies were seriously mismatched. In fact, once the exchange rate was pegged, wages should have been frozen, perhaps under the political umbrella of a temporary price freeze.

One may argue that once the exchange rate has been fixed and once nominal wages have been frozen, there is no need to freeze prices, provided both aggregate demand and monopoly profit margins are kept under control. The argument is perfectly correct from the technical point of view, except that it can be reversed. If prices, exchange rates, aggregate demand and monopoly profit margins are properly controlled, there is no need to freeze wages. In fact, redundancy may be necessary to make a price-stabilization package politically palatable, since no part may be willing to behave as a Nash player in the absence of a guarantee that others will follow the same pattern. In fact, the political marketing of income policies is largely dependent on its visible content, that is, on price freezes. One should recognize, however, that price freezes may become the poison pill of stabilization packages. In fact, once prices are frozen, inflation falls to zero by definition. Yet the risk remains that price inflation may be replaced by queue inflation, not necessarily a better achievement in terms of social welfare, nor a stable arrangement capable of preventing a swap of future for present inflation.

7. STAGGERED WAGE SETTING AND INFLATIONARY INERTIA

Let us now examine the complications created by staggered wage setting. In the following discussion while labor is assumed to homogeneous, workers are uniformly distributed among a continuum of classes, one for each real number $0 \leq x < 1$. Nominal wages of class x are reset at time $x + n$, for every positive, zero or negative integer n. This is to say that individual nominal wages move by steps of time length $T = 1$ (the step curve $S(\tau)$ in Figure 1), but that adjustment dates for different classes are uniformly spread over time.

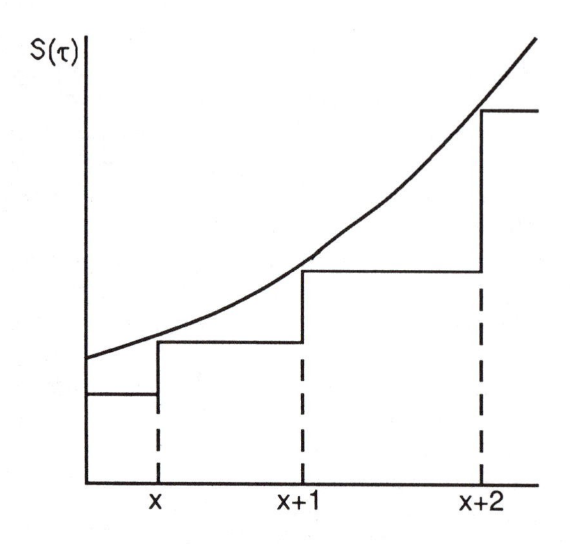

FIGURE 1 The step curve $S(\tau)$.

Under the above assumptions, the average nominal wage at time t will be given by the shaded area in Figure 2:

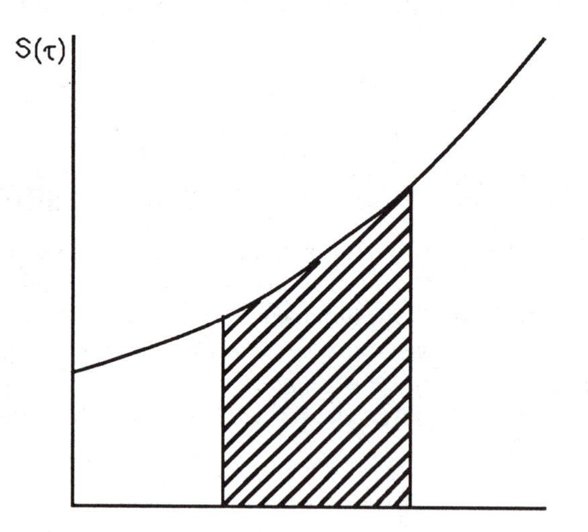

FIGURE 2 The average nominal wage at time t.

$$W(t) = \int_{t-1}^{t} S(\tau)d\tau\,, \tag{7.1}$$

$S(\tau)$ indicating the individual nominal wage of the class with resetting data $\tau(t-1 \leq \tau < t)$. To make sure that the integral on the right side of Eq. (7.1) does exist, we shall assume $S(\tau)$ to be a continuous function of τ.

Let us further assume that the price level $Q(t)$ is determined by multiplying unit labor costs by a constant mark-up factor:

$$Q(t) = (1+m)bW(t)$$

where m is the profit margin and b the labor input per unit output. In the following discussion we shall overlook supply shocks as well as both cyclical and long-term changes in productivity, thus treating b as a constant. This makes the average real wage $W(t)/Q(t)$ a constant:

$$W(t) = zQ(t) \tag{7.2}$$

where $z^{-1} = (1+m)b$.

Staggered wage setting in an inflationary economy introduces two different real wage concepts, the peak and the average. The peak $k(\tau) = S(\tau)/Q(\tau)$ (OP in Figure 3) is the individual purchasing power of the class with resetting date τ, immediately after the nominal wage increase. The average,

$$z(\tau) = k(\tau)\int_{\tau}^{\tau+1} \frac{Q(\tau)}{Q(\sigma)}\,d\sigma \tag{7.3}$$

indicated by the shaded area in Figure 3, is the worker's average real wage in the period where his nominal wage remains fixed. For example, if $Q(\sigma) = Q(0)e^{\pi\tau}$, that is, if the inflation rate $Q'(\sigma)/Q(\sigma)$ is a constant π, the peak/average ratio will be given by

$$\frac{k(\tau)}{z(\tau)} = \frac{\pi}{1-e^{-\pi}}\,, \tag{7.4}$$

an increasing function of π that, for small inflation rates, can be approximated by

$$\frac{k(\tau)}{z(\tau)} = 1 + \frac{\pi}{2}\,.$$

What really matters for both employers and employees is the average real wage $z(\tau)$. Yet the only objective element in each wage contract is the real wage peak $k(\tau)$. That the peak should be set so as to achieve a target average $z(\tau)$ seems pretty obvious. Under perfect foresight one may plausibly assume that $k(\tau)$ is calculated so

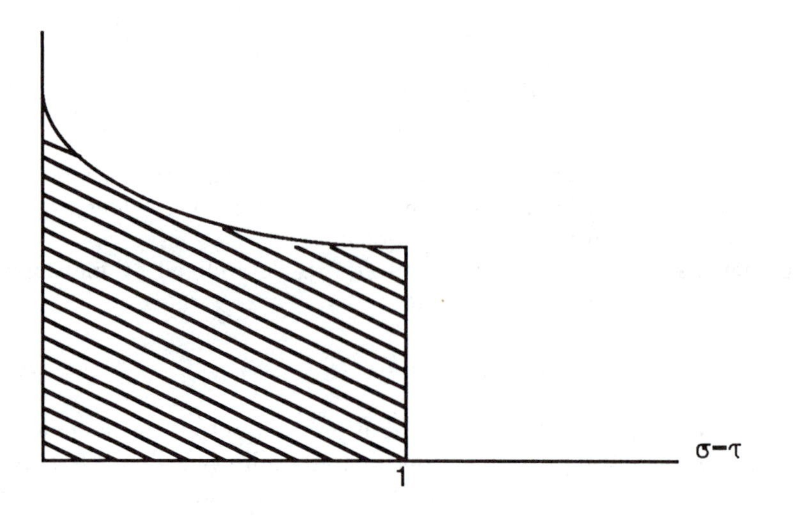

FIGURE 3 In staggered wage setting, the worker's average real wage in the period when his nominal wage remains fixed.

as to make $z(\tau) = z$ in the case of full employment, unemployment forcing workers to accept $z(\tau) < z$ and excess demand for labor leading to $z(\tau) > z$. Complications arise when strategic interdependence is brought on stage. In fact, the optimum $k(\tau)$ depends on both the nominal output path and the cost-of-living curve in the time interval $[\tau, \tau + 1]$. Even if the former is credibly announced by the government, the problem remains that the cost-of-living path will depend on how other labor groups will react in terms of nominal wage increases. In fact, according to Eq. (7.1) and (7.2)

$$zQ'(t) = W'(t) = S(t) - S(t-1)\,. \tag{7.5}$$

Let us first explore the mathematics of staggered wage setting. Three basic theorems are proved in the Appendix:

THEOREM 1 (the constant inflation-rate theorem): if, for $t \geq 0$, $Q(t) = Q(0)e^{\pi t}$, where π is a positive constant, then $z(\tau)$ converges to z and the peak/average ratio to:

$$\lim_{\tau \to \infty} \frac{k(\tau)}{z(\tau)} = \frac{\pi}{1 - e^{-\pi}}$$

THEOREM 2 (the peak/average ratio theorem): if, for all $t \geq 0$, $k(t) = rz$, r indicating a constant greater than one, then the inflation rate converges to π, where

$$r = \frac{\pi}{1 - e^{-\pi}}$$

THEOREM 3 (the non-neutrality theorem): if the inflation rate decreases for $t-1 \leq \tau \leq t+1$, then $z(\tau) > z$ for some $t-1 \leq \tau \leq t$.

The first theorem explains why chronic inflation eventually leads to formal or informal indexation. If prices increase at approximately stable rates, real wage peaks will be recomposed at fixed time intervals. In practice this is achieved by periodic nominal wage increases for cost-of-living escalation:

$$S(\tau) = S(\tau - 1)\frac{Q(\tau)}{Q(\tau - 1)}$$

making $k(\tau) = k(\tau - 1)$.

The second theorem proves the converse. Since indexation periodically resets real wage peaks above the real average the economy can afford to pay, inflation becomes predominantly inertial, being no longer the result of too much money chasing a few goods. The problem at this stage is no longer to explain why prices actually increase, but to explain how inflation can accelerate or decelerate.

Excess demand for labor is one possible source of inflation acceleration by making $k(\tau) > k(\tau - 1)$, namely, by substituting indexation-plus for simple cost-of-living escalation. Increased inflationary expectations can do the same, either because economic agents anticipate more expansive monetary policies or because each labor group expects other labor groups to move from simple indexation to indexation-plus. Adverse supply shocks, lowering the equilibrium real-wage average z, are another cause of inflation acceleration in an indexed economy. Recomposing the same real-wage peaks now means increasing the coefficient r in Theorem 2, thus lifting the equilibrium inflation rate. Finally, if labor unions manage to reduce the wage-adjustment interval, say, from twelve to six months, what was previously the annual inflation rate will become the six-month cost-of-living increase, as occurred in Brazil in late 1979. This, once again, is a consequence of the peak/average ratio theorem, where π is the equilibrium inflation rate for the wage-resetting time interval.

One might argue that the foregoing analysis overstates the problem of inflationary inertia by ignoring that the length of nominal wage contracts is a decreasing function of the inflation rate. And that, once this fact is taken into consideration, a wage escalation clause is not a serious obstacle to fight inflation. A virtuous circle can be put in motion by a slight decrease in the inflation rate, which in turn will increase the length of the nominal wage contracts, further lowering the inflation rate and so on. The essence of the argument is that wage indexation at constant time intervals is not a convincing arrangement, since it makes average real wages a decreasing function of the inflation rate, a superior scheme being the trigger-point system where wages are adjusted at fixed-inflation-rate thresholds.

Both trigger-point and fixed-time-interval indexation schemes are found in inflationary economies, and one may argue that the second arrangement is nothing but a practical implementation of some implicit trigger-point, since wage adjustment intervals usually decrease as the inflation rate jumps up. Two points, however, should be stressed. First, once annual inflation rates reach a high two-digit figure or more, trigger points become highly impractical, since nominal wages can be adjusted monthly, quarterly or every six months, but not at time intervals such as two months, seventeen days and three hours. Second, the response of the nominal wage contract length to changes in the inflation rate tends to be highly asymmetric. If inflation accelerates, employers and employees may naturally agree on shorter indexation periods, leading to further inflation acceleration, and paving the road to hyperinflation. Yet, once inflation falls, strategic interdependence complications emerge again. No labor union will take the lead in accepting an extension of the indexation interval without the guarantee that other labor unions will do the same. Summing up, the implicit trigger-point argument shows that inertia is an obstacle to inflation deceleration, but not to inflation acceleration.

Turning back to our staggered-wage-setting model, let us assume that for $t \leq 0$ inflation runs at a constant rate π with equilibrium wage indexation:

$$Q(\tau) = Q(0)e^{\pi\tau} \qquad (\tau \leq 0), \tag{7.6}$$

$$S(\tau) = z\,\frac{\pi}{1-e^{-\pi}}\,Q(\tau) \qquad (\tau \leq 0). \tag{7.7}$$

At time $t = 0$, the government announces a stabilization program eventually intended to achieve zero inflation and full employment. This requires some spontaneous or managed de-indexation scheme where all peak/average ratios in nominal wage contracts are eventually reduced to one. Because of both strategic interdependence and lack of synchronization of wage adjustment dates, de-indexation cannot escape three complications.

First, in the absence of income policies, the government can, at best, determine the nominal output path, but not the price path, which is entirely dependent on nominal wage increases. Why then, in the absence of unemployment, any labor group should take the lead in accepting lower real-wage peaks, can only be explained by imprudent playing that only yields perfect foresight if the other players are equally imprudent. In a word, staggered wage setting with high inflation rates makes a strong case in favor of income policies.

Second, even if an ideal policy mix leads to perfect foresight, the price path for $t \geq 0$ must be chosen so as to make $z(\tau) \geq z$ for all labor groups. In fact, under perfect foresight, workers will only accept $z(\tau) < z$ in case of unemployment. The neutrality dream, a declining inflation rate path with $z(\tau) = z$ for all, is a mathematical impossibility according to Theorem 3. Yet, as will be shown below, perfect foresight price paths that yield $z(\tau) > z$ during the whole transition phase do exist. The problem is that they may be rather complicated, and that simpler paths may imbalance the real-wage structure.

Third, unless nominal wages can be cut, gradualism is inevitable, some residual price increase being necessary to reduce from peak to average the purchasing power of the labor groups adjusted just before the announcement of the stabilization program. Prices can only be stabilized at a level $Q(f)$ such that

$$\frac{Q(f)}{Q(0)} \geq \frac{\pi}{1-e^{-\pi}}\,. \tag{7.8}$$

As an example, if the inflation rate runs at 300% a year before the announcement of the stabilization plan and if wages are adjusted every six months (which makes $\pi = \ln 2 = 0.6931$), the residual price increase must be at least

$$\frac{Q(f)-Q(0)}{Q(0)} - 38.63\%\,.$$

Even if normal wages can be cut, a shock treatment intended to make $Q(t) = Q(0)$ is an awkward policy choice. In fact, according to Eq. (7.5), this would require

$$zQ'(t) = W'(t) = S(t) - S(t-1) = 0 \quad \text{for all } t \geq 0$$

which, with stable prices, would perpetuate a disequilibrium wage structure, where nominal wages of class x $(0 < x \leq 1)$ would be frozen at

$$S(x) = \frac{\pi e^{(x-1)}}{1-e^{-\pi}}\, z\, Q(0)$$

standing below the equilibrium level $zQ(0)$ for $e^{x} < \pi/e^{\pi} - 1$, and above $zQ(0)$ for higher values of x.

An attractive incomes-policy arrangement is the D-day proposal of former Brazilian Minister of Finance Octavio Gouveia de Bulhões. Nominal wages would be adjusted at the contractual dates for price increases before the announcement of the stabilization plan, but not for the residual inflation following that announcement. In our model, this would imply $S(t) = S(0)$ for all $t \geq 0$. Prices, according to Eqs. (7.2) and (7.5) would move up so that

$$Q'(t) = \frac{\pi}{1-e^{-\pi}}(1 - e^{\pi(t-1)})Q(0) \quad (0 \leq t \leq 1)$$

namely, along the curve

$$Q(t) = \frac{Q(0)}{1-e^{-\pi}}(1 + \pi t - e^{\pi(t-1)}) \quad (0 \leq t \leq 1)\,,$$

$$Q(t) = \frac{\pi}{1-e^{-\pi}}Q(0) \qquad (t \geq 1)\,.$$

The fact that the inflation gradually declines from π to zero without requiring nominal wage cuts and that prices stabilize with $z(\tau) = z$ for all labor classes after $t = 1$, only displays part of the attractiveness of the D-day proposal. A still more interesting conclusion is that in the transition period, namely, for $-1 < \tau < 1$, the average purchasing power in every individual wage cycle is greater than the equilibrium level z. For contracts signed before the D-day and that mature after the beginning of the stabilization program, the conclusion is obvious. Such contracts would yield an average purchasing power z if the inflation rate was kept equal to π. Since the inflation rate declines after $t = 0$, it follows that $z(\tau) > 1$ for $-1 < \tau \leq 0$.

For $0 < \tau < 1$ and since $s(\tau) = k(\tau)Q(\tau) = S(0)$,

$$z(\tau) = S(0) \int_{\tau}^{\tau+1} \frac{1}{Q(\sigma)} \, d\sigma \, .$$

Hence,

$$z'(\tau) = S(0) \left\{ \frac{1}{Q(\tau+1)} - \frac{1}{Q(\tau)} \right\} .$$

Since $Q(\tau+1) > Q(\tau)$ for $0 < \tau < 1$, it follows that $z(\tau)$ is a decreasing function of τ for $0 < \tau < 1$. Since $z(1) = z$, one concludes that $z(\tau) > 1$ for $0 < \tau < 1$. The possibility of having $z(\tau) > 1$ in all individual wage cycles while, at any time, the average real wage of all workers stands at $W(t)/Q(t) = z$ should not be taken as a paradox. In fact, we are dealing with overlapping wage cycles, and the fact that $z(\tau) > 1$ for $-1 < \tau < 1$ does not imply that the purchasing power of all labor groups is greater than z in the transition period $0 \leq t \leq 1$.

Assuming perfect foresight, the D-day price path might be understood as a rational-expectations response to the announcement of a policy that would gradually reduce the rate of expansion of nominal output from τ to zero. In fact, once strategic interdependence is taken into account, one can hardly believe that markets would spontaneously produce any arrangement such as the D-day wage-adjustment rule. It may be hard to implement, even as an incomes-policy decision, since labor unions are not expected to understand the mathematics of staggered wage setting. It may be politically much easier to dampen inflation through a fractional indexation arrangement

$$\frac{S(t) - S(t-1)}{S(t-1)} = a \, \frac{Q(t) - Q(t-1)}{Q(t-1)} \qquad (0 < a < 1)$$

technically inferior to the D-day proposal, but that can be understood without much intellectual sophistication.

The foregoing analysis describes how staggered wage setting complicates anti-inflationary policies, the same conclusions applying to staggered price setting. The central problem is that, since adjustment dates are not synchronized, instant wage-price stabilization becomes virtually impossible. In fact, at any instant some wages and prices are too high, some too low, equilibrium being reached across time by continuous rotation of relative positions. Of course, a tempting proposal is to immediately realign all wages and prices to equilibrium, cutting some and increasing others, depending on how recent the last adjustment was. Yet in most cases this will be found illegal, since it violates outstanding contracts. Moreover, it would involve nominal wage cuts, which politicians consider a nightmare. Gradualism then becomes inevitable, the D-day proposal indicating the quickest possible achievement.

Still, there is one escape route for a shock treatment, that of monetary reform. In fact, a currency change is not necessarily limited to a cut of a certain number of zeros in the country's monetary unit. It may well include a number of rules that establish how contracts should be translated from the old currency into the new one. Recent experiences such as the Austral Plan of Argentina and the Brazilian Cruzado Plan indicate how an imaginative use of such translation rules can lead to immediate relative wage and price synchronization.

8. THE ROLE OF MONETARY REFORMS

Attempts to stop a big inflation are often coupled with monetary reforms for two well-known reasons: first, as a political move to signal a new age of price stability; second, as a practical action to increase the purchasing power of the currency unit, which is usually achieved by cutting three, six or even twelve zeros in the old money bills. Needless to say, changing names and cutting zeros is a fruitless exercise unless fiscal austerity is imposed and the inflationary expectations abated. As in the case of income policies, the list of failed monetary reforms is too long to be ignored.

How a monetary reform can erase the inflationary psychology is an intriguing question, since it involves a contradictory policy answer. Incomes policies may be essential to break inertia, but economic agents must also be convinced that the Central Bank will stop fueling inflation by money printing. Yet once the expected rate of inflation falls dramatically, fiat money must be generously printed for a period of time to meet the increase in money demand. The classic rule of thumb adopted by the German stabilization program of November, 1923, and borrowed by the Argentine Austral Plan, i.e., prohibiting money creation to finance public-sector deficits, has a strong appeal, since it suggests that the money supply will be properly kept under control in the long run. How to bridge short-term problems however, when renewed confidence in the domestic currency (based on the assumption that it will be no longer generously printed) requires a lot of additional printing, remains a complicated policy issue? Policy-makers must walk on a sharp knife-edge, avoiding so quick a reliquefaction that might re-ignite inflationary expectations, as well as one so slow that it would lead to recession caused by a credit crunch.

The best solution might well be to set targets for total financial assets held by the public (M_4), allowing M_1 to expand as long as other financial assets decline. Yet, since M_4 statistics do not include foreign currency held by the public, the rule must be amended in case there is a significant shift from dollar assets to domestic currency securities. Moreover, when the inflation rate falls from 400% or 2,000% a year to virtually zero, there is no guarantee that the demand for financial assets will be kept unchanged in real terms. Hence, besides tracking M_4 as a basic guideline, the Central Bank should keep a close eye on what is actually happening to nominal GNP.

Monetary reforms can also perform a very important function, that of dissipating the Fischer-Taylor inertia. As previously discussed, a serious obstacle to overnight price stabilization is that outstanding non-indexed contracts reflect old inflationary expectation. Honoring such contracts when the inflation rates fall dramatically implies unanticipated wealth transfers between contracting parties that may lead to bankruptcies and recession. Of course, high inflation rates limit non-indexed contracts to short maturities, but even so the transfer effect may be sizeable. Let us assume, for instance, that the inflation rate falls abruptly from 15% a month to zero, as occurred in Brazil on February 28, 1986. In the absence of a monetary reform, a promissory note maturing three months after the stabilization day would cost the borrower a 52% unanticipated real surcharge. As to staggered wage-setting, the obstacles it creates to immediate price stabilization were extensively discussed in the preceding section.

Monetary reforms can solve all these problems created by unanticipated stabilization and by lack of synchronization by establishing appropriate translation rules. In fact, once a new currency is substituted for the old one, the law must state how outstanding contracts should be rewritten in the new monetary unit. The imaginative innovation of the Argentine Austral Plan, borrowed by the Brazilian Cruzado Plan, was to recognize that there is no legal or economic reason why the translation rule for currency bills should hold for other obligations and contracts. In the case of non-indexed liabilities, the natural way to eliminate the effects of unanticipated stabilization is to establish conversion rates declining by a daily factor corresponding to the old inflation rate. As an example, the Cruzado reform immediately converted cruzeiro bills and demand deposits into cruzados by a cut of three zeros. Conversion rates for cruzeiro-denominated liabilities with future maturities were set to decline by a daily factor, starting February 28, of 1.0045. Thus, a promissory note of one-million cruzeiros maturing April 30, 1986 was worth 770.73 cruzados, not 1,000 cruzados. This specific translation rule was borrowed from the Argentine Austral Plan. Neither Argentina nor Israel has to face problems of staggered wage setting, since in both cases wage adjustment dates were already synchronized. Brazil did, and the true innovation of the Cruzado Plan was to convert wages from cruzeiros into cruzados by computing their average purchasing power of the last six months with an increase of 8%. This actually meant reducing the nominal wages of the labor groups adjusted in January and February, 1986, and in all cases increasing wages substantially below the past inflation rate. The 8% bonus was intended to offset the inflationary effect of a number of price increases which were to be enforced on February 28, 1986, so that the plan was to synchronize all real wages at their average level of the previous six months, since the length of nominal wage contracts was exactly six months. At the last moment the government decided not to enforce these sectorial price increases, so that average real wages were actually increased by 8%. This may have been the original sin of the Cruzado Plan.

APPENDIX

THE MATHEMATICS OF STAGGERED WAGE SETTING

We shall now prove the three basic theorems on the mathematics of staggered wage setting mentioned in the preceding text. Key concepts and relations are provided by the following equations:

$$W(t) \int_{t-1}^{t} S(\tau)d\tau\,, \tag{A.1}$$

$$W(t) = zQ(t)\,, \tag{A.2}$$

$$k(\tau) = \frac{S(\tau)}{Q(\tau)}\,, \tag{A.3}$$

$$z(\tau) = S(\tau)\int_{\tau}^{\tau+1} \frac{1}{Q(\sigma)}d\sigma\,. \tag{A.4}$$

From Eq. (A.4) it follows that, if $Q'(\sigma)/Q(\sigma) = \pi$ is a constant, then $Q(\sigma) = Q(\tau)e^{\pi(\sigma-\tau)}$, and

$$z(\tau) = k(\tau)\frac{q - e^{-\pi}}{\pi}\,. \tag{A.5}$$

THEOREM 1 (the constant inflation-rate theorem): if, for $t \geq 0$, $Q(t) = Q(0)e^{\pi t}$, π indicating a positive constant, then

$$\lim_{t\to\infty} z(t) = z$$

and

$$\lim_{t\to\infty} \frac{k(t)}{z(t)} = \frac{\pi}{1 - e^{-\pi}}$$

PROOF Taking derivatives in both sides of Eq. (A.1):

$$W'(t) = zQ'(t) = S(t) - S(t-1)\,.$$

Dividing by Eq. (A.2),

$$\pi = \frac{Q'(t)}{Q(t)} = \frac{W'(t)}{W(t)} = \frac{S(t) - S(t-1)}{zQ(t)} = \frac{1}{z}\left\{ k(t) - \frac{Q(t-1)}{Q(t)}\, k(t-1) \right\}$$

or, since $Q(t) = Q(t-1)e^{\pi}$,

$$k(t) - e^{-\pi} k(t-1) = \pi z.$$

Since $o < e^{-\pi} < 1$, any solution of this difference equation converges to

$$\lim_{T \to \infty} k(t) = \frac{z\pi}{1 - e^{-\pi}}\,.$$

It follows from Eq. (A.5) that

$$\lim_{t \to \infty} z(t) = z$$

and, as a consequence,

$$\lim_{t \to \infty} \frac{k(t)}{z(t)} = \frac{\pi}{1 - e^{-\pi}}\,.$$

Before dealing with the peak/average ratio theorem, let us prove:

LEMMA 1.: if $|\, r \,| < 1$, any solution of the differential-difference equation:

$$X'(t) = r\big(X(t) - X(t-1)\big) \qquad (A.6)$$

converges to a constant X.

PROOF For every integer n, let us define:

$$v_n = \max\left\{ \mid X'(t) \mid \mid n-1 \leq t \leq n \right\}.$$

If $\mid X'(\tau) \mid = v_n$, then Eq. (A.6), combined with the Lagrange theorem, implies:

$$v_n = \mid X'(\tau) \mid = \mid rX'(\tau - \theta) \mid$$

where $0 < \theta < 1$. Since $\mid X(\tau - \theta) \mid \leq \max[v_n, v_{n-1}]$ and since $\mid r \mid < 1$, this implies $v_n \leq \mid r \mid \max[v_n, v_{n-1}]$, or equivalently

$$v_n \leq \mid r \mid v_{n-1} \,. \tag{A.7}$$

It follows that

$$\lim_{t\to\infty} \mid X'(t) \mid \leq \lim_{n\to\infty} v_n \leq \lim_{n\to\infty} v_1 \mid r^{n-1} \mid = 0.$$

According to Eq. (A.7), for any $0 \leq s < 1$

$$\sum_{n=1}^{\infty} \mid X'(s+n) \mid \leq \sum_{n=1}^{\infty} v_n \leq v_1 \frac{1}{1- \mid r \mid} \,.$$

This implies that the series

$$f(s) = \sum_{n=1}^{\infty} X'(s+n)$$

converges, since it is absolutely convergent. From Eq. (A.6) it follows that

$$\lim_{n\to\infty} X(s+n) = X(s) + \frac{f(s)}{r} \,.$$

To complete the proof of the lemma, it remains to prove that if $0 < s < s' < 1$, then $X(s') + (f(s'))/r = X(s) + (f(s))/r$, or equivalently that

$$\lim_{n\to\infty} [X(s'+n) - X(s+n)] = 0 \,.$$

Using once again Lagrange's theorem

$$X(s'+n) - X(s+n) = (s'-s)X'(\alpha+n) \quad (s < \alpha < s')$$

which yields

$$\lim_{n\to\infty} \mid X(s'+n) - X(s+n) \mid \leq \lim_{h\to\infty} v_n = 0 \,.$$

THEOREM 2 (the peak/average ratio theorem): if, for all $t \geq 0$, $k(t) = rz$, r indicating a constant greater than one, then the inflation rate $Q'(t)/Q(t)$ converges to r, where

$$r = \frac{\pi}{1 - e^{-\pi}}.$$

PROOF $S(\tau) = rzQ(\tau)$, for $\tau \geq 0$. Hence, by Eqs. (A.1) and (A.2), for $t \geq 1$

$$Q(t) = r \int_{t-1}^{t} Q(\tau)d\tau. \tag{A.8}$$

Making

$$X(t) = Q(t)e^{-\pi t}, \tag{A.9}$$

Eq. (A.8) is equivalent to

$$X(t) = r \int_{t-1}^{t} e^{\pi(\tau - t)} X(\tau)d\tau. \tag{A.10}$$

From Eq. (A.9)

$$\frac{Q'(t)}{Q(t)} = \pi + \frac{X'(t)}{X(t)}.$$

Hence, to prove the theorem, it suffices to prove that $X(t)$ converges to a positive constant. Taking derivatives in both sides of Eq. (A.10):

$$X'(t) = re^{-\pi}(X(t) - X(t-1)) = \frac{\pi}{e^{\pi} - 1}(X(t) - X(t-1)). \tag{A.11}$$

Since $re^{-\pi} = \pi/(e^{\pi} - 1) < 1$, it follows from Lemma 1 that $X(t)$ converges to a constant X. To complete the proof, we just need to show that $X > 0$.
For any integer n let us define

$$m(n) = \min\left\{X(\tau) \mid n - 1 \leq \tau \leq n\right\}.$$

Since the price index $Q(t)$ is always positive, $m(n) > 0$. Let us assume that $X(t) = m(n)$, where $n - 1 \leq t \leq n$. If $t = n$, Eq. (A.10) yields

$$X(n) = m(n) = r \int_{n-1}^{n} e^{\pi(\tau - n)} X(\tau)d\tau \geq m(n)r \int_{n-1}^{n} e^{\pi(\tau - n)}d\tau = m(n),$$

the equality requiring $X(\tau) = X(n) = m(n) > 0$ for all $n - 1 \leq \tau \leq n$. In this case, Eq. (A.11) yields $X(t) = m(n) > 0$ for all $t \geq n - 1$. If $X(t) = m(n)$ and $n - 1 \leq t \leq n$, Eq. (A.10) leads to

$$X(t) = m(n) \geq rm(n-1) \int_{t-1}^{n-1} e^{\pi(\tau - t)}d\tau + rm(n) \int_{n-1}^{t} e^{\pi(\tau - t)}d\tau$$

which implies

$$m(n) \geq m(n-1).$$

In either case, $m(n)$ is a non-decreasing sequence of positive real numbers, which proves that $X > 0$.

THEOREM 3 (the non-neutrality theorem): if the inflation rate decreases for $t-1 \leq \tau \leq t+1$, then $z(\tau) > z$ for some $t-1 \leq \tau \leq t$.

PROOF We shall prove that the assumption of $z(\tau) \leq z$, for all $t-1 \leq \tau \leq t$, is inconsistent with that of a declining inflation rate. In fact, to say that the inflation rate decreases for $t-1 \leq \tau \leq t+1$, is equivalent to saying that, in that interval, the function

$$q(\sigma) = \ln Q(\sigma)$$

is strictly concave. Hence, if $m = q'(t)$, the curve lies below the tangent at the point $(t; q(t))$, namely

$$q(\sigma) - q(t) \leq m(\sigma - t)\,,$$

the equality holding only when $\sigma = t$. It follows that

$$Q(\sigma) \leq Q)t\, e^{m(\sigma-t)}$$

and as a consequence that, for $t-1 \leq \tau \leq t$

$$\int_{\tau}^{\tau+1} \frac{1}{Q(\sigma)} d\sigma > \frac{1}{Q(t)} \frac{1-e^{-m}}{m} e^{m(t-\tau)}\,.$$

Let us assume that $z(\tau) \leq z$ for all $t-1 \leq \tau \leq t$. It follows, from Eqs. (A.2) and (A.4), that

$$S(\tau) < \frac{m}{1-e^{-m}} W(t) e^{m(\tau-t)}\,. \qquad (A.12)$$

Introducing Eq. (A.12) in Eq. (A.1) yields the contradiction $W(t) < W(t)$.

REFERENCES

1. Arida, P., and A. Lara Resende (1985), "Inertial Inflation and Monetary Reform in Brazil," *Inflation and Indexation: Argentina, Brazil and Israel*, Ed. John Williamson (Washington, D.C.: Institute for International Economics).
2. Dornbusch, R., and M. H. Simonsen (1987), "Inflation Stabilization with Incomes Policy Support," *The Group of Thirty.*
3. Fellner, W. (1976), *Towards a Reconstruction of Macroeconomics* (Mercer, PA: American Enterprise Institute).
4. Fischer, S. (1977), "Long-Term Contracts, Rational Expectations and the Optimal Money Supply Rule," *Journal of Political Economy* February.
5. Jones, A. J. (1980), *Game Theory: Mathematical Models of Conflict* (New York: Halstead Press).
6. Keynes, J. M. (1972), *A Tract on Monetary Reform, Collected Writings of John M. Keynes* (Cambridge, MA: University of Princeton Press), vol. 4; and Keynes, J. M. (1973), *The General Theory of Employment Interest and Money, Collected Writings of John M. Keynes* (Cambridge, MA: University of Princeton Press), vol. 7.
7. Lopes, F. (1986), *O Choque Heterodoxo* (Editora Campus).
8. Lucas, R. (1976), "Econometric Policy Evaluation: A Critique," *The Phillips Curve and Labor Markets*, Eds. K. Brunner and A. Meltzer (Amsterdam: North Holland).
9. Lucas, Robert, and Thomas Sargent (1979), "After Keynesian Macroeconomics," *Federal Reserve Bank of Minneapolis Quarterly Review* **3**.
10. Lucas, Robert, and Thomas Sargent, eds. (1981), *Rational Expectations and Econometric Practice* (Minneapolis: University of Minnesota Press).
11. Modiano, E. (1986), *Da Inflação ao Cruzado* (Editora Campus).
12. Modigliani, F. (1977), "The Monetarist Controversy, or Should We Forsake Stabilization Policies," *American Economic Review, Papers and Proceedings.*
13. Sargent, T. (1979), *Macroeconomic Theory* (San Diego: Academic Press).
14. Sargent, T. (1985), *Rational Expectations and Inflation* (New York: Harper and Row).
15. Schelling, T. (1982), *Micromotives and Macrobehavior* (New York: Norton).
16. Simonsen, M. H. (1970), *Inflação, Gradualismo versus Tratamento de Choque* (APEC).
17. Simonsen, M. H. (1985), "Contrato Salariais Justapostos e Política Anti-Inflacionária," *Revista de Econometria* Novembro.
18. Simonsen, M. H. (1986a), *Keynes versus Expectativas Racionais* (FGV).
19. Simonsen, M. H. (1986b), "Rational Expectations, Income Policies and Game Theory," *Revista de Econmetria* Novembro.

20. Taylor, J. B. (1978), "Staggered Wage Setting in a Macro Model," *American Economic Review, Papers and Proceedings.*
21. Tobin, J. (1981a), "Diagnosing Inflation: A Taxonomy," *Development in an Inflationary World*, Eds. M. J. Flanders and A. Razin (New York: Academic Press).

III. Working Group Summaries

JACK GUENTHER, JOHN HOLLAND, STUART KAUFFMAN, TIM KEHOE, TOM SARGENT, & EUGENIA SINGER
September 17, 1987

Report of Work Group A: Techniques and Webs

Though the rubric "webs" was used to distinguish this subgroup from the others ("patterns" and "cycles"), the subgroup also devoted substantial time to questions of adaptation and evolution in economics.

There were two broad themes:

The first theme centered on the two-nation *overlapping generations* model from economic theory. We started by looking for a well-established model from economics that, even in its simplest form, raises some of the central quandaries of the global economy (such as the ability of one nation to implicitly tax another by "printing money," or any of the several questions about international debt raised by Mario Simonsen in his lecture). The model also had to be easily extendible in several dimensions (number of countries, number of commodities, number of generations, arbitrage procedures, etc.). The two-nation *overlapping generations* model goes a long way toward meeting all of these criteria. We were able to identify several ways of modifying the model that would allow comparisons between the standard *equilibrium* solutions (which require perfect one-step *foresight* in the agents) and solutions that would arise if adaptive rule-based agents were to replace the agents with foresight. In particular we were able to identify counterparts of the "prisoner's dilemma" model studied by Axelrod (in this context, the possibility that both nations print money, each attempting to implicitly tax the other, until the money is worthless). This opens the possibility of employing a genetic algorithm to explore the space of agent strategies, until solutions stable against incursions of other unfavorable strategies are attained (the *evolutionary stable strategies*—counterparts of the strategy TIT FOR TAT in Axelrod's work). We also identified some aspects of the n-country version that might act much as spin glasses (such as the transformation of an initially more-or-less homogeneous group of countries into clusters of trading partners, co-ordinated lenders, etc.).

The second theme was more web-like, and more abstract. It began with considerations of co-evolution of technologies (such as the introduction of automobiles giving rise to requirements for roads, gas stations, special steels, etc.) and continued with discussions of walks over rugged adaptive landscapes. The analogies with well-established biological concepts and interactions are quite appealing, but there is much work to be done in recasting these models in ways that are both precise and in contact with economic theory. Turing's work on reaction-diffusion equations may provide a way of initiating this program (suggesting ways in which a web may develop local self-sustaining clusters of technology). There are also some possibilities of using a canonical representation of nonlinear utility functions (in terms of the average utility of available components or building blocks) to discuss the expected evolution of the web in terms of new combinations of building blocks. Some of the discussion centered on ways of adding dimensions to the rugged landscape as new technologies are injected. It was generally agreed that approaches along the lines of this second theme are in need of worked examples.

Both themes contribute to the broad objective of a deeper understanding of the (price, commodity, etc.) trajectories induced on an economy by continuing technological innovations *and* the interaction of a *variety* of limited agents. We would hope to form several computer "cartoons" of the world economy (simulated dynamic models) that both develop the theoretical points (such as those presented by Simonsen) and illustrate them.

KENNETH J. ARROW, WILLIAM BROCK, DOYNE FARMER, DAVID PINES, DAVID RUELLE, JOSÉ SCHEINKMAN, MARIO SIMONSEN, & LARRY SUMMERS
September 17, 1987

Report of Work Group B: Economic Cycles

The general topics for discussion and questions addressed to this group were the following:

1. Relation of deterministic to stochastic models. Role of foresight is removing positive Lyapunov coefficients (stability, chaos, etc.). Meaning of predictability in deterministic systems. Can there be deterministic models without perfect foresight?

2. Securities markets: economic theory, martingales, excess volatility, explanation of trading volumes. Does deterministic nonlinear analysis help? Psychological components (overconfidence; non-Bayesian expectation formation).

3. Business cycles: stochastic (monetary or real) models versus deterministic non-linear; large numbers of variables with leads and lags. Are there significant changes in regime, e.g., two world wars, Great Depression, emergence of business cycles.

A list of 4 discussion topics was decided upon early. Here they are, with some comments.

1. *Drafting guidelines for the analysis of economics time series.* The people most interested in this are William Brock and José Scheinkman. This is a technical topic, and was not discussed extensively. But it was noted that the analysis of

transients in time series (without stationarity assumption) is possible, although not much work has been done on this. It was also noted (David Pines) that the theory of a large number of incomplete time series might supplement the theory of the few high-grade time series which have been analyzed up to now.

2. *What is the connection between speculation and volume, and the problem of excess volatility?* Larry Summers proposed an explanation which seemed to satisfy everybody. The idea is that the influx of information is the driving force. Information is interpreted differently by different traders, leading to an increased trading of shares, and an increased volatility. Note that the stock market is itself a source of information; the process is thus self-sustaining.

3. *Dynamical models of the economy: stochasticity and nonlinearity.* An aspect of this is problem 4 below. This general subject led to long discussions involving in particular Kenneth Arrow and Larry Summers. One point is that the economy seems to have remarkable stability properties: external accidents either have very little effect, or the effects (as for wars) are important but wiped out rather rapidly. This would suggest that there are robust laws for economic time evolution, and that it should be relatively simple to find and state these laws. Looking at things more carefully, however, one finds that successive adaptations of the economy to successive crises do not correspond to the same phenomenon: the crises are different and their resolution is different. This remark makes one less hopeful about finding simple laws for economic evolution.

4. *The cloudy crystal ball model.* "Chaos" in nonlinear time evolutions makes the prediction of the future difficult. Kenneth Arrow therefore suggested we look at the economic problem of matching the price of prediction and its expected dividends. Consider the discounted utility

$$U(x,\mu) = \int e^{-\delta t} u(f^t{}_\mu x) dt$$

where $(x,t) \rightarrow f^t{}_\mu x$ is a smooth time evolution, and m is some control parameter. It is readily seen that U is a continuous function of x, μ, but the derivative with respect to x appears to diverge when the discount rate δ is less than the largest characteristic exponent of the dynamical system $(f^t{}_\mu)$. It can nevertheless be shown (David Ruelle—homework) that U is a Hölder continuous function of x and μ. Let μ be adjusted to make U maximum at x_0. Then a small error on x_0 or m can lead to a relatively large loss on U.

Another topic which was discussed was to find a simple toy model explaining fluctuations in the stock market in terms of nonlinear dynamics. The idea proposed by Larry Summers is that there are "noisy traders" and "sophisticated traders" which share the market. They have different (deterministic) strategies, and one obtains in this manner rather easily a chaotic time evolution. But it must be admitted that a really convincing model has not yet been obtained.

P. W. ANDERSON, W. B. ARTHUR, E. BAUM, M. BOLDRIN, N. PACKARD, J. SCHEINKMAN, & L. SUMMERS
September 17, 1987

Report of Work Group C: Patterns

The general topics for discussion and questions addressed to this group were the following:

1. Economic development as a complex system, as a pattern problem. Are there many equivalent states? Can one model the diffusion of development as a reaction-diffusion front? Political input as damping/enhancing factor in development.
2. Regional economic differences. Is there a tendency to polarization? Do reaction models offer a parallel? Is this paralleled by international economic differences?
3. Review of possibilities for lock-in. Do such methods as annealing suggest policies or spontaneous actions of the economic system for removing harmful lock-in?

Aside from a great deal of very interesting and enlightening general conversation, which at this point is lost except for Michele Boldrin's notes (at the end of this chapter), the group came up with three rather specific projects, at least two of which will undoubtedly be followed up by collaborative work at peoples' institutions.

1. Eric Baum, Bill Brock, and Larry Summers will attempt to use neural-net learning algorithms on sets of economic data, especially macroeconomic data.

Perhaps more interesting than the crude problem of prediction is the system's capabilities for generalization and feature extraction.

2. Michele Boldrin, José Scheinkman and Norman Packard have evolved a two-country model with internal structure and an increasing return feature which may serve as a zero-order model for bistability effects—a kind of McCulloch-Pitts neuron or Hebb synapse of the economic field. The variable e embodying the various effects of "learning by doing," infrastructure, education, etc., is introduced in a way which is suggestive for many other problems.

3. Brian Arthur and Phil Anderson discussed a cooperative "nucleation-like" model which might, according to Brian, be very applicable to known features of pattern of fertility change.

Both 2 and 3 follow the general topic of "patterns" set in the list of topics for working groups. Both in this group and in group B, there was a general feeling among some of the natural scientists and a few of the economists that the role of inhomogeneity—the observed coexistence of different types of economy in the same world or even country—is a question which present fashion in economic theory underplays.

The group as a whole endorses the concept of a network, but would prefer to see the word "global" given less controlling force in our title; perhaps, "Fundamentals of Economic Theory with Emphasis on the Global Economy."

MICHELE BOLDRIN'S NOTES

Anderson: We need a theory of economic development as, it seems, we do not have one. We need to be able to explain how countries grew and how they interact dynamically.

Boldrin: The distinction between growth (increasing quantitites within a given economic order) and development/evolution (qualitative changes in the economic structure). We have some decent theories for the first phenomenon but none for the second. It is the second one which is of concern here.

Summers: Not sure we agree with the previous remarks. The problem is that technological innovation is the driving force and we have no economic explanation of its determinants, at least at the macroeconomic level. In particular, there are no formal analyses of the phenomena of pattern formation, unequal growth, virtuous cycles vs. vicious ones in the development of national economies, etc. There is a verbal tradition, mostly based on historical evidence, but no theory.

Arthur: Suggests that his own research (as well as others' contributions along the same line) may be of potential usefulness to answer some of these problems. For example, models of competitive exclusion that he has developed, so far, at a micro-level may well be modified to explain macro phenomena. He believes there are techniques in chemistry, physics and biology that may be used by economists.

Anderson: It seems that what we have is very similar to what Kauffman talked about in his speech: rugged landscapes, spin glasses, etc. There are various classes of irregular landscapes, the two most important are: a) totally random and very irregular; random field models have very simple properties (linear max); and b) spin glasses, the next simplest set-up. Here we have interactions that are not totally random and come in pairs. When all pairs are interactive, statistical techniques may help to understand the outcomes (see the lecture given by Palmer). An important feature of this "landscapes" is that, very often, you end up in deep local minima from which it is difficult to escape.

Baum: The idea of universal computation: with an appropriately used computer, you can forecast. Economists seem to have a lot of troubles in forecasting. One is led to ask: is there a computability property for the economy? This is clearly preliminary to any tentative computation. Given the difficulties we have been facing, one may well conjecture an impossibility theorem for economic systems: the economy, simply, is not computable.

Summers: He believes that our problems in forecasting do not depend on the absence of large enough computers. There are problems with our models and techniques. Four reasons that may explain our commonly recognized failure:

1. We use "too simple" models (i.e., representative agents, perfect foresight, etc.).
2. We have no idea about what the FED and other politically motivated entities are going to do and this strongly affects the real outcomes.
3. The equilibria (of the real world) are multiple; therefore, actual behaviors are very sensitive to initial conditions and some shocks imply big changes.
4. We simply do not have *the* good model. There is something like the "true model," but economists seem not to have found it yet.

Boldrin: Some comments on the reasons listed above. It is clear that 1 is a special case of 4. All in all, we have the right model and the information on the initial conditions/driving forces (1,2,4), or the real economy is properly described by a formalism which, essentially, has a non-predictable dynamics. Which way do we take, Laplace or Poincaré? If nonlinear phenomenon are perceived as relevant, clearly one has to lean toward a Poincaré perspective. Computability issue and church thesis: should we come up with a "proof" that the economic system is not computer-computable; we also must conclude that some of the operating rules of an economy are not of the logical-aritmomorphic class (Georgescu-Roegen has written something on this).

Summers: There is a practical, everyday side of the forecasting macro-data activity, i.e., you have to use your nose and the New York Times. Look at historical data, look at where you are today, and try to understand what may "reasonably" happen. You may like it or not, but that's the state of the art!

Anderson: In forecasting macro variables, economists always use "representative" entities. But aggregation distorts: the system is not a unit, is not a household. An unemployed white yuppie is different from an unemployed unskilled black. This is important.

Summers: Agrees, but reports that the profession has been moving in the other direction.

Boldrin: Report some very recent researches "against the mainstream." His own on multisector growth models. Woodford model where workers are different from entrepreneurs. The old Cambridge (U.K.) tradition. All this leads to nonlinearities and irregular dynamics.

Arthur: Recalls that if the underlying dynamics is nonstationary, then a neuron-network method would probably perform better from the pure forecasting viewpoint.

Baum: Explain the logic of the neuron-net technique. Discussion of its applicability to economic data. It does. This will lead to Summers-Baum and Brock collaboration.

Arthur: There is a theory of economic development. It is essentially "wordy": history more than theory. Approximately, economists see LDC countries as "growing" in a framework where a vector of state is given (say, prices of tradeable goods) and all they can do is to react to it and grow, if possible. Quantitative approach. There are some prescribed stages that one has to go through and only one itinerary is recognized as possible. There are other theories, instead, that allow for transients and where different patterns may be possible. Biological approach: admit that there are many interacting species, instead of only one. His own research on this area (see papers). Locational patterns: could another Singapore be somewhere else? How do spin glasses relate to this?

Summers/Boldrin/Arthur: Possibility of multiple Nash equilibria is a choice-of-location game with external effect. Similar to the VHS vs. Beta video game. Introduce cost of movements and repeat the game, i.e., introduce a dynamics: where will you go? This is not clear. Anyhow, these kind of mechanisms will not explain LDC's but, much better, industry agglomeration with different patterns of equilibrium. In this case, externalities are important; big-push theory follows when you talk about regional policies for development. There are many examples of failure of such a policy (e.g., South of Italy). On the other hand, you have the "new cities movement" in U.K. the big policy question is: how do you estimate the true level of positive feedback?

Anderson: We are still short of a dynamic story. What we want is countries' interaction that leads to patterns of development that may depend on initial conditions and where it is possible that every country grows, as well as some that do and some others that do not. This remark prompted a brief discussion after which Boldrin and Scheinkman said they could build a simple competitive-equilibrium dynamic model with such features.

IV. Final Plenary Discussion

RICHARD PALMER
Department of Physics, Duke University, Durham NC 27706

Final Plenary Discussion

INTRODUCTION

This is a report of a discussion which formed the last session of the conference on *Evolutionary Paths of the Global Ecomony*, on September 18th, 1987 in Santa Fe. A few days earlier I had jotted down a few nagging general questions about the approach of the economists. Three of these questions eventually became a framework for the final discussion. Here I try to reiterate the questions, comment on them, and give a sense of the discussion.

Several disclaimers are necessary. First, although I formalized the questions, I am not solely responsible for them. Indeed, my intent was to capture what I perceived as the sense of several of the participants from the natural sciences. Second, I am totally naive about economics; all that I now know was learned during the Santa Fe meeting. I make neither claims nor apologies for this, but do mean it seriously. Third, my reporting of the discussion is sketchy at best and probably libelous at worst. Partly for this reason, but also for brevity, I have made no attempt toward attribution of particular comments, and speak in terms of supposedly homogeneous "economists" and "natural scientists." I hope that my colleagues from both fields will forgive me.

QUESTION 1

Why do economists downplay or ignore the role of psychological, sociological, and political forces in economic systems?

COMMENTS

This question came up again and again during the meeting. Answers from the economists included: "such forces really *aren't* important"; "they're important, but too hard to treat"; "they're being studied in certain specific cases"; and "they're automatically included through their economic effects." None of these answers was entirely satisfying to the natural scientists, and their variability concerned some of us. The importance of such forces seemed self-evident and not easily dismissed. Moreover, several of us felt, in individual discussion, that it should not be too hard to build such factors into models.

A related issue was the role of *positive* feedback in economic systems. The forces under discussion would naturally lead to positive feedback in many cases. However, positive feedback is apparently frowned upon by some economic practitioners because of its conflict with a belief in a basically stable economy. Brian Arthur's examples of positive feedback mechanisms seemed unremarkable to the physical scientists, but apparently generate strong opposition in some quarters.

DISCUSSION

Many examples of the influence of psychological, sociological and political forces were suggested, including moral hazard in insurance, oil shocks, the influence of labor laws and unions, political influences in trade tariffs, ties between cities for sociological reasons, and mass psychology in the stock market. The last example is particularly apt at the time of writing (October 30, 1987), when the phrases used frequently by market analysts and commentators include *consumer confidence*, *national buoyancy*, and *panic*.

It seems that there are several areas in which psychological, sociological and political forces *are* allowed for, or are even the focus of inquiry. Not all of these lie in the domain of economics, however. One can find, for instance, the inclusion of economic factors in political and sociological theories, perhaps more readily than the reverse. It was suggested that it would be worth inviting some people from these related fields to any follow-up meeting at the Santa Fe Institute. Specifically mentioned were researchers in the "mathematical theory of social networks," and mathematically trained political scientists. Experts in "public choice theory" might also be appropriate.

The view that non-economic forces are not in fact usually very important was again expressed. They are normally regarded as explaining only minor effects, or perhaps as tipping the balance in a marginal case, whereas the leading terms are seen as purely economic. Some natural scientists still found this hard to accept.

There was a little discussion on *how* to include psychological, sociological and political forces in economic models. One could certainly modify a utility function to allow for such factors, but there is no obvious *a priori* method. Game theory is a conventional tool in studying political forces. It is not clear to what extent rational expectations theory already includes, or can be adapted to include, non-economic factors. Perfect foresight can in principle allow for all possible influences, but may not do so as currently implemented.

QUESTION 2

Rational Expectations (RE) theory with infinite foresight appears obviously wrong. Why is it so well accepted?

COMMENTS

Perhaps "obviously wrong" is too strong, but most of the natural scientists were certainly disturbed by the theory. Predicting the future, or even possible futures, is fraught with danger in almost any system. There seems to be too much emphasis on determinism in RE theory, especially since even deterministic dynamics does not necessarily allow long-term prediction because of the divergence of trajectories. Similarly, any requirement for a *unique* answer appears groundless in such cases. Multiple equilibria are seen as natural (and perhaps even expected) by the natural scientists but not by the economists.

An interesting case is the sensitive dependence of the derived equilibrium on final conditions in the distant future, exhibited for example in the Overlapping Generations model described by Tim Kehoe. This can be seen as an example of multiple equilibria, not distinguishable with present information, and as such should not be problematic. Regarding it literally leads to nonsense.

Besides determinism and uniqueness, some of the natural scientists were concerned by the assumption of rationality in human behavior. This ties in, of course, with question 1 above. On the individual level it would be hard to claim that behavior is totally rational or perfectly chosen; indeed, a current topic in psychiatry is "self-defeating" behavior. But perhaps this does not matter on the aggregate level where the mistakes of one individual can mean enhanced profit for another? Nevertheless, if the the assumption is crucial, it deserves careful examination. Perhaps one could form rational theories of the irrationality of others?

DISCUSSION

The RE theory was defended in various ways. It is admitted to have some faults, but they are not as severe as first perceived. Some of the limitations were discussed earlier in the meeting, especially by Larry Summers. There is no doubt that examples can be constructed in which there are either no answers or only nonsensical answers.

When problems arise they may actually be signifying poor or inconsistent assumptions. This may be the case in the Overlapping Generations model, which is not accepted by all. In that model, the problems disappear if additional mechanisms that span several generations are added, such as corporations living longer than individuals. In general in RE theories, one needs to be able to average over *many* lifetimes, or *many* plays of the game.

Exogenous shocks (or stochastic noise) can also be used to make models more realistic. They can, for example, be used to allow for the possibility of unforseen technological innovation. In general one can find ways of repairing RE models when they display problems.

Several ways of improving the RE models were suggested. One was to add a modicum of irrationality. Although this makes some sense and has been tried, in practice it is not liked because of its ambiguity; there are too many different ways of being irrational, whereas rationality is presumed unique. This demonstrates again the desire for unique deterministic answers. In fact, many economists are hoping for a theory that is much *more* deterministic than RE.

Another approach discussed was working with probability distributions for each possible future, instead of with expected values. The distributions should presumably become more and more broad (fuzzy) further into the future, giving an effective smooth cutoff to predictability. For each possible future thus described, the usual analysis would give a probability distribution for optimum behavior (e.g., consumption), which could then be made self-consistent just as expectation values are in the conventional RE theory. The theory of "fuzzy sets" might be useful.

Finally, there was some discussion as to whether learning models could replace or supplement RE theory. One can ask whether an economic system itself learns by its mistakes in a cumulative way, as opposed to local optimization (*cf.* tâtonnement). Most models do not include a true learning process. An adaptive rule-based system, such as those described by John Holland, might possibly provide a very different modeling approach in which learning is natural.

QUESTION 3

Can a system with a fixed number of variables adequately model innovation?

COMMENTS

Natural scientists are aware of fundamental differences between open and closed systems. A system in which the number of degrees of freedom N is fixed, albeit large, is not the same as one with N itself a free dynamic variable. In Stuart Kauffman's talk, various results were derived for adaptive walks on rugged landscapes in a large *fixed* number of dimensions N. For example, progressive improvement of an optimum became harder and harder, and as $N \to \infty$ the local optima fell towards the mean of the whole domain. While highly interesting in themselves, there is no compelling reason to believe that these results also apply to an open system where N is allowed to increase with development. Indeed, there are areas of technology (e.g., computers) in which improvement seems to accelerate as new dimensions are introduced.

It seems possible that some economic models may have fallen into the same trap. Generally one seems to start with and retain a fixed set of N variables describing the system. New commodities that have not yet been invented have to be included implicitly, but with parameters that are initially 0, ∞, or irrelevant. But it is not at all clear that this is equivalent to having a truly open set of variables. One may need instead to introduce budding or bifurcation processes to generate new variables stochastically when there is a strong enough incentive or need for invention. This was also discussed by Kauffman.

DISCUSSION

It was pointed out that some work *has* been done in this area. Innovation has been studied particularly in the context of international trade theories, where the introduction of new commodities plays a large role. An exogenously given stream of innovations is an ingredient of some such theories. Similar ideas are used in some models of patents.

There was also some discussion on modeling bifurcation or budding processes. One clearly needs a cost of introducing a new node in the interacting network of commodities. The venture capital that can be raised for a given innovation is obviously related to this, but also involves the probable degree of success.

V. Summaries and Perspectives

P. W. ANDERSON
Santa Fe Institute, 1120 Canyon Road, Santa Fe, NM 87501, and Joseph Henry Laboratories of Physics, Princeton University, Jadwin Hall, Post Office Box 708, Princeton, NJ 08544

A Physicist Looks at Economics: An Overview of the Workshop

1. CONCEPTS OF "EQUILIBRIUM" AND DYNAMIC CHAOS

From William Brock's summary and José Scheinkman's and Thomas Sargent's discussion of the concept of the Arrow-Debreu theory, we learned that even theories which appeal to the concept of "equilibrium" do not necessarily avoid the apparently random fluctuations in the course of time which are characteristic of driven dynamical systems in physics. In physics these are called "non-equilibrium" systems; a liberal education on the various meanings of the word "equilibrium" was a bracing experience for all.

It has been speculated for a few years that such fluctuations were the source of much of the cyclic behavior of the overall economy and would therefore not result from random, unpredictable "shocks," but would be the consequence of some set of deterministic, nonlinear, integro-differential equations describing overall human economic behavior. Such equations often exhibit what is called chaotic dynamics, involving orbits which depend sensitively on initial conditions and which therefore diverge with positive Lyapunov exponents. These technical phrases will be explained, I am sure, in David Ruelle's and other contributions. The conditions which make it possible for this kind of theory to happen, for instance, are that the future is discounted at a sufficiently large rate or finite lifetime effects are included. Thus, one of the central questions underlying the workshop—what can serve as a driving mechanism for chaotic behavior?—was answered at least in part. Intuitively, it is very reasonable that the presence of agents willing to trade present gain for greater future loss, or of simply uninformed or improvident agents, or of inept strategies, all could invalidate the notion of an overall sum of utilities as a Lyapunov function which always is increased in every transaction (since each party to a transaction must feel he is gaining by it). If there is an overall function which always increases, transactions can only drive the economy towards a local maximum possible value of all utilities, or "fixed point." The effects of inadequate foresight probably do account for much dynamic behavior. On the other hand, major regime changes—such as occurred in 1980, possibly in 1930, and certainly in 1815, are clearly not accounted for by such ideas, nor are the rise and fall of national hegemonies.

In this area of macroeconomic dynamics, the major focus was on the idea that whatever the underlying dynamics behind macroeconomics, it might well be modeled as a deterministic, driven, dynamical system with a small finite number of exogenous control variables and endogenous non-decaying degrees of freedom (it is heartening that external "shocks" can be incorporated as extra degrees of freedom, an idea which I felt had not been clearly expressed in the literature I read.) This behavior is common in physical systems which are not "driven" too hard, so that most of the variables do relax towards a single, finite, dimensional region, called an attractor, which hopefully can be discovered empirically. Brock, Scheinkman and to some extent other economists showed that such attractors do exist in narrow regions of detailed economic data such as market averages, but these have not yet been successfully abstracted from the more limited sequence of data on major macroeconomic factors such as GNP or G World P. David Ruelle and Doyne Farmer discussed promising new mathematical techniques for seeking such attractors. A major question is whether the enhancement of the data set with more kinds of parallel, related measurements such as interest rates, demographics, inventories and the like can be incorporated meaningfully, and if so under what limitations. Clearly, for instance, if the entire data set is too brief or too sparse in time, endogenous variables with low or high frequencies cannot be studied. Nonetheless, it seems hopeful that if one includes some of the exogenous driving forces as part of the data, the resulting attractor could somehow be made lower in dimension. Scaling of different kinds of data will be a problem.

A perhaps minor but, for the physicists, an interesting related question was the kind of anomaly discussed by Tim Kehoe in his "overlapping generations" model, which is fairly common in economic modeling—which he sees as a sensitive dependence of the present on projection into the future. A physicist is reminded of the nineteenth century arguments about Boltzmann's theorem and irreversibility which, among a certain kind of purist, unfortunately persist into the present. In physics we have it much easier: time is symmetric and large fluctuations die off both ways, as Boltzmann observed, because equilibrium states are also most probable ones, whereas in economics "most probable" and "equilibrium" do not mean the same thing. Nonetheless, Tim seems to have built a system which reverts to a most probable state—can he not borrow our concepts to deal with it, by building some kind of probability measure of initial states?

2. PREDICTION AND DATA MASSAGING BY UNORTHODOX (OR ORTHODOX) MEANS

All of us, I hope, realize that for the few decades of our lifetimes the attempt to completely substitute machine intelligence for human will be a mistake. Nonetheless, "learning machines" can often be a considerable aid to the human intelligence, in that they can sort out great masses of data which our senses are not well equipped for classifying, and identify features in them or generalize from them.

Two basic approaches are worth pursuing. The more appealing is, of course, the attempt to build a machine which can learn more or less in the same way that biological learning takes place: by evolution or by neural networks. The former is the basis of John Holland's scheme, in which a program for performing some task is steadily improved by a random evolutionary process based on some evaluation of performance, operating on the individual statements of the program. Eric Baum described his own neural-net scheme based on the multi-layer perceptron with a back propagation learning algorithm, which is not atypical of present neural net ideas. The weights in a network of "neurons" are altered to optimize the "good" output and minimize the bad. Baum's scheme is more evidently a means for finding features in the data, which then can activate complex responses; but one wonders if Holland's scheme is not based on a similar multi-layer sorting process, looking for features or regularities, with possibly more or a different kind of built-in flexibility.

When thought of in this way, the "learning machine" ideas seem related to the kinds of more orthodox data-handling scheme which Norman Packard described, based on doing an optimal separation in Hamming information space. This in turn goes back to the multidimensional scaling algorithms and ideas used by psychologists to sort data into relevant variables when even the scales of different variables are not known.

I was glad to see at least one collaboration begun (Eric Baum with Larry Summers and William Brock) on attempting prediction with the neural net machine. It is clearly better to have the unbiased machine classifying and abstracting one's data than the fallible and often politically biased economist. Nonetheless, one may still fall into the "Phillips curve" kind of trap where time and history alter relationships, and past experience is no reliable guide. I would wonder if this kind of scheme might be combined with some analytical knowledge, building in some known or hypothesized relationships before letting pure empiricism take over.

Karl-Heinz Winkler delighted our eyes with a demonstration of what really fast computing and good date display can do in hydrodynamics, but I felt that most participants see economics as a data-limited, poorly determined system where most of the interesting things involve poor sets of data in very high-dimensional spaces rather than hard equations integrated carefully in few dimensions as one needs for hydrodynamics. Display and detail are not yet problems.

The discussions often touched on the question of prediction by relatively orthodox autoregression or input-output matrix models—for instance, during Larry Summers', Buz Brock's, and Tim Kehoe's presentations. The economists were uniformly rather skeptical of most such models and, significantly, for my point of the last paragraph but one, better impressed by the brute force autoregression methods with *no* built-in theory of Sims and Litterman, than with apparently more sophisticated schemes.

Brock emphasized, and later contributions repeated, that the best autoregression fit for GNP in terms of a purely random stochastic process is a two-parameter random walk, with total persistence of fluctuations ($\lambda = 1$). It is not clear to me whether this fact is obvious or profound: it seems to the outsider to be a rather mysterious preference for marginal stability, with no manifest cause, but it does not seem to surprise economists.[1]

3. WEBS AND PATTERNS: TOWARD THEORIES OF GROWTH AND CHANGE

Five minutes into the workshop, Brian Arthur intrigued us with his opinion that theories of growth, while the least fashionable of specialties in economics at the moment, would come to be dominant for the next twenty-five years. Whether or not one agreed with this provocative statement, it was clear that this area lay at the heart of our subject matter, yet that it was the least amenable to standard economic methodology. The presentations here were those of Arthur, Stu Kauffman, and from the methodological side Richard Palmer and the brief discussion by Norman Packard of pattern theory.

[1]Note added as a result of discussions with M. Jaric: a characteristic of many physical dynamic systems is the so-called "$1/f$ noise": fluctuations increasing in magnitude proportionally to their period. All too many theories of this phenomenon exist, many of them true, no doubt, for particular cases. This led me to wonder whether a spectral index has been measured for the business cycle or other economic indicators: does the economy have $1/f$ noise, and is this the meaning of the marginal stability result? Or is it $1/f^2$, like true Brownian motion? Second, while $1/f$ noise arises in some cases of few-dimensional dynamical chaos, it is more characteristic of systems with very many degrees of freedom. Some recent investigations associate it with a "scale-free" dynamical system, one in which perturbations at all possible scales are equally close to dynamical instability. A characteristic model which shows $1/f$ noise is sand pouring down a slope with avalanches constantly occurring at all the possible scales.

Brian pointed out to us the many regimes in which the standard assumptions of equilibrium theory, such as diminishing returns, concave utility functions, etc., manifestly break down and drive positive feedback behavior and exponential explosions. These include economies of scale and large startup costs, learning-by-doing, infrastructure buildup, education and all the other goods of economic development, and simple fashion and social conformity. Web-like effects involving the intricate interlocking of technologies are also of this type. One can identify many situations where inhomogeneous patterns of development seem to have been driven by these mechanisms—e.g., Silicon Valley, or Detroit.

Stu Kauffman's concern was with the complex economic interrelationships of modern technology, involving not just the invention of new goods but the dependency of these on each other—e.g., the hi-fi on the broadcast network, the supply of records or tapes, but also on a complex of technologies and distribution networks for quite technically complicated electronic devices, high-tech speakers, or earphone parts, etc. (An amusing semantic problem was Stu's use of "new needs" which sent economists up the wall—I have been careful to say "new goods"!) The analogy is to the growth of a whole interrelated organism or ecology in biology, which once started in a given direction is often locked in, in the same way that a complex of technologies becomes locked in.

Economics differs from biology in that foresight and game-playing exist. For instance, a whole collection of actors can suddenly and simultaneously switch strategies. Also, government action and regulation exist to correct or control lock-in. Nonetheless, Stu's is an attractive model of development, especially of the original breakthroughs of the agricultural and industrial revolutions, seen as an explosion caused by the natural catalysis of many interacting developments. One can also see, in any particular country like England, that the way that explosion took place can lock the economic system into a relatively inefficient mode because of the many essential interrelationships. Whatever one may say about the "web" 's effect on macroeconomics, one can accept that it is an excellent metaphor for the effects of technology in the economy.

Stu's presentation emphasized the biological point that improvement at a given level of complexity becomes increasingly hard with time, and that often the only way out is an increase in the complexity level—certainly a good metaphor not only for the course of evolution but for the social and economic history of man.

Given all these stimulating ideas, what is one to do with them? From Galbraith to most of our physical scientists, outsiders protest the homogenization of economics by economists, and on the other hand, the economists develop more and more single-actor or single-product models. Arthur has produced a number of inhomogeneous models, and yet another arose as a brief collaboration with me, basically a nucleation model of social effects on human fertility as the transition from agricultural society to industrial passes through a country. Other models were discussed both in groups A and C attempting to deal with regional or social fragmentation of economies—the numbers of examples of backward regions or ghettoized subeconomies existing side by side with fully developed industrial societies.

It strikes me as I look back that we outsiders were far too easygoing. There are just too many examples of situations where the rich get richer and the poor starve: regional hegemonies as in Europe, tribal hegemonies in Africa, socioeconomic ones in the Latin American model with, often, outside assistance, purely economic ones in the U.S. It should not be a forbidden subject to observe that differences are usually partially enforced by sociopolitical action, usually, but not always non-violent; and it should be an interesting question whether in fact the purely *economic* interests of the hegemony are not harmed by closing out large groups from the economic engine. Every society needs prosperous consumers. The effect of the near-revolution that was the Great Depression in the U.S. was a great reduction of economic difference coupled with a very considerable reduction in the amplitude of dips in the economic cycle—look at Larry Summers' data. I will later discuss means of including the political dimension and political feedback in economic modeling, but I think that here, already, we are ducking some of the basic questions John Reed posed to us, by avoiding almost deliberately any brush with the political dimension.

A second interesting question was posed by group C; namely, could one model intercountry interaction (or even interactions between two regions within a country) in such a way as to permit the model to exhibit all of the various observed relationships? For example, could one model encompass countries which continue to dominate others economically, countries which contrive to pass others and take over the high-tech, high-profit sector (Japan *vis-à-vis* Britain, e.g., or Korea vs. Japan)? This discussion led to the Boldrin-Scheinkman-Packard model described elsewhere in this volume. We have hopes for this as a basic model of nonlinear positive feedback effects allowing multistability of economic regimes. Particularly interesting is the "e-factor" embodying the positive feedback in a way suitable for "learning by doing," but which in fact can represent, for example, education, infrastructure, or research improving overall background technology. The key is to model the e-factor as an enhancement of productivity automatically growing with the total overall production.

4. THE GLOBAL ECONOMY PROPER: PATTERNS OF CHANGE, DEBT, AND POLITICS

A presentation by Hollis Chenery of his data on growth patterns of developing countries set the stage for much of our discussion. The crude picture was of a general similarity in the time-scale of the process of countries' passing through pre-industrial to industrial states. There are few but notable exceptions and differences. Hollis has presented his own data elsewhere; I would suggest the caveat that growth does not really seem to be uniform and inevitable once one has reached a certain stage. Israel and the U.K. peak out and reverse. Many other countries' growth rates have been phenomenally higher than standard for some period, then reverted to lower or to stagnation. What Chenery can and has done for us is to establish a baseline for the "standard developing country."

Mario Simonsen presented a fascinating and very deep discussion of the debt crisis of the 80's. He showed how a regime which he called, mildly, "good arithmetic," could induce a developing country to borrow well ahead of its capacity to repay when the "arithmetic" changes from "good" to "unpleasant." "Good" vs. "unpleasant" is a matter of rate of growth of exports, $\mathbf{x}$, being greater or less than interest rate $\mathbf{i}$; if less, the ratio of debt to exports begins to rise rather than fall, and inevitably debt repayment stifles investment and cripples the developing nations' ability to further expand exports. That is, whether debt service can be financed out of export growth depends on $\mathbf{i} - \mathbf{x}$, and in turn, on this depends whether real growth can happen at all, and whether standards of living can rise. The oil shock led, it appears, inevitably to a decrease in net exports and an increase in $\mathbf{i}$ (see later) which tipped the balance towards unpleasant arithmetic without the lenders realizing the danger. In fact, in a first "pump-priming" stage the lenders actually stepped up their loans. Mario detailed the then unavoidable slide into undeclared default and regaled us with a description of the game-playing strategy between the banks and the LDC's, leaving the LDC's by no means paying enough to service their debts, but paying enough to ward off sanctions.

It is easy enough to analyze the past; could one have seen it coming in the future? That was the question posed to us by John Reed, and I have to say that from a purely economic point of view I would feel, probably agreeing with a large majority, the answer is no. What is involved here is what we learned to call a "change of regime" of which at least three were identified fairly clearly in our discussions.

The first regime change occurred at the end of the Napoleonic Wars, at which time the "business cycle" seems to have come into existence; it is more or less associated with the rise of an industrialized capitalistic world economy. Could one speculate that this was the first point at which overcapitalization became possible?

A second regime change—surely not the only one in the previous hundred years I suspect—took place in the years of the Great Depression, though one cannot help but wonder if the entire period 1914–44 was not a single series of events which left the world in a different economic regime, just as the corresponding 30 years of war and revolution ending in 1815 did so. Certainly 1914–44, or in the narrower economic sense 1920-35, left the world's economically powerful nations with regimes which, while not necessarily democratic in the one-person, one-vote sense, all took as one of their primary goals the economic well-being of the populace as a whole, and did not shy away from direct intervention in the economic process in pursuit of that goal. The result seems to have been a decrease in the virulence of economic swings.

What happened to cause the regime change of the 1980's? Certainly many economists ascribe it to the reverberations from the monetary changes of the early '70's. Personally, as a non-economist I suppose that I tend to feel with many of the non-economists that the manipulation of tokens of exchange can't really make that much difference. What did seem to change radically was political attitude: the majority of the dominant world economies first went through a period of severe inflation and then switched political directions so as to reduce inflation and create positive, real interest rates.

5. A PERSONAL MODEL FOR A "REGIME CHANGE"

At this point I would like to end my report on the workshop *per se*, and indulge in a digression into my own thoughts as stimulated by those final hours of discussions. I think that one may be able to model the forces behind such an event as a global response to a unique (but not necessarily isolated or unparalleled) occurrence: the Oil Shock. What I mean by not being isolated, of course, is that the oil shock is our *first* clear example of resource restriction, but not the only one yet to occur (e.g., ocean front property, irreplaceable art objects) or potentially the worst (e.g., the coming CO_2 crisis). As a starter let me build a simple model for inflation, one which many economists must know—I don't even know if it is taught, but certainly several people I have talked to understood it. We imagine a country in which we have a spectrum of actors—I shall think of them as W's (workers), M's (managers and entrepreneurs) and O's (widows and orphans—the elderly, the poor, the school children as educated by local government, etc.). They share in the national income:

$$I_W + I_M + I_O = I_{TOT} \quad .$$

Through their unceasing efforts and the beneficence of nature, I_{TOT}, which is produced jointly by the efforts of W and M and the savings of O, M and W, increases at a rate r per year, the lion's share of which, $(r + \epsilon) \cdot (I_W + I_M)$, is appropriated by W and M who have the political and economic power; but a fraction of which, $rI_O - \epsilon\,(I_W + I_M)$, trickles down to O, since O is after all the parents, relatives, former members, and children of W and M, and has some political influence. $\epsilon\,(I_W + I_M/I_O)$ cannot exceed r because O's income is denominated in fixed monetary units and their political and moral power is surely equal to keeping a fixed *monetary* income, if only because much of O's income is supplied by fixed pension funds, statutory entitlements, etc.

Now the oil shock happens. As a result, r become negative:

$$r \rightarrow r - \Delta \qquad |\,\Delta\,| > |\,r\,|$$

the economy as a whole must take an actual living standard decrease. Thus, even $\epsilon = 0$ cannot maintain the monetary income I_O. What is worse, it cannot maintain the real income $I_W + I_M$, unless we set $\epsilon = \Delta$, in which case I_O drops even more precipitously.

But in fact W and M has the bulk of real *political* power, and they can choose to change the monetary units in such a way that I_O remains constant in monetary units while the real income of these dominant groups also remains constant. The corresponding rate of inflation is, I believe,

$$| \Delta - r | \times \frac{I_W + I_M}{I_O}$$

and increases to infinity with the decrease of the fraction of *non-indexed* income, making the perfect socialistic democracies the most liable to inflation (but, of course, with the fewest real losers, in so far as all income tends to be indexed.)

The above is a relatively simple and well-documented example of the interaction of politics and economics. It does not, however, explore the full range of possibilities, in that it assumes only two unchanging political groupings, the productive sector $W + M$ involving management and organized labor, and the rest of the "general public," O. In fact, both sectors are composed of the same people in varying fractions, depending on the level of "shock" Δ. Slightly more realistic would be to assign a "swing" group S whose income—more accurately, their utility function—is made up partly of fixed monetary income "I_O" and partly of true earnings "I_W." This is the standard situation of professional and lower management groups, and increasingly in the modern state even of workers, who have pension rights and savings as well as earnings. For these groups, the attitude towards inflation will depend severely on the amount of loss, Δ, and the resulting degree of inflation, and in fact when $\Delta > r$, they too will experience a loss in real income and some resulting level of discomfort.

In fact, there will be a continuum, and as Δ increases relative to r, a larger and larger fraction will suffer real losses from inflation which counterbalance the rise in monetary earnings. Eventually, public opinion, and then the entire political process, will focus on inflation as an issue, and the necessary measures will be taken to even out—at least, for all but a relatively small sector such as the genuine poor—the real income losses due to Δ. I would conjecture that one could put the above in the form of a relatively clear politico-economic model which would describe the fact that not just the U.S., but England and several continental European countries swung away from inflationary policies in the decade after the oil shocks, in spite of the evident damage to large and powerful groups; the defection of large parts of labor from the liberal coalition was a consequence of their job and pension security, and their ownership of capital, as well as of declining productivity growth which reduced r at the same time as the oil shock increased Δ. The whole regime change was a response to obvious economic motivation on the part of a large swing group of voters.

With this incursion into regions where economic angels clearly fear to tread, I will close. The experience of the workshop was one I shall continue to value and hope to repeat in one form or another, and I wish to thank all the participants as well as our indispensable financial supporters in Citicorp and the Sage Foundation.

KENNETH J. ARROW
Department of Economics, Stanford University, Stanford, CA 94305

Workshop on the Economy as an Evolving Complex System: Summary

The last thirty years have seen a steady development of dynamic analysis in economics, both in theoretical development and in empirical implementation. The theoretical equations were frequently nonlinear in form, but the empirical methods have largely employed the tools of linear stochastic analysis. The assumptions of randomness were made partly because clearly no simple system fit the data very well and partly because of the clear subjective perception that the future was uncertain. The theoretical analysis of dynamic systems was largely devoted to the determination of their steady states or solutions which exhibited steady (exponential) growth. In stochastic models, the equivalent of a steady-state solution is a stationary or invariant probability distribution.

The economic theory underlying the dynamic systems studied tended to emphasize amplitude-reducing behavior (negative feedback). A brief summary of the central theory of general competitive equilibrium (GCE) will bring out these points. (It should be noted, however, that many modifications have been studied, but they have not had the coherence and generality of what I have called the GCE model.) The system is one of private ownership. There are two kinds of economic agents, firms and households. Associated with each firm is a production possibility set, defined by the technology available to it; the set consists of all vectors defining possible combination of inputs and outputs. It is assumed in the GCE model that the production possibility set does not contain incentives to go to extremes. More precisely, it is assumed that any weighted average of two possible production vectors is also a possible production vector, i.e., that the production possibility set is a convex set. In particular, a production process can be run on arbitrarily small scale in the model, so that there are no advantages to large scale or to specialization. Each household owns a set of assets (most especially, labor-time) which it can sell on the market and has preferences over possible vectors defining consumption and willingness to work, and each is small on the scale of the national economy. Suppose a set of prices is announced, one for each commodity. Then each firm chooses that production vector from its production possibility set which maximizes profit. This vector defines its supply of or demand for each commodity as a function of the prices; because of the assumptions on the production possibility set, these functions are continuous. Analogously, each consumer sells his or her assets at the market price, yielding an income, and purchases a consumption vector which is more preferred among those that can be afforded at the given prices and with the given income. For each commodity, call the excess of the sum of the demands by some consumers or producers over the supplies offered by others the *excess demand*. Under the assumptions made, the excess demand is a *continuous* function of all prices. *Equilibrium* is defined as a vector of prices, one for each commodity, such that the excess demand for every commodity is zero, i.e., the supply of each commodity equals the demand for it. A major result is that, under the stated conditions, *an equilibrium set of prices exists.*

So far, the presentation has taken no account of time. A straightforward reinterpretation of the GCE model is all that is required. A commodity has to be identified not only by the usual characteristics but also by the date at which it is available. We can imagine, for the moment, that markets exist today for deliveries of commodities at any date in the future. Then there will be an equilibrium for the system in which supply and demand are equated for all periods of time. A consideration of the meaning of this system shows that it automatically introduces borrowing and lending, and interest rates appear in a natural way. If, at the prices prevailing now and in the future, I would prefer to consume less than my current income today and consume more in the future, I will lend today for repayment in the future. Thus, on the hypothesis that markets exist for all commodities at all times, the GCE model implies a time path of *equilibrium dynamics*.

The hypothesis that all markets for all future times exist today is, of course, unrealistic, but it is equivalent to the assumption that all individuals correctly anticipate all future prices, the so-called *rational-expectations hypothesis*. Furthermore, much recent analysis has gone into models which modify the hypothesis of complete markets by assuming that some markets for future commodities do not exist today and are not replaceable by expectations. In the *overlapping generations* model, there is envisaged a sequence of generations, with the young of one generation living simultaneously with the old of the preceding one. There is one natural restriction; individuals cannot enter into contracts before they are born. It turns out that this creates great differences in outcomes, particularly in models where there is no limit to time.

The equilibrium dynamics of these models is characterized by difference equations (or differential equations in continuous-time models). The original tendency of the research was to search for solutions which were constant in time. There was a generally held point of view, which indeed goes back to the origins of economics as a systematic discipline, that solutions which were not constant would tend to the constant solution or *steady state*. But more recent research, partly influenced by the new methods of nonlinear dynamic analysis developed in physics and ecology, has demonstrated that there are solutions to the same equations with cycles and even with chaotic behavior. The multiplicity of solutions is itself an embarrassment, since it suggests that economic theory, even if accurate, does not yield a unique pattern of dynamic behavior and hence its predictions are far from sharp.

It may be added that among the accomplishments of the last thirty years has been the systematic incorporation not only of time, but also of uncertainty into the GCE model and its relatives. Commodities are distinguished not merely by their usual characteristics and their date, but also by the unfolding of the events which are uncertain today but will be resolved by the date at which the contract is to be realized. Corresponding to the difference equations which define the dynamic equilibrium in the case of certainty are a set of stochastic difference equations. The search for steady states was replaced by that for stationary distributions. The analogues to cycles and chaotic behavior have not yet been studied.

Though this is not the place to go into the detailed developments of equilibrium dynamics under uncertainty, there is one point that must be emphasized. The uncertainty is, at equilibrium, perceived by all the economic agents (households and firms). They know that they don't know some aspects of the future. This is manifest in subjective knowledge; it is also manifest in certain kinds of economic behavior, particularly diversification in investments, which would not occur if the agent were certain of the future.

The discussion above has concerned the equilibrium dynamics, the movement of prices which serve to equilibrate supply and demand at each point of time. But from the time the concept of equilibrium was clearly formulated (in the 1870's), the question arose: how would an economy with decentralized knowledge get to an equilibrium. Here again the assumptions were developed in a way which appeared to be amplitude-reducing. Start with an arbitrary set of prices. There will be excess demands for some commodities and excess supplies (negative excess demands) for others. It was assumed that the price of any commodity would move in the same direction as the excess demand (this process is called by the French term, *tâtonnement*). Thus, a scarcity (positive excess demand) would lead to higher prices which would both lower demand and increase supply, thereby offsetting the initial perturbation. However, it has become recognized that the multiple connections within the economy might cause instability even though each reaction was stabilizing in its first-order effects. To give an intuitive feel, consider a world with just two commodities, bread and butter. At the initial prices, suppose that the demand for bread exceeds the supply, while the supply of butter exceeds the demand for it. The price of bread rises, while the price of butter falls. But the demand for bread certainly increases when the price of butter falls, and it can happen that the net effect is to increase the demand for bread, thereby amplifying the initial deviation.

The general perspective of mainstream (so-called neoclassical) economic theory has certainly had some empirical successes. From a very broad perspective, there is a considerable degree of self-maintenance or homeostasis. What would appear to be major shocks, especially wars, are absorbed, not immediately of course, but in the sense that after a relatively few years the economy appears to be about where it would be if the shock had not occurred. Similarly, some proportions in the economy, for example, the ratio between capital and output or the share of wages in national income, are broadly constant over time in specific countries.

But it is clear that many empirical phenomena are not covered well by either the theoretical or the empirical analyses based on linear stochastic systems, sometimes not by either. The presence and persistence of cyclical fluctuations in the economy as a whole of irregular timing and amplitude are not consistent with a view that an economy tends to return to equilibrium states after any disturbance. The persistence of unemployment undermines the assumption that prices and wages work to reduce imbalances between supply and demand. Equilibrium theorytheoryequilibrium would tend to suggest that as technology spreads throughout the world, the per capita national incomes would tend to converge, but any such tendency is very weak indeed. Similarly, different ethnic and class groups and economic regions within a country show only fitful tendencies to converge balanced by the equal likelihood of divergence. Instead of stochastic steady states, we observe that volatility tends to vary greatly over time, quiescent eras with little period-by-period fluctuations alternating with eras of rapid fluctuation. The securities markets have always shown great volatility, while the international financial markets, on which currencies are exchanged, have shown virtual disorganization since exchange rates were allowed to float, although most economists would have regarded free-exchange rate movements as an aid to stabilization.

These empirical results have given impetus to the closer study of dynamic models and the emphasis on application of new results on nonlinear dynamic models. They have also given rise to criticism of the models themselves, and this tradition goes back far; it suffices to mention the alternative theories of J. M. Keynes.

Many though not all of these themes are reflected in the presentations at this workshop. Equilibrium dynamics in different forms has been expounded by Scheinkman, Sargent and, to some extent, Brock. They have emphasize the underlying perspective that individuals make their optimizing decisions on the basis of correct predictions of future prices and their own future decisions, with suitable generalizations when the uncertainty of the future is recognized.

Scheinkman showed how the general model gives rise to a theory of the pricing of assets, which get their present value from the future values of their uses and themselves. He also showed how the theory could be altered, though with considerable increase in complication, when the set of markets for transferring wealth from the present to the future is not complete. Sargent went over similar ground, but with much more specific hypotheses about the nature of production and the preferences of households, and derived in that case a system of *linear* stochastic difference equations.

As already observed, the standard equilibrium dynamics, though it appears to have many negative feedback elements, is nevertheless capable of exhibiting behavior other than convergence to a steady state. In fact, though, it is not hard to find in the real world positive feedback elements. The presentation by Arthur exhibited such phenomena as specialization due to economies, learning by doing (the improvement in efficiency attributable to extended production), and the economies of agglomeration (geographical coexistence of different firms). He developed models of their implications, in particular that the observed distribution of economic activity might be determined by the history, in particular, by the actual outcomes of a random distribution.

To orient the discussions, Summers presented a number of empirical generalizations about economic activity, stressing the characteristics of business cycles, the failure of national economies to converge, and the tendency of nationals to invest at home in spite of the existence of an international capital market. Chenery presented by telephone the typical characteristics of countries in varying stages of development. The generalizations partly accorded with the outcomes of equilibrium dynamics, but in considerable measure disagreed.

The empirical tools of nonlinear dynamics analysis were applied to several economic time series by Brock. Measures of the correlation dimensions of a number of economic series were made by Brock and others. Unfortunately, economic time series are typically and necessarily very short by the standards of the physical sciences, and reliable estimates are correspondingly hard to find. Strong evidence that the correlation dimension was low could be found only for a few series, most notably stock market prices.

As indicated earlier, the competitive equilibrium is characterized as the solution to a system of nonlinear equations (and inequalities). Kehoe reported on the methods of solution of this and similar systems, developed originally by Herbert Scarf and by the mathematician, Stephen Smale. The problem is essentially equivalent to finding the fixed point of a transformation, and by now there is a considerable set of variants on the basic method of solution. The algorithm is complex. Though the matter was not much discussed at the Workshop, the complexity of the algorithm may be reflected in corresponding difficulties of achievement of equilibrium in the real world. Fixed-point algorithms necessarily require more information than the methods of adjustment of prices to excess demands discussed above; it is not sufficient to adjust each price solely according to information about the corresponding market. In the rational-expectations framework, where actual markets for future goods are replaced by (correct) expectations of future prices, it is implicitly assumed that firms and households are capable of computing the solution of the general equilibrium model. Its complexity might be held as an argument against the realism of that assumption.

Kehoe also expounded his work and that of others on the equilibrium dynamics of overlapping generations models. These models appear very realistic in some ways, but their solutions have very peculiar features. If the horizon of the model (the terminal period) is finite, it turns out that there is a very strong dependence of present economic conditions, prices, quantities and so forth, on the terminal conditions, which may be very far in the future. This paradox does not hold for the chief rival of the overlapping generations model, where each individual is assumed to live forever. The last absurd-sounding assumption could be made to sound more reasonable by interpreting it to mean that individuals care about and make provisions for their descendants, though with decreasing weight on those further away in time.

Mario Simonsen has introduced to us a very different approach to a different problem, bargaining over the resolution of the debt problem of Brazil and similar countries. Simple models based on game theory were used, not to model the strategies of the different agents (the banks, the countries involved) in realistic detail, but to give insight into the nature of the issues, including the limitations of fully rational analysis.

New problems were raised and new perspectives unfolded in the discussions between economists and natural scientists. The tension between chaotic behavior and perfect foresight was observed. Start with an equilibrium dynamics of a standard type derived from the hypothesis that future prices are predicted perfectly. Suppose that the solution to the difference equations characterizing the solution exhibits chaotic behavior. Is it realistic to assume that the future, even though deterministic, is in fact predictable? Clearly, part of the lessons drawn by natural scientists, especially meteorologists, from nonlinear dynamics is precisely the opposite; chaotic behavior implies that small errors of observation in the starting position may lead to virtually total unpredictability after some period of time. This creates no difficulties of consistency when the predictor is not part of the system being predicted. But when the predictors are the economic agents being examined, there is a fundamental inconsistency. This epistemological antinomy is reinforced by the empirical observation that actual behavior of prices of assets such as securities could never reasonably have been predicted; if it had, there would have been much more buying or selling at earlier stages. This problem was dubbed "the cloudy crystal ball," and was the subject of much informal discussion in Working Group B. If deterministic models permitting chaotic behavior are to replace stochastic equilibrium dynamics, in whole or in part, there has to be a theory in which chaotic solutions cause the economic agents being studied to recognize the limitations on their predictive ability and act accordingly.

A new perspective not yet absorbed by economists is that of adaptive learning models, such as those advanced by Holland and Kauffman. These have arisen by analogy with and as an attempt to understand biological evolution through natural selection. As Kauffman suggests, it is tempting to apply these models to the emergence of new ideas as well as new species and particularly to technological innovations. The analogy between evolution and technological progress has been almost a commonplace since the time of Darwin (who was himself influenced by earlier economic thought, especially that of Malthus), but has in fact not led to much. In the last decade, Richard Nelson and Sidney Winter have made a much more elaborate attempt, typically with difference equations in which rewards lead to expansion and failures to contraction. The European students of the economics of innovation have taken up the Nelson-Winter model, somewhat mixed with terminology derived from Prigogine's work. It seems very possible that the more precise and imaginative learning algorithms discussed, particularly in Working Group A, will enable us to achieve at least a qualitative feel for the possible future of innovations. Innovations, almost by definition, are one of the least analyzed parts of economics, in spite of the verifiable fact that they have contributed more to per capita economic growth than any other factor. An understanding of possible futures is not only interesting in itself, but will help to allocate resources today better.

In the last few year, there has also been interest in learning in the context of rational-expectations models. The hypothesis that individuals know the future or its probability distribution *ab initio* makes little sense, and the usual assumption has been that these distributions are learned by repeated exposure. However, the environment in which the learning takes place is influenced by the fact that other agents are also learning. There has not been a coherent account of whether this learning is possible except in a few special cases, and even a consistent account of a learning process is hard to arrive at. Bray, Sargent, Marcet, Grandmont and Evans have all proposed learning schemes and studied the stability or instability of competitive equilibria under them. Arrow and Hahn have been seeking to relate similar ideas to the tatonnement process discussed above.

There are many more aspects of recent work in natural sciences which have potential analogies in economics: self-organizing systems, the building and erosion of links in network formation, possibilities of nucleation. In turn, economic phenomena and analysis may turn out to inspire new and different kinds of concepts in the natural sciences. Indeed, the use of supply-and-demand analysis in Holland's work hints at even further possibilities. The identity of predictors and predictands which we have observed in dynamic economics may not be without analogy in biology. We look forward to mutual systems-building.

VI. Research Papers

MICHELE BOLDRIN† & JOSÉ A. SCHEINKMAN‡
†University of California, Los Angeles, and ‡ Goldman, Sachs & Co., New York and the University of Chicago

Learning-By-Doing, International Trade and Growth: A Note

1. INTRODUCTION

The research effort that underlies the simple model we are presenting was motivated by a few (to us) undisputable facts. The first, and most evident, is that across the world different countries have been growing at very different speeds during, say, the last fifty years. In particular, if we want to interpret their behavior in terms of steady-state growth rates, we have to conclude that they are on different steady states: some countries grow very fast, some others at a slower pace, and a few do not seem to grow at all. A second fact is, maybe, less "theory free" but, in our opinion, equally compelling: such differences in the rate of development do not seem to be explainable in terms of differences in natural resources, capital stocks, technologies and tastes. In particular, if we define "technology" as a list of available blueprints describing how to combine inputs to obtain outputs and "labor force" as some measure of the existing population, then it should be easy to see that, with any suitable definitions of taste and natural resources, there exist countries that are similar in any respect (or at least were similar when their development processes started), but that have been growing very differently. An easy way out is always available: to claim that tastes are indeed different, that some countries are inhabited by people with a disutility of work and/or high discount rate so that they do not work, do not save and consequently do not grow. But this seems nothing more than a trick.

Finally, it is also a fact that countries growing at different rates end up producing distinct sets of goods: they may or may not completely specialize, but it is certain that the product mix of the fast-growing nations will typically contain a larger portion of high-technology, advanced, non-primary goods than the one of the slow-growing countries. In short, if we aggregate goods in "low tech" and "high tech," then the process of development seems to imply a specialization in the second group for those countries that exhibit high rates of growth. Our question is: can we build a model that accommodates the three qualitative facts listed above and that does it in a parsimonious way, i.e., without introducing a plethora of special assumptions on preferences, market structures, trading constraints, etc.? The answer is positive, at least to a first approximation.

We consider a world with two countries, two produced goods and a finite number of inputs, exogenously supplied in fixed quantities. The consumers in each country are assumed to satisfy the standard neoclassical hypothesis and the production of each good is organized competitively in the presence of many identical firms. The output of each single firm in an industry is a function of the amounts of inputs it hires and of the average level of "expertise" in the country. The amount of the latter factor that we denote by θ, may be increased only through learning-by-doing (see Arrow, 1962). More precisely, the rate of growth of "expertise" in a country depends on the share of its work force allocated to the production of each good. We specify that the first good is a "high technology" (industrial) product whereas the second one is a "low technology" (agricultural) commodity. It is then natural to assume that a certain effort allocated to the production of the industrial good will have a larger positive effect on the growth rate of θ than the same amount allocated to production in the agricultural good. The idea here is that by producing potatoes, one may get some increase in overall expertise, but not as much as when producing computers. We also assume that as θ increases, its productivity in the industrial sector increases relative to the agricultural sector.

Finally, we assume that except perhaps for the initial values of θ, the two countries are identical.

At each point in time, a competitive firm takes as given prices and its production possibilities in making its input-output decisions. Each firm's decision, in turn, affects the future production possibilities of all firms in the same country, but given the presence of many producers in a country, the individual firms correctly ignore the impact of their decision on their own future production costs. It is the presence of this *externality* that makes our model not entirely conventional.

It is clear that in such a framework, a small difference in the initial levels of θ may be magnified by the dynamical process. In fact, the country with larger θ at the beginning will have some comparative advantage in producing the industrial good which in turn will reinforce such advantage as the learning-by-doing mechanism is stronger in this sector. We will observe then two different growth rates in the two countries and, if some steady state exists to which they converge, it will be an asymmetric position in which one country is richer than the other (i.e., has a higher level of θ), produces a larger proportion of the industrial commodity, and pays its factors of production higher returns as their marginal productivity is in fact higher in the rich than in the poor country. All of this simply follows in competitive equilibrium as a consequence of the initial difference in expertise, everything else being identical.

The remark that externalities may affect the dynamic evolution of comparative advantages was previously made by Krugman (1985) and Lucas (1985). Both dealt with a linear technology and with industry-specific knowledge. As a consequence, the results they obtained are similar to the ones in Section 3 below.

Finally, in contrast with Krugman (1985) and Lucas (1985) our learning-by-doing is not an industry-specific mechanism, i.e., the variable θ measures the overall level of expertise for the country as a whole. We have made this choice partly because we believe that this type of externality actually spills over across industries and partly because it leads to simpler mathematics without any loss of explanatory power.

One may also argue that a growth mechanism driven only by learning-by-doing does not look very attractive. In particular, the assumption that all of the "expertise/technical knowledge" is disembodied seems to be rather odd. We believe that this is a serious issue to which a more detailed analysis should be dedicated. It is our conjecture that the appropriate route is to embody the advancement in expertise and/or knowledge in the capital goods and allowing accumulation of such goods. The embodiment may or may not be full, as human capital and pure expertise factors ought to be considered. Nevertheless, we believe capital accumulation to be an essential instrument through which progress in productive ability and efficiency of an economic system are transferred over time. The rest of the present note is organized into three other sections and some brief conclusions. In the next section, we present a general formalized version of the world economy we have in mind, solve for a competitive equilibrium, and briefly describe the dynamic process for expertise and the associated competitive growth path. In the third section, we present a simple linear model where such a dynamic is realized, albeit in a very extreme form. Finally, Section 4 contains an analysis of the conditions under which the general model of Section 2 produces the asymmetric outcomes we have in mind.

2. THE GENERAL FRAMEWORK

Consider a world with two countries and two goods: let $i = 1, 2$ denote the countries, and x and y denote the goods. Each country is inhabited by a large number of identical, infinite-lived agents, maximizing their lifetime discounted utility from consumption and supplying a fixed quantity of labor in each period. The latter together with a finite number of productive resources that are inelastically supplied in fixed quantities during each period, will be denoted by the vector z^i. Let θ^i denote the quantity of "expertise" in country i. Notice that all variables depend on time, but we suppress the t-variables as we are using a continuous time setup.

TECHNOLOGY

At each time t, the firms in country i have the production functions

$$x^i = \alpha_x(\theta^i)F_x(z_x^i) \tag{2.1a}$$

in the x-producing industry, and

$$y^i = \alpha_y(\theta^i)F_y(z_y^i) \tag{2.1b}$$

in the y-producing industry. Here z_j^i, for $j = x, y$, denotes the amount of inputs employed in sector j, with $z_x^i + z_y^i = z^i$, constant over time.

We assume:

(T.1) For $j = x, y$, $F_j : \mathbf{R}_+^m \to \mathbf{R}_+$ is an increasing, homogeneous degree-one and concave function, which is C^2 on the interior of its domain.

Note that under (T.1), Eqs. (2.1a) and (2.1b) also define the industry's production function.

We also assume:

(T.2) $\alpha_j : \mathbf{R}_+ \to \mathbf{R}_+$, for $j = x, y$, is smooth almost everywhere on $\mathbf{R}_+$ and $\lim_{\theta \to \infty} \alpha_j(\theta) \leq A_j$, where A_j is a finite number. Also α_x is strictly increasing and α_y non-decreasing.

PREFERENCES

The representative agent in each country maximizes his period-by-period utility function $u(c_x^i, c_y^i)$, which amounts to intertemporal maximization as no savings are allowed and the learning-by-doing mechanism works as a pure external effect. Of his utility function we assume:

(U.1) $u : \mathbf{R}_+^2 \to \mathbf{R}_+$ is strictly concave, homothetic and of class C^2. Also it satisfies

$$\lim_{c_j \to 0} \frac{\partial u(c_x, c_y)}{\partial c_j} = +\infty, \qquad j = x, y.$$

Consumers in country i own the total amount of resources z^i and lend them out to the firms at a price π^i which will be, in equilibrium, a function of $(\theta^1, \theta^2) = \theta$. Write $M^i(\theta)$ for the income that they so receive and $P(\theta)$ for the price of the good x in terms of the good y. By solving the problems

$$\max\ u(c_x^i, c_y^i), \qquad \text{s. t. } c_y^i + P(\theta)c_x^i \leq M^i(\theta) \tag{2.2}$$

for $i = 1, 2$, we get the four demand functions

$$c_j^i = d_j\big(P(\theta), M^i(\theta)\big), \qquad i = 1, 2; \quad j = x, y. \tag{2.3}$$

COMPETITIVE EQUILIBRIUM

On the supply side, for given $\theta = (\theta^1, \theta^2)$, each country allocates z^i competitively across sectors, taking $P(\theta)$ and π^i as given. The maximum problems that are solved are:

$$\max: \; P(\theta)x^i - \langle z_x^i, \pi^i(\theta)\rangle \qquad \text{s. t. } x^i \le \alpha_x(\theta^i)F_x(z_x^i) \tag{2.4a}$$
$$\max: \; y^i - \langle z_y^i, \pi^i(\theta)\rangle \qquad \text{s. t. } y^i \le \alpha_y(\theta^i)F_y(z_y^i) \tag{2.4b}$$

for $i = 1, 2$.

Once again we will have a vector of factor-demand correspondences in each country and related supply correspondences for the output that will depend, parametrically, on θ:

$$z_x^i \in z_x(P(\theta), \pi^i(\theta)) \tag{2.5a}$$
$$z_y^i \in z_y(P(\theta), \pi^i(\theta)) \tag{2.5b}$$
$$x^i \in \alpha_x(\theta^i)F_x\left[z_x(P(\theta), \pi^i(\theta))\right] \tag{2.6a}$$
$$y^i \in \alpha_y(\theta^i)F_y\left[z_y(P(\theta), \pi^i(\theta))\right] \tag{2.6b}$$

for $i = 1, 2$.

A competitive equilibrium at a certain time (for given θ) in the world economy is then defined by price functions: $\{P(\theta), \pi^1(\theta), \pi^2(\theta)\}$ such that

$$d_x(P(\theta), M^1(\theta)) + d_x(P(\theta), M^2(\theta)) \tag{2.7a}$$
$$\in \left\{\alpha_x(\theta^1)F_x\left[z_x(P(\theta), \pi^1(\theta))\right] + \alpha_x(\theta^2)F_x\left[z_x(P(\theta), \pi^2(\theta))\right]\right\},$$
$$d_y(P(\theta), M^1(\theta)) + d_y(P(\theta), M^2(\theta)) \tag{2.7b}$$
$$\in \left\{\alpha_y(\theta^1)F_y\left[z_y(P(\theta), \pi^1(\theta))\right] + \alpha_y(\theta^2)F_y\left[z_y(P(\theta), \pi^2(\theta))\right]\right\},$$
$$z^1 \in \left\{z_x(P(\theta), \pi^1(\theta)) + z_y(P(\theta), \pi^1(\theta))\right\}, \tag{2.7c}$$
$$z^2 \in \left\{z_x(P(\theta), \pi^2(\theta)) + z_y(P(\theta), \pi^2(\theta))\right\}. \tag{2.7d}$$

Budget constraints and normalization of the price of y at one will make either Eq. (2.7a) or (2.7b) redundant. Existence of an equilibrium is a trivial result under our assumptions; uniqueness is also easy to prove as we are in fact facing a "Hicksian" economy (see Arrow-Hahn (1971), p. 220). In equilibrium, the quantities $x^1(\theta)$, $x^2(\theta)$, $y^1(\theta)$, and $y^2(\theta)$ of goods will be produced and consumed in the two countries.

We remark that at each time t, θ^1, θ^2 are fixed and the competitive equilibrium described above will be "instantaneously" Pareto optimal, i.e., it will maximize the welfare at time t of the representative consumer subject to the production possibilities. All deviations from Pareto optimality are of a dynamic nature. Our learning-by-doing hypothesis states that the time variations of θ^i are determined by:

$$\dot{\theta}^1 = E\big(z_x^1(\theta), z_y^1(\theta)\big) \quad (2.8a)$$
$$\dot{\theta}^2 = E\big(z_x^2(\theta), z_y^2(\theta)\big) \quad (2.8b)$$

with E increasing in both of its arguments. What can be said about the dynamical system of Eqs. (2.8)? Given the level of generality we have kept so far, it seems highly improbable to prove anything specific about the patterns of evolution of our model economy. We try to show in Section 4 that, indeed, with a couple of additional assumptions, we may deduce a picture of the state space that fits with the one we have in mind. But, first, we like to turn to a simple example where the desired conclusions follow almost trivially.

3. THE SKELETON OF THE MODEL: A RICARDIAN ECONOMY

We begin by discussing a very simplified model that formalizes, in an extreme form, our basic intuition. We specify the two production functions to be linear in the exogenously supplied factor (labor) as in Lucas (1985, sect. V). Set:

$$x^i = \alpha_x(\theta^i)\ell_x^i, \quad (3.1a)$$
$$y^i = \alpha_y(\theta^i)\ell_y^i. \quad (3.1b)$$

Assume that only labor is used in production and normalize units so that $\ell_x^i + \ell_y^i = 1$ in both countries. For the sake of the example, let's take a logarithmic utility function in both countries:

$$u(c_x^i, c_y^i) = \beta \ell n c_x^i + (1-\beta)\ell n c_y^i \quad (3.2)$$

with $\beta \in (0,1)$. This will give, upon maximization under budget constraint, the demand functions:

$$c_x^i = \frac{\beta M^i}{P_x} \quad (3.3a)$$
$$c_y^i = \frac{(1-\beta)M^i}{P_y} \quad (3.3b)$$

where M^i is the total income of country i (at given θ), and P_x and P_y are the two prices. Denote with W^i the wage in country $i = 1, 2$ (this also will depend on θ).

Maximization of profits on the part of the firms under the simple linear technology of Eqs. (3.1) yields the supply rules:

$$x^i(\theta^i, W^i, P_x) \begin{cases} = 0 & \text{if } \dfrac{W^i}{P_x} > \alpha_x(\theta^i) \\ \in (0,\infty) & \text{if } \dfrac{W^i}{P_x} = \alpha_x(\theta^i) \\ = \infty & \text{otherwise} \end{cases}, \tag{3.4a}$$

$$y^i(\theta^i, W^i, P_y) \begin{cases} = 0 & \text{if } \dfrac{W^i}{P_y} > \alpha_y(\theta^i) \\ \in (0,\infty) & \text{if } \dfrac{W^i}{P_y} = \alpha_y(\theta^i) \\ = \infty & \text{otherwise} \end{cases}. \tag{3.4b}$$

Labor market clearing will imply that in each country, the equilibrium wage will be:

$$W^i(\theta^i, P_x, P_y) = \max\left\{P_x\alpha_x(\theta^i), P_y\alpha_y(\theta^i)\right\}. \tag{3.5}$$

Remember that, in equilibrium, P_x and P_y also will depend on θ. It is clear from Eq. (3.5) that factor-price equalization does not need to hold in our model. The wages in the two countries will in general be different.

For given $\theta = (\theta^1, \theta^2)$ the instantaneous competitive equilibrium at time t is Pareto efficient even if (as noted in Section 2) the whole path described by such Competitive Equilibria is not a Social Optimum. In any case, for given θ, total income at time t for country i is

$$M^i(\theta^i, P_x, P_y) = \max\left\{P_x\alpha_x(\theta^i), P_y\alpha_y(\theta^i)\right\}. \tag{3.6}$$

The competitive equilibrium prices and quantities can finally be found by solving the international market-clearing conditions:

$$\frac{\beta}{P_x}\left[M^1(\theta^1, P_x, P_y) + M^2(\theta^2, P_x, P_y)\right] = x^1(\theta^1, W^1, P_x) + x^2(\theta^2, W^2, P_x) \tag{3.7a}$$

$$\frac{(1-\beta)}{P_y}\left[M^1(\theta^1, P_x, P_y) + M^2(\theta^2, P_x, P_y)\right] = y^1(\theta^1, W^1, P_y) + y^2(\theta^2, W^2, P_y) \tag{3.7b}$$

where Eqs. (3.4), (3.5) and (3.6) have to be used.

In order to describe the time evolution induced by the solution of Eqs. (3.7), we need to specify the learning-by-doing mechanism. Assume it is:

$$\dot{\theta}^i = f(\ell_x^i) + g(\ell_y^i) - \gamma\theta_i \tag{3.8}$$

where $f \geq 0$, $f' > 0$, $g \geq 0$, $g' > 0$ and bounded above, and $\gamma > 0$.

To simplify the discussion, we have excluded intersectoral influences. The "depreciation" factor γ may raise some doubts; we claim that expertise and knowledge depreciates. People die or forget what they have learned, machines wear out and are destroyed, etc. It would be easier to argue this point if "expertise" was embodied in the factors of production, but, as we said, it is also very difficult and we prefer at this stage to settle for less. Let's consider Eqs. (3.7) and (3.8), and draw a phase-plan for the dynamical system of Eq. (3.8) in the (θ^1, θ^2) space (see Figure 1).

To begin with, let's show that the diagonal is an invariant set for the associated flow. Take $\theta^1(t_0) = \theta^2(t_0) = \theta_0$ as an initial condition. The symmetric solution to Eq. (3.7) must be of the form:

$$P^* = \frac{P_y^*}{P_x^*} = \frac{\alpha_x(\theta_0)}{\alpha_y(\theta_0)} \tag{3.9a}$$

$$M^1 = M^2 = P_x^* \alpha_x(\theta_0) = P_y^* \alpha_y(\theta_0) \tag{3.9b}$$

$$x^1 = \beta \alpha_x(\theta_0), \quad x^2 = \beta \alpha_x(\theta_0) \tag{3.9c}$$

$$y^1 = (1-\beta)\alpha_y(\theta_0), \quad y^2 = (1-\beta)\alpha_y(\theta_0) \tag{3.9d}$$

and, by substituting into Eq. (3.8), we conclude

$$\dot{\theta}^i \left(t \mid \theta^1(t_0) = \theta_0\right) = \dot{\theta}^2 \left(t \mid \theta^2(t_0) = \theta_0\right)$$

for all t. Hence, $\theta^1(t) = \theta^2(t)$ forever. Moreover, there exists a unique, attracting stationary state at $(\overline{\theta}, \overline{\theta})$ where $\overline{\theta} : [f(\beta) + g(1-\beta)]/\gamma$. This is point A in Figure 1.

Now let's consider an asymmetric initial condition, say, $\theta^1(t_0) > \theta^2(t_0)$ and assume x is the industrial good. Our basic intuition on the different speeds of learning-by-doing requires:

ASSUMPTION i. For every $\ell \in (0,1] : f(\ell) > g(\ell)$.

ASSUMPTION ii. The function $\alpha : \mathbf{R}_+ \to \mathbf{R}_+$, defined as $\alpha(\theta) = \alpha_x(\theta)/\alpha_y(\theta)$ is increasing.

Assumption i guarantees that, when initial conditions are different, comparative advantage will be important. Linearity of the technology implies that, for $\theta^1(t_0) > \theta^2(t_0)$, either country 1 specializes in production of good x or country 2 specializes in production of good y or both. The analysis may become rather complicated if we seek a complete description of each single case. Given the illustrative purposes of this example, we consider the situation in which both countries fully specialize. Notice, anyhow, that the qualitative conclusions will not be affected in the general case. Competitive equilibrium quantities and prices at any time $t \geq t_0$ will therefore be:

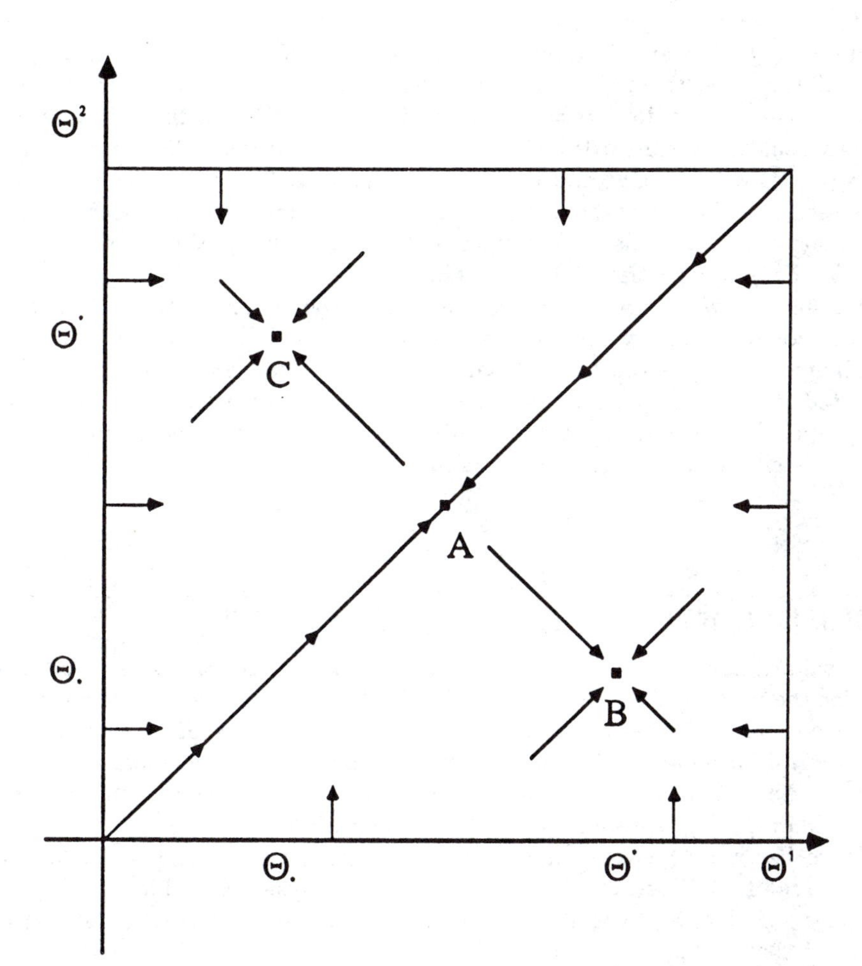

FIGURE 1 Phase-plan for the dynamical system of Eq. (3.8).

$$P^* = \frac{P_y^*}{P_x^*} = \frac{1-\beta}{\beta}\,\frac{\alpha_x(\theta^1)}{\alpha_y(\theta^2)} \tag{3.10a}$$

$$M^1 = P_x^*\alpha_x(\theta^1); \quad M^2 = P_y^*\alpha_y(\theta^2) \tag{3.10b}$$

$$x^1 = \alpha_x(\theta^1); \quad x^2 = 0 \tag{3.10c}$$

$$y^1 = 0; \quad y^2 = \alpha_y(\theta^2) \tag{3.10d}$$

and the two countries will grow according to:

$$\dot{\theta}^1 = f(1) - \gamma\theta^1 \tag{3.11a}$$

$$\dot{\theta}^2 = g(1) - \gamma\theta^2 \tag{3.11b}$$

Denote $\theta^* = f(1)/\gamma$ and $\theta_* = g(1)/\gamma$, then $\theta^* > \theta_*$ because of Assumption i and $\theta^* > \overline{\theta}$ if $f(1) > f(\beta) + g(1-\beta)$. It is also immediate to see that, in fact, the two new steady-state positions $B = (\theta^*, \theta_*)$ and $B' = (\theta_*, \theta^*)$ are the mirror image of each other and are locally attractive. The basin of attraction of B is given by all the points (θ^1, θ^2) with $\theta^1 > \theta^2$ and that of B' is the other half of the positive orthant.

Notice also that, because of the crude simplifications we have been using, two other stationary positions in fact exist at $(\overline{\theta}, 0)$ and $(0, \overline{\theta})$. When one of the two countries has no expertise at time t_0 (say, $\theta^1(t_0) > 0$ and $\theta^2(t_0) = 0$), then the other country reverses to autarky and moves to the steady-state level $\overline{\theta}$, whereas the poor country "disappears" from the scene.

Figure 1 contains the bulk of our argument: in the presence of externalities generated by learning-by-doing mechanism and with differential products, free trade and competitive behavior tend to magnify small differences in the initial conditions and may easily lead to huge disparities in the long run.

4. THE DYNAMICS OF THE GENERAL MODEL

As we said at the end of Section 2, we need a little more structure to be able to consider the vector-field of Eq. (2.8). We will do it here in order to show that the conclusions we have reached in the previous section may indeed persist under a more general formulation. The analysis will not be, in any case, exhaustive nor will we try to be rigorous in all our assertions. A formal treatment of the problem in terms of "Proposition-proof" requires additional work.

From Section 3 we keep the preferences' specification. The technology is as described in (T.1)–(T.2). In order to simplify the analysis, we find it more attractive to change variables and consider the dynamic processes in terms of new variables α^i and ω^i that will soon be defined.

Let a path for $\theta^1(t), \theta^2(t)$ be given. This will induce a path $\alpha_x(\theta^i(t))$ and $\alpha_y(\theta^i(t))$, for $i = 1, 2$. Set $\alpha^i(t) = \alpha_x(\theta^i(t))$ and define $\alpha_y^i = h(\alpha^i)$. This is always possible as we are considering monotone functions. Also set $\omega^i = x^i/\alpha^i = F_x(z_x^i)$, i.e., ω^i is an "aggregate index" of the amount of resources country i invests in the production of good x. For each ω^i let

$$T(\omega^i) = \sup\left\{F_y(z_y^i), \quad \text{s. t. } \omega^i = F_x(z_x^i), \quad z_x^i + z_y^i = 1\right\}$$

i.e., T describes the "Production Possibility Frontier" (PPF) when $\alpha_x^i = \alpha_y^i = 1$, a level which may very well not correspond to any θ^i. We will directly assume T to be strictly concave and differentiable, but this could be derived from a slight strengthening of our assumptions in Section 2. For $i = 1, 2$ given α^i (or equivalently θ^i), we may now write $y^i = h(\alpha^i)T(\omega^i)$ with $h(v) = \alpha_y(\alpha_x^{-1}(v))$. As we observed above, the "instantaneous" competitive equilibrium is Pareto efficient and, hence, we must have for interior solutions:

$$\frac{P_x}{P_y} = -\frac{dy^i}{dx^i}\,|_T = -\frac{h(\alpha^i)}{\alpha^i}\,T'(\omega^i) = p \tag{4.1}$$

for both countries. We will in fact, for simplicity, assume in this section that interiority prevails. This would follow from Assumptions 1 and 2 of Section 2 if we further impose the classical "Inada Conditions."

We need now to redefine our dynamical system. As we have chosen α^1, α^2 as the two-state variables, we write

$$\dot{\alpha}^i = \phi(\omega^i) - \gamma\alpha^i, \qquad i = 1, 2. \tag{4.2}$$

Note that each ω^i depends on both α^1 and α^2 so that the new dynamical system is not decoupled. Moreover we are not, implicitly, making expertise sector specific: ω^i also determines the amount of resources used in the y-producing sector so that the form of Eq. (4.2) amounts to nothing more than a renormalization. The idea that x is the advanced sector is then conveyed in the new framework by the assumptions:

ASSUMPTION 1. ϕ is positive and increasing.

ASSUMPTION 2. $h(\alpha)/\alpha$ is decreasing.

In order to avoid unbounded growth (not an undesirable feature, but not our concern here), we also impose:

ASSUMPTION 3. ϕ is bounded and $h(\alpha)/\alpha$ is also bounded.

Once again let's consider what happens when the initial conditions are on the diagonal. Clearly if $\alpha^1 = \alpha^2$, then also $\omega^1 = \omega^2$. Moreover, because of the homothetic nature of the utility function, it is possible to show that the level of the ω^i's are constant over time and equal to the (unique) solution to the fixed-point problem: $T(\omega)/\omega = -T'(\omega) = -T'(\omega)(1-\beta)/\beta$, (uniqueness here follows from the concavity of T). Call this value ω^*; the dynamics on the diagonal is then

$$\dot{\alpha}^i = \phi(\omega^*) - \gamma\alpha^i, \qquad i = 1, 2. \tag{4.3}$$

so that a rest point will exist and all the orbits on the diagonal will converge to it. This is point A in Figure 2.

Before moving ahead and considering asymmetric initial conditions, let's pause and outline our strategy. We want once again to show that A is a saddle point for the vector field of Eq. (4.2) and that such a vector field points inward on the boundaries of some appropriate square $[0, \overline{\alpha}] \times [0, \overline{\alpha}]$ in the (α^1, α^2) plane. If this is the case, then standard Hopf-Poincaré degree arguments will guarantee that an odd number of equilibria (rest points) for Eq. (4.2) will exist.

Given the general nature of the functions ϕ, F_x, F_y, α_x and α_y, we have no method for computing the number of such equilibria and their dynamic stability. But this will not affect our qualitative argument: we may have other saddle points on each side of the diagonal, or a unique attracting cycle, or even sinks and sources and limit cycles (either stable or unstable); *in any case,* the competitive equilibrium paths will share the common feature of being asymmetric either because they converge to an asymmetric rest point or because they cycle along a closed curve that (being all on one side of the 45° line) will exhibit average levels of the α's (θ's)

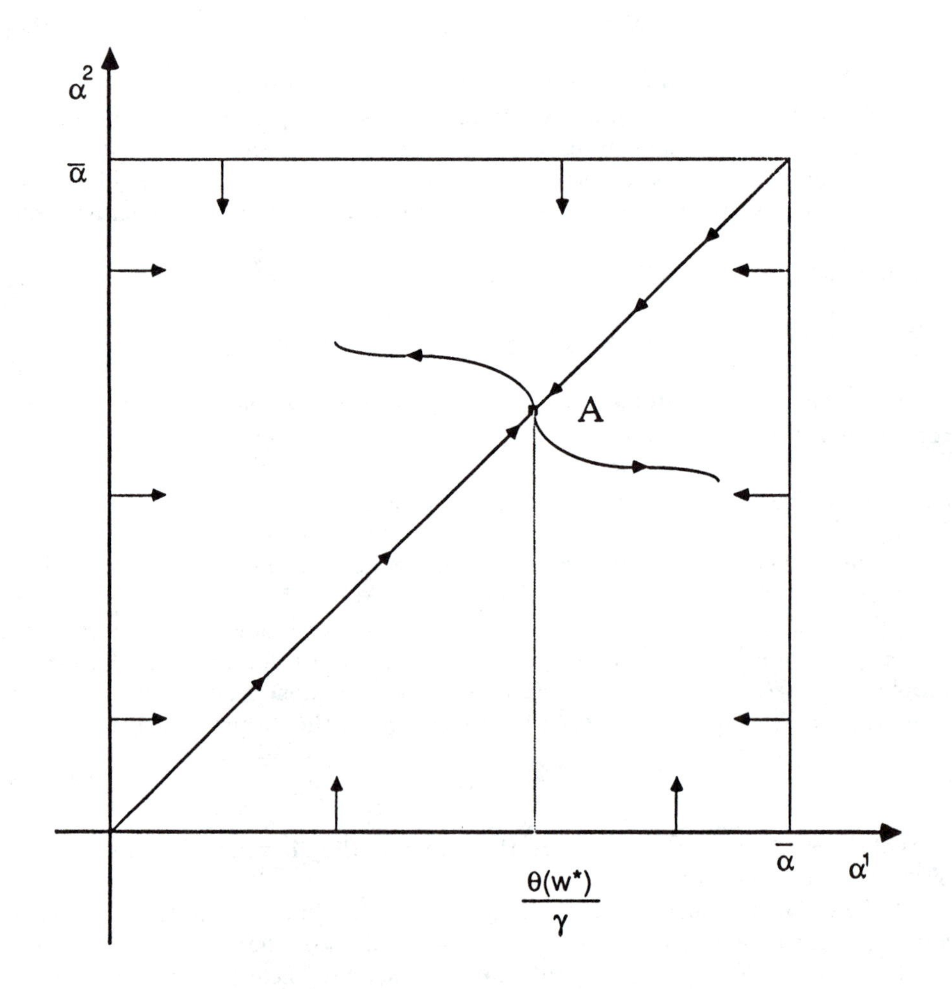

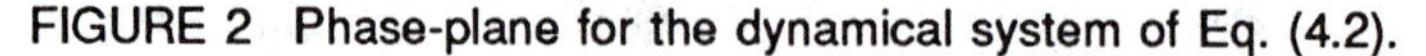
FIGURE 2 Phase-plane for the dynamical system of Eq. (4.2).

that are different across countries. And this qualitative behavior is exactly what we have in mind.

Set $\overline{\omega} = F_x(z^i)$ (remember that $z^1 = z^2$), then

$$\overline{\alpha} = \frac{\phi(\overline{\omega})}{\gamma} \tag{4.4}$$

is the maximum sustainable level of α for both countries and, clearly, the vector field of Eq. (4.2) points inward from any point of the type $(\overline{\alpha}, \alpha^2)$ and $(\alpha^1, \overline{\alpha})$ for $\alpha^i \in [0, \overline{\alpha}]$ (see Figure 2). Pointing inward from the other side requires further assumptions. In fact, we may have a situation in which a point of the type $(\phi(\omega^*)/\gamma, 0)$ (respectively $(0, \phi(\omega^*)/\gamma)$ will attract all the trajectories starting in an ϵ-neighborhood of the horizontal axis (respectively, vertical). Notice that such a feature is not necessarily harmful to our argument, as such a rest point is, indeed, an equilibrium where one country is much richer than the other. Nevertheless, we may like a model with less extreme predictions. It is not difficult to see what is required to guarantee our result. One sufficient condition is the technology to be such that you can produce something even when your expertise is zero and that you, in fact, choose to do so. This may be obtained either because some resources are always allocated to the production of good x, so that $\omega(\alpha^1, \alpha^2) > 0$ everywhere (i.e., total specialization never occurs) and/or because $\phi(0) > 0$, i.e., that even if all the resources are employed in the production of y, some (gross) expertise is acquired that can be used in sector x. Then, as long as $\alpha^i < \phi(0)/\gamma$, $\dot{\alpha}^i > 0$ is obtained. Another more subtle argument can be developed by using the "Inada" conditions mentioned above, together with the boundedness of $h(\alpha)/\alpha$, to ensure that no matter how small α^i is, provided $\alpha^j \leq \overline{\alpha}$ for $j \neq i$, $\omega(\alpha^i, \alpha^j) > \epsilon$. For then, if $\alpha^i < \phi(\epsilon)/\gamma$, $\dot{\alpha}^i > 0$.

We will simply assume the first case to be realized:

ASSUMPTION 4. Either $\phi(0) > 0$ or $\omega(\alpha^1, \alpha^2) > 0 \; \forall (\alpha^1, \alpha^2) \in \mathbf{R}^2$ and $\phi(\omega) > 0$ for $\omega \in (0, \overline{\omega}]$.

This understood, we may proceed to the last step. Consider the case in which $\alpha^1(t_0) > \alpha^2(t_0)$, then as $h(\alpha)/\alpha$ is decreasing and T is concave, we have $\omega^1 < \omega^2$ at t_0. This is our basic comparative advantages intuition and it follows from Eq. (4.1). Therefore, $\alpha^1(t_0) > \alpha^2(t_0)$ implies that $\phi(\omega^1) > \phi(\omega^2)$ at that point. Next we observe that, under our hypothesis on h and T, the conditions for applying the implicit function theorem hold in a neighborhood of $(\phi(\omega^*)/\gamma, \phi(\omega^*)/\gamma) = (\alpha^*, \alpha^*)$. (They, in fact, hold everywhere on $(0, \overline{\alpha}] \times (0, \overline{\alpha}]$ which is why we can define the dynamical system of Eq. (4.2).) Then write $\omega^i = f_i(\alpha^1, \alpha^2)$, for $i = 1, 2$. Since $\omega^i = \omega^*$ on the diagonal, we have that:

$$\frac{\partial f_i(\alpha^*, \alpha^*)}{\partial \alpha^1} + \frac{\partial f_i(\alpha^*, \alpha^*)}{\partial \alpha^2} = 0, \qquad i = 1, 2. \tag{4.5}$$

From Eqs. (4.1) and (4.3) we have that $\alpha^1 > \alpha^2$ implies $\omega^1 > \omega^2$. This and Eq. (4.5) yields:

$$\frac{\partial f_i}{\partial \alpha^i} \geq 0, \quad \frac{\partial f_i}{\partial \alpha_j} \leq 0, \qquad i \neq j. \tag{4.6}$$

To exclude that the derivatives in Eq. (4.6) are both zero, we need to consider again the Competitive Equilibrium condition of Eq. (4.1). Write it as

$$F(\alpha^1, \alpha^2, \omega^1, \omega^2) = \frac{h(\alpha^1)}{\alpha^1} T'(\omega^1) - \frac{h(\alpha^2)}{\alpha^2} T'(\omega^2) = 0 \tag{4.7}$$

and use the implicit function theorem to compute $\partial f_i / \partial \alpha^i = \partial \omega^i / \partial \alpha^i$. As we have assumed $h(\alpha)/\alpha$ to be decreasing, this is nonzero; therefore, both derivatives in Eq. (4.6) are nonzero in a neighborhood of (α^*, α^*) and equal in modulus.

To check under which conditions (α^*, α^*) is a saddle, we need only to linearize Eq. (4.2) around the symmetric equilibrium. The Jacobian computed, there is

$$\gamma \left\{ \gamma - \phi'(\omega^*) \left[\frac{\partial f_1(\alpha^*, \alpha^*)}{\partial \alpha_1} + \frac{\partial f_2(\alpha^*, \alpha^*)}{\partial \alpha_2} \right] \right\} \tag{4.8}$$

As one root is certainly negative, we need Eq. (4.8) to be negative. Notice that the symmetry of the equilibrium can be used to simplify Eq. (4.8) so that our necessary and sufficient condition reads:

$$\gamma < 2\phi'(\omega^*) \frac{\partial f_1(\alpha^*, \alpha^*)}{\partial \alpha_1}. \tag{4.9}$$

It is easy to construct examples verifying Eq. (4.9) since one may, for instance, increase $\phi'(\omega^*)$ without altering either α^* or γ. If acquired expertise depreciates too fast with respect to its rate of self-reproduction, any divergent path is bound to snap back eventually. When the learning-by-doing mechanism is of some relevance (why bother otherwise?), then asymmetric equilibria are the logical outcome of our simple model. Notice, finally, that the positive synergies occurring from trade are reflected in Eq. (4.9) by the fact that the term on the right side sums up the effects from both countries. This suggests that in a general n-countries, m-commodities world, the asymmetric effects are more likely to dominate.

Finally, we show that in fact income of the most productive country will be the largest independent of tastes. Let M^i be the income of country i, i.e.,

$$M^i = p_x \alpha^i \omega^i + p_y h(\alpha^i) T(\omega^i).$$

From Eq. (4.1) we have that

$$\frac{M^1}{M^2} = \frac{h(\alpha^1)}{h(\alpha^2)} \frac{T(\omega^1) - T'(\omega^1)\omega^1}{T(\omega^2) - T'(\omega^2)\omega^2}.$$

If $\alpha^1 > \alpha^2$, then as observed above $\omega^1 > \omega^2$. Further, by the strict concavity of T:

$$T(\omega^1) - T'(\omega^1)\omega^1 > T(\omega^2) - T'(\omega^1)\omega^2 \geq T(\omega^2) - T'(\omega^2)\omega^2.$$

Since $h(\alpha^1) \geq h(\alpha^2)$, we have that $M^1 > M_2$.

5. CONCLUSIONS

We refrain from deriving too many implications from such a simple model and, in particular, to discuss the kind of government policies—production subsidies, import tariffs—that could ameliorate the dynamic inefficiencies. Better insights should come from a more articulated analysis that we are already developing. We only point out a few remarkable limits of this exercise, limits we hope to be able to overcome in the near future.

1. One may want more definite predictions. This will require the choice of "reasonable" functional forms and, very likely, the use of numerical simulations.
2. The shortcomings of the log utility functions outlined at the end of Section 2 must be eliminated by the choice of more sophisticated and more flexible specification of the utility function. The linearity of the Engel curve with respect to income is an especially disturbing limitation.
3. The notion of expertise/knowledge must be analyzed more deeply and cannot be relied upon as the only "engine" of growth. This will amount to an explicit consideration of the capital accumulation process which will yield, in turn, a real intertemporal optimizing framework. The notion of human capital and the ways in which "social knowledge" is embodied in "objects" and transferred over time is an important, related topic.
4. Finally, we may want to allow borrowing-lending to occur across the two countries. This will enable intertemporal consumption smoothing and may therefore affect the temporal pattern of demand and prices. The dynamical system to be considered in this case is a three-dimensional one and is not *a priori* clear that the same simple conclusions will replicate.

ACKNOWLEDGMENTS

The research reported here was mostly done while we both attended the Global Economy workshop at the Santa Fe Institute. We benefited from discussing the topic with almost every participant of the workshop, but we are especially thankful to Phil Anderson, Ken Arrow, Brian Arthur, Buz Brock, David Pines, David Ruelle and Larry Summers. We particularly thank Norman Packard for several discussions and for having helped us to work out a numerical example. The NSF supported the second author's research through Grant SES-8420930 to the University of Chicago.

REFERENCES

1. Arrow, K. J. (1962), "The Economic Implications of Learning-by-Doing," *R. Ec. St.* **29**, 155–73.
2. Arrow, K. J., and F. Hahn (1971), *General Competitive Analysis* (New York: North Holland).
3. Krugman, P. (1985), "The Narrow Moving Band, the Dutch Disease and the Competitive Consequences of Mrs. Thatcher: Notes on Trade in the Presence of Dynamic Scale Economics," *mimeo, MIT working paper.*
4. Lucas, Jr., R. E. (1985), "On the Mechanics of Economic Development," *Marshall Lectures delivered at the University of Cambridge, revised version August 1987, mimeo, University of Chicago.*

J.-P. ECKMANN,* S. OLIFFSON KAMPHORST,* D. RUELLE† &
J. SCHEINKMAN‡
*Département de Physique Théorique, Université de Genève CH-1211 Genève 4, Switzerland; †I.H.E.S., 91440 Bures-sur-Yvette, France; and ‡Department of Economics, University of Chicago, 1126 East 59th Street, Chicago, IL 60637, U.S.A.

Lyapunov Exponents for Stock Returns

Recently, J. Scheinkman and B. Le Baron[3] have analyzed a time series of weekly stock returns from the point of view of nonlinear dynamics. In particular, using the Grassberger-Procaccia algorithm, they found the data to be compatible with a deterministic dynamical system of low dimension (around 6) perturbed by a moderate amount of noise. In the present note, we pursue the analysis of the same time series, and obtain an estimate of Lyapunov exponents, i.e., rates of exponential growth of small perturbations to initial conditions.

First, however, we analyze the data for temporal homogeneity by producing a recurrence plot[1] (see Figure 1) to obtain this plot the weekly returns time series is first used to generate a six-dimensional trajectory by the time-delay method. The points of trajectory are indexed by week. Plotting the index of a point horizontally form left to right, the indices of the other trajectory points closest to it (neighbors) are plotted vertically from bottom to top. This produces the gray square with ascending diagonal (called recurrence plot) at the top of Figure 1. (The diagonal corresponds to the fact that each trajectory point is in a neighborhood of itself.) The graph at the bottom of Figure 1 gives the radius of the sphere in six dimensions containing a required minimum number of neighbors. The recurrence plot, apart form a short exceptional period near the beginning of the time series, appears reasonably uniform (i.e., gray, without clustering around the diagonal). This seems

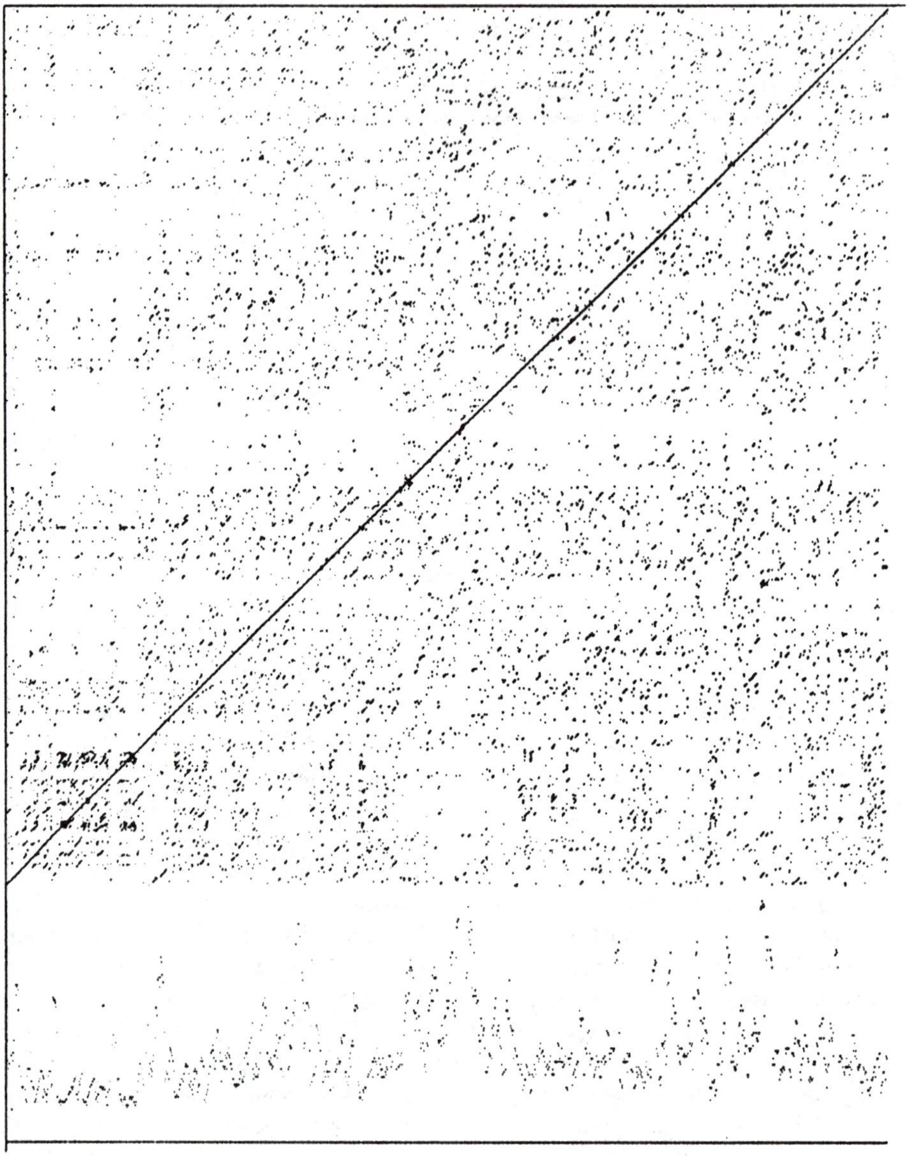

FIGURE 1 Analysis for temporal homogeneity via a recurrence plot.

to indicate that there is not much drift—in the more than twenty years studied—for the behavior of weekly returns over a few weeks.

Having satisfied ourselves with the temporal homogeneity of the time series, we proceed to apply an algorithm for the estimation of Lyapunov exponents.[2] The results are given in Figure 2, where embedding dimension is plotted horizontally, and Lyapunov exponents vertically. It is known that negative exponents or exponents near zero cannot be trusted.[2] As the figure indicates, however, there seems to be a positive exponent ~.15 per week and another one ~.10 per week. This gives at least a rough estimate of how unpredictable the evolution is, for the series of weekly returns which we have analyzed.

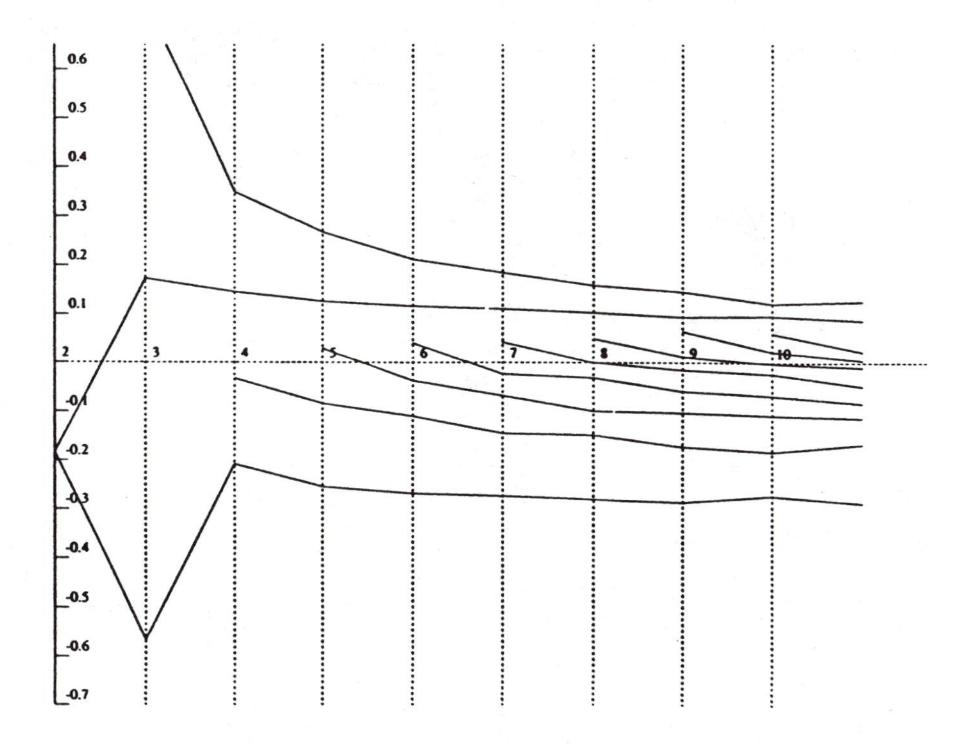

FIGURE 2 Results of applying an algorithm for the estimation of Lyapunov exponents.

REFERENCES

1. Eckmann, J.-P., S. Oliffson Kamphorst and D. Ruelle, "Recurrence Plots of Dynamical Systems," *Europhys. Lett.*, to appear.
2. Eckmann, J.-P., S. Oliffson Kamphorst, D. Ruelle and S. Ciliberto (1986), "Lyapunov Exponents from Time Series," *Phys. Rev.* **A34**, 4971–4979.
3. Scheinkman, J. A., and B. Le Baron, "Nonlinear Dynamics and Stock Returns," *J. Business,* to appear.

VII. Appendix

GEORGE A. COWAN† & ROBERT MCCORMICK ADAMS‡
†Los Alamos National Laboratory, Los Alamos, NM 87545 and Santa Fe Institute, 1120 Canyon Road, Santa Fe, NM 87501; and ‡Smithsonian Institution, 1000 Jefferson Drive, Washington, DC 20560

Summary of Meeting on "International Finance as a Complex System" at the Rancho Encantado, Tesuque, New Mexico, August 6–7, 1986

The Santa Fe Institute, an educational and research organization dedicated to promoting interdisciplinary efforts on problems related to highly complex systems in science and society, sponsored a meeting on "International Finance as a Complex System" at the Rancho Encantado, August 6 and 7, 1986. The discussion was organized by Robert McCormick Adams, Secretary, Smithsonian Institution, and Vice-Chairman of the Institute's Board of Trustees, and George A. Cowan, President of the Institute, in response to a suggestion to Adams by John Reed, Chairman of the Board of Citicorp. Support for this meeting was provided by a grant from the Public Service Company of New Mexico and by the Santa Fe Institute.

The major purpose of the meeting was to examine the potential application of some recent developments in the natural sciences, mathematics, and computer technology to certain problems in economics, more specifically to the better understanding and management of world capital flow and debt. The participants included a contingent from Citicorp headed by Chairman Reed together with Eugenia Singer, Byron Knief, and Victor Menezes; the Chairman and Chief Executive Officer of Public Service Company of New Mexico; Jerry Geist; Nobel Laureate Philip Anderson from Princeton University, computer scientist John Holland from the University of Michigan, and Larry Smarr, Director of the National Center for Supercomputing Applications at the University of Illinois, all members of the Board of Advisors of the Institute; Kenneth Rogoff, Professor of Economics at the Social Systems Research Institute, University of Wisconsin; Doyne Farmer, a physicist specializing in the behavior of nonlinear dynamical systems at the Los Alamos National Laboratory; Carl Kaysen, Professor of Political Economy at the Massachusetts Institute of Technology and member of the Institute Board; David Pines of the University of Illinois and Chairman of the Board of the Institute; Robert McCormick Adams; and George A. Cowan. Adams chaired the meeting.

The discussion began with a review by John Reed and members of the Citicorp staff of the development of a globally interconnected economic system in the period dating from 1850 to the present, with emphasis on the increasing volatility and difficulties which have characterized the system in the past several decades. They provided an impressively broad and consistent view of current conditions and their antecedents. The presentation rested on an earlier finding that existing simplifying models of the contemporary global economic system are of limited accuracy or predictive utility. In particular, such models have not been found to meet the needs of financial managers faced with large risks and uncertainties and a continuing responsibility for real-time decisions. The result is that key institutions find themselves with limited analytical tools as they deal with a dynamic system that seems to be moving in indeterminate directions and at an accelerating pace.

It was recalled that the two world wars and the inter-war period saw a series of severe breakdowns in a long-term, uneven but essentially worldwide growth in per capita GNP. Parallel efforts to cope with that series of shocks led, by the end of World War II, to a recognition that a genuinely and unprecedentedly interconnected system already was emerging. The new arrangements necessary for it, as the need was then understood, were formalized in the Bretton Woods agreements. Emphasis was consensually shifted away from isolationism and protectionism as national policies and toward the mutual support of institutions like the World Bank, the IMF, GATT, etc. Recognition of the impending entry of an enlarged and changing set of major actors was a further important contribution. On the other hand, due allowance may not have been given to the significance, diversity, and capacity for independent action of entirely unforeseen numbers of less-developed countries that would also presently appear as actors on the world economic stage.

After an ensuing interval of decades that was fairly stable, at least in financial terms, the system was again beset by new shocks and mounting conditions of turbulence: serious inflation in many countries, high unemployment, rapidly rising prices for energy, etc. The decision to shift from fixed to floating exchange rates, viewed rather narrowly at the time as a palliative to fairly specific imbalances, turned out instead to be a profoundly important new step with new destabilizing consequences of its own. It was perhaps inevitable that a brake was eventually applied to this seemingly runaway process, symbolized by the appointment of Paul Volcker in 1979 and to a considerable degree embodied in subsequent steps for which he was chiefly responsible. But the cumulative effect of the actions taken, and of their abruptness, was to alter fundamentally the economic environment to which the developing countries had to adapt.

In the new environment, the LDC's are beset by a significant decline in the world's real rate of economic growth (from 3.7% per year in the period 1973–1980 to 2.3% after 1980). Declining growth translates to declining prices and markets for raw materials and other products entering world trade to balance their needed capital and other imports. Developing countries make a minor contribution to world GDP (roughly $12.8 trillion), but a very significant contribution to total world trade (roughly $1.9 trillion). Responding to what were, in effect, negative interest rates (due to the high rate of inflation), LDC's had quite naturally been willing to tolerate large deficits in their current accounts and paid for imports with borrowings during the '70's. This contributed substantially to an immense rapid growth in exports, particularly stimulating demand for commodities of which a large proportion enters world trade, e.g., oil 52%, autos 43%, wheat 21%.

With the abrupt reversal in interest rates since 1980, however, LDC's have had to cut imports and expand exports even under adverse terms of trade. Many now manage to run net surpluses, but the aggregate effect is to keep world trade at essentially stagnant levels. In turn, that aggravates their difficulties in managing the debt, let alone resuming growth in GNP. As confidence in their capabilities wanes, large-scale capital flow to LDC's has virtually ceased (beyond steps that only serve to keep their interest payments current). Instead, other industrialized countries now direct major capital flows to the U.S. (the cost of capital in Japan, for example, is roughly half that in the U.S.). Although our deficits continue to mount, our rates of return on capital are very good. Continuing investment in what might be characterized as our "lifestyle" (e.g., $230 billion annually in auto and mortgage financing alone) accompanies, and directly contributes to, an ongoing deterioration in the economic condition and prospects of the LDC's. Apart from the suffering involved, long-continued trends of this kind will cause the whole system to become progressively more tension-ridden and fragile.

If the current situation continues, borrowers must contemplate a period of at least twenty years of flat economic growth while the debt load is reduced to a manageable size through cash flow repayment. An alternative method of adjustment is to swap debt for equity in the resources of the debtor nations. This appears to be an attractive solution to creditors, but the overall consequences to LDC's and to our domestic economy cannot be readily assessed. The use of economic models to predict such consequences is hampered by the inability of current models to deal with large-scale adjustments and adaptations, e.g., to deal with the dynamics of a rapidly changing world.

Mrs. Singer presented a review of global economic models, some of which attempt to incorporate dynamics. These include Project Link which interconnects the national economic models of thirteen OECD countries, the U.S.S.R., LDC countries, eastern Europe, and the PRC. The model contains 4500 equations and 6000 variables. Other global models which largely assume general equilibrium are the Federal Reserve Multi-Country Model, the World Bank Global Development Model, the Whalley Trade Model, the Global Optimization Model, and various forms of input-output models. Major deficiencies are evident in the way all of these models treat or ignore systems dynamics.

An "ideal" model would examine the development process in more detail; incorporate more explicit statements about social and political structures; include such endogenous variables as capital flows, exchange rates, interest rates, and inflation rates; appropriately hybridize general equilibrium and global optimization considerations; and treat major LDC and industrial countries individually.

In the following discussion, Rogoff noted that dynamics is not an overlooked subject in economics but rather one in which economists are deeply interested. However, their successes with very large systems have been few as compared to analyses of component parts. He felt that it would not be profitable to further enrich the FRB model with dynamics, that its problems lie elsewhere. But there are many issues in macro-economic modeling, rational-expectations theory, and game theory which could be strengthened by contributions from the kinds of scientific expertise represented at this meeting.

In view of some of the participants, our present ability to plan economic actions to minimize further unfavorable developments is not very encouraging. In the broadest sense, as a physicist observed, it distorts analysis to isolate issues of monetary policy as an enclosed causal framework. Even when the field is expanded to include trends and balances in world trade, shifts in interest rates, and the like, we are left with a somewhat artificially confined group of interactions. While it is more manageable in terms of traditional analysis and computation, this limited system can sustain prediction only in the form of short-term linear projections. Virtually excluded is the possibility of unexpected outcomes resulting from what have been defined *a priori* as exogenous forces.

Considerations of international rivalries and national security are good examples, and a number of specific cases were cited during the discussion. "Confidence" is a volatile, not well-understood entity that ebbs and flows in ways not closely responsive to utility-maximizing rationality, sometimes simultaneously favoring and disfavoring countries with considerable similarities in their overall economic position. Another difficulty, also linked to limitations on the complexity that economic managers have been prepared to take into account, has arisen from a prevailing tendency to make decisions after viewing each country's prospects individually. Necessary as such a view is, this means that there has been little effort to consider the world economic system as a whole.

Similar issues arise if consideration is given to adding new lines of approach in order to take "exogenous" forces in other countries more fully into account. Rates of return on capital and prospects of political instability, for example, cannot be factored neatly into a single set of equations. Different and unfamiliar conditions of uncertainty are obtained, and the appropriate experts are rarely to be found in the same circles as bankers and ministers of economics. Loans once made are irrevocable, so that there is little disposition to reconsider critically the decisions leading to them. Political constraints on choices of strategy, and even on individual decisions, are frequent but rarely acknowledged. For these and other reasons, there is little disposition among bankers to examine closely the quality of their foreign assets, one participant observed. Most banks persistently, and perhaps even consciously, "keep their heads under a pillow." Yet this posture is obviously apparent to investors and by no means cost-free, as a comparison of price-earnings ratios for bank stocks with other equities quickly shows.

Still another challenge to improved predictability involves the nonlinearity of many forms of response. A half-dozen years or so ago, "stagflation" had been widely accepted as something approaching an indefinite condition of near-normality. It did not seem possible that the world collectively might be willing to slow its growth, including the acceptance in a number of European countries of levels of quasi-permanent unemployment that were then regarded as politically intolerable, in order to stem inflation. A nonlinear response, a turning point, on the part of very broad segments of the public was necessary for inflation to be perceived and resolutely dealt with as the central problem. Yet it is not clear, even in retrospect, how that turning point was reached.

Drawing these numerous observations together (while necessarily ignoring many others), we see that the world financial system is complex, multi-centered, subject to severe disturbances or even breakdowns, and poorly understood in its more dynamic aspects. It is a system loosely coupled in some respects and tightly coupled in others, harboring the real possibility that changes in the direction of tighter coupling could alter (perhaps inducing magnified oscillations in) the system's behavior. It is an open rather than closed system, without clear boundaries, exposed to nonlinear responses that presently defy even retrodiction. Hence, it depends on a heavy discounting of the future rather than on efforts to take the future more fully and accurately into account.

As mentioned at the outset, the essential question before the meeting was whether there was some likelihood that new potentials for modeling and computation, growing largely out of advances in the physical sciences, mathematics, and computer technology, might be at least marginally helpful toward this latter end. The discussion of this question was begun with some examples of the techniques and methodologies presently under development.

Smarr reviewed the rapidly increasing ability of graphics software on supercomputers to present movies of the dynamical features of very larges complex systems. With the aid of such graphics, the human eye grasps and the brain more readily interprets the dynamic structures of these systems, including features which can be totally overlooked when scanning static charts and numerical listings.

Holland discussed examples of self-learning systems on computers, describing in some detail a computerized system for managing a gas pipeline network which has previously required the extensive experience and judgment of highly trained individuals. The principles involved in constructing artificial intelligence algorithms of this type which learn from experience may be useful in dealing with inputs from the stochastic external environments involved in adaptive economics.

Farmer described the mathematical treatment of nonlinear dynamic systems which has, in recent years, led to general recognition of the importance of chaos and attractors in such systems. It was agreed that the data base to which Citicorp has access could be made available for searches for similar patterns of behavior in economics. Rogoff observed that this subject is already of interest in economics and that a recent meeting of a prestigious econometrics society opened with a keynote talk on the subject of chaos given by Jean-Pierre Grandmont.

Anderson continued in a similar vein on the subject of the current treatment of complex systems with very large numbers of independent variables and the development of new conceptual frameworks such as the spin-glass Hamiltonians. He pointed out that in the social sciences, one must include an important new factor, "tire-tracks," meaning memory of past performance and the ability to change the trajectory based on previous experience and preferred outcomes (selection rules). Research on such systems must better identify major driving forces, of which technological change must be one of the most important, and the principal dissipative forces, which are not so readily identifiable nor easily rank-ordered in importance. He expressed cautious optimism concerning our ability to incorporate these forces.

Reed summarized his sense of the meeting by remarking that, on the demand side, it was clearly worth an immense amount of effort to achieve a better understanding of the forces which shape the development of a global economy. A major weakness is in our understanding of the adaptive or dynamic forces. In the short space of time available, it was not possible for him to judge whether the time is ripe for a greatly expanded effort along the disciplinary lines represented at this meeting and that the question should be examined further. Kaysen remarked that perhaps an examination of next steps should be undertaken under auspices more clearly representative of the economics side of the problem. Cowan expressed a concern that, under other auspices, the momentum to undertake a broadly interdisciplinary approach might be weakened. Pines stated that the Institute should be prepared to organize a meeting, perhaps two weeks in length, with an appropriate breadth of representation to discuss the content of a mutually supportive program incorporating the kinds of science and computer technology represented at the meeting.

Geist expressed a deep conviction concerning the great importance of the subject, particularly emphasizing the problem of presenting data and arguments for action to public leaders with little technical background. He was impressed by the potential of the graphics techniques described by Smarr for conveying information on highly complex subjects to lay people. There was general agreement that this aspect could constitute a major contribution.

Adams summarized the consensus recommendation for further action. He noted that a cautious formulation of goals is appropriate, since such a venture would take us across wholly uncharted terrain. However, there was a general disposition among participants to regard the venture as entirely credible and decidedly worth pursuing. He promised prompt follow-up review and planning which will lead to a more thorough identification of the options for useful research.

Index

A and B games, 210
activity analysis matrix, 155, 160
adaptive dynamics, 112
adaptive expectations, 10
adaptive nonlinear network, 118
adaptive walk, 128
adoption market, 21
advantage
    comparative, 286
aggregate excess demand function, 148, 160
algorithm, 179
    back-propagation learning, 33, 41-45
    learning, 249
    rule-generating, 122
    Scarf's, 153
    simplicial, 152
allocation processes, 18
annealed average, 188
annealing, 16, 187
    average, 188
    schedule, 187
    simulated, 187
anticipation, 118
Argentine Austral Plan, 233-234
asymptotic states, 9
attractors, 13
autoregression
    orthodox, 268

back-propagation learning algorithm, 33, 41-45
bifurcation, 65, 257
    period-doubling, 65
Brazilian Cruzado Plan, 234
broken ergodicity, 181
Brouwer's fixed-point theorem, 148, 150-151, 154
    Hirsch's proof of, 154
budget constraint, 148
budget constraints, 152
building blocks, 118, 122
business cycles, 200

capital accumulation, 287
central limit theory, 77
chaos, 102, 196, 248
chaos theory, 77, 86
chaotic dynamical systems, 100
    nonlinear model, 100
chaotic dynamics, 149
chaotic paths, 50
chaotic trajectories, 164
characteristic exponent, 198
chemical kinetics, 9
class struggle, 54
classifier systems, 121
coadaptation, 135
coevolution, 199
comparative advantage, 282
    dynamic evolution, 287
comparative statics analysis, 147
competition, 121
competitive equilibrium, 289
competitive exclusion, 11
competitive firm, 286
complex dynamics theory, 77
computation of equilibria, 152
configuration space, 181
configurations, 180
confirmation, 122
conformity effects, 24
constant inflation-rate theorem, 228, 235
constant returns technology, 155
constant-sum game, 211
consumer, 148
consumer's demand function, 148
coordination effects, 10
correlation dimension, 87, 89, 92, 95
cost function, 180
cumulative causation, 10

data massaging, 263
    by orthodox means, 267
    by unorthodox means, 267
decreasing returns technology, 155

default hierarchy, 121
detailed balance condition, 186
deterministic chaos, 77
detrending, 196
development
    specialization, 286
disequilibrium adjustment process, 149
divergence of trajectories, 259
dynamic chaos
    concept of, 261
dynamic network, 171
    with random couplings, 171
dynamical model
    open, 169
dynamical systems model, 170
    conventional, 170
dynamical systems theory, 170
dynamical systems, 9
    chaotic, 100

ecology, 199
economic cycles, 247
economic development, 10
economic equilibria, 147
economic profit, 156
economic system, 16, 254
    political forces, 258
    psychological forces, 258
    sociological forces, 258
economic theory, 9, 250
economic webs, 125-126, 246
    evolution, 140
    growth, 137, 141
    growth constraints, 141
economy
    Richardian, 290
effects
    conformity, 24
    coordination, 10
    learning, 10
    synergistic, 17
    threshold, 10
emergent structures, 9, 118
employment, 17
endowment, 148, 158
entropy, 185
equilibria, 157, 163-164
    computation of, 152
    computing existence of, 163
    economic, 147
    multiple, 10
equilibria (cont'd.)
    multiplicity of, 157, 164
equilibrium, 148, 152, 156-157, 199, 245, 276
    competitive, 289
    concept of, 265
    continuity of, 157
    existence of, 148
    inferior in economics, 16
    Nash, 205, 208
    regular, 159
        index of, 159
    with transfer payments, 152
equilibrium dynamics, 277
equilibrium theory, 78, 91, 269, 278
excess demand function, 149, 151, 158
    continuously differentiable, 158
exchange economy, 148
exit, 13, 16
expansion dynamics, 169
expectations, 26
expertise, 286
externality, 286

feedforward model, 33
financial theory, 77, 84
fixed-point index theorem, 158
fluctuations, 11
forecasting, 99
free disposal, 156
free energy, 185
free trade, 199
frustration, 177
function
    aggregate excess demand, 148, 160
    consumer's demand, 148
    cost, 180
    excess demand, 149, 151, 158
    implicit function theorem, 158
    inverse function theorem, 157
    Lyapunov, 35, 80-81, 266
    partition, 184
    transfer, 152, 157
fuzzy sets, 260

Game Theory, 205
game of half of the average, 206, 212
games
    A and B, 210
    constant-sum, 211
    half of the average, 206, 212

games (cont'd.)
  prisoner's dilemma, 211
general competitive equilibrium theory, 275
global economy proper, 270
global Newton's method, 154
golf course, 182
goods
  agricultural, 286
  high-technology, 286
  low-technology, 286
graph bipartitioning, 179
ground state, 185
growth theory, 268

high-technology good, 286
Hopf bifurcation, 51-52
Hopfield model, 33-40
human capital, 286
hyperplane transform, 121
hypotheses
  learning-by-doing, 290
  martingale, 92
  rational expectations, 205, 276

implicit function theorem, 158
income policies, 222
increasing returns, 10
index, 159
industrial organization, 9, 12
inertial inflation, 218
infinitely lived agents, 63
inflationary inertia, 205, 225
information theory, 43
instances, 180
internal energy, 185
internal models, 119
international trade, 285
international trade theory, 9, 11
intertemporal duality theory, 83
intertemporal models, 162
inverse function theorem, 157

Kaldor's business-cycle model, 51
Keynesian model, 51

Lagrange multipliers, 152
Lagrange's theorem, 237
landscape, 181
laws
  Strong Law of Large Numbers, 18
  Walras', 148
learning algorithms, 249
learning effects, 10
learning machines, 263
learning-by-doing, 285-286
learning-by-doing hypothesis, 290
limit cycle, 149
limited agents, 123
linear production technology, 155
local uniqueness, 157
location by spin-off, 22
lock-in, 10, 183
low-technology good, 286
Lyapunov coefficients, 247
Lyapunov exponents, 99, 102, 262, 303
Lyapunov function, 35, 80-81, 266

macro-dynamics, 12
market shares, 24, 11
Markov process, 85, 120
Markov stochastic shocks, 80
martingale hypothesis, 92
mean field theory, 189
meta-dynamic, 170
metric, 181
microstate, 184
model, 25
  Anderson's personal, 272
  dynamical systems, 170
  feedforward, 33
  Hopfield, 33-39
  internal, 119
  intertemporal model, 162
  Kaldor's business-cycle, 51
  Keynesian, 51
  morphogenesis, 143
  NK, 133
  nonlinear, 99-100, 104, 248
  open dynamical, 174
  overlapping generations, 245, 58-62
  123, 259, 276
  prey-predator, 54
  prisoner's dilemma, 123, 245
  random link, 189
  regime change, 272
  stochastic, 14, 248
  trial-and-error, 207
  Turing's, 143
  open dynamical, 169
monopoly, 21
Monte Carlo simulation, 186
multi-arm bandit, 26

multiple equilibria, 10
mutual causal processes, 10

Nash equilibrium, 205, 208
network externalities, 12
neural networks, 106, 112-113
Newton, 154
    global method, 154
NK model, 133
noise, 196
non-convexity, 10
non-neutrality theorem, 229, 239
nonlinear dynamics, 49, 195, 197, 248
nonlinear interactions, 118
nonlinear model, 100, 104, 248
nonlinear modeling, 99
nonlinear physical system, 13
NP-complete, 179
nuclear reactors, 15
numeraire, 149

open, 261
open dynamical model, 174
optimization, 42, 127, 131, 151, 177
order, 11
order parameter, 189
orthodox autoregression, 268
overlapping generations model, 58-62, 123, 163, 245, 259, 276

Pareto efficient, 79, 151, 157, 162-163, 291
Pareto optimal, 79, 289
partition function, 184
path-dependence, 17, 11, 23
pattern theory, 268
patterns of change, 270
patterns of debt, 270
patterns of politics, 270
patterns, 254, 268
peak/average ratio theorem, 229, 238
perfect foresight, 163
period-doubling bifurcation, 65
periodic cycles, 50
perpetual novelty, 120
phase-locking, 13
policy, 26
positive feedback, 258, 9
potential barrier, 13
PPF, 294
prediction, 201, 267
    by orthodox means, 267
prediction (cont'd.)
    by unorthodox means, 267
prey-predator model, 54
price domain, 148
principle of strategic interdependence, 209
prisoner's dilemma model, 123, 211, 245
production possibility frontier, 294
public choice theory, 258

quasi-homomorphisms, 119
quenched average, 188

random link model, 189
randomness, 100, 177
rational expectations, 208
rational expectations hypothesis, 205, 276
rational expectations theory, 259
rationality assumption, 56
rationality, 259
recombination, 122
recontracting processes, 23
recurrence plot, 301
regional economics, 9
regular economy, 157
regular equilibrium, 159
    index of, 159
regular production economies, 159
regular value, 154
replica method, 190
requisite variety, 27
retraction, 154
Richardian economy, 290
rugged landscapes, 261
rule-generating algorithms, 122

Sard's theorem, 154
Scarf's algorithm, 153
selection problem, 12
selectional advantage, 11
selective pressures, 120
self-reinforcement, 10
self-reinforcing mechanisms, 9
sequential decisions, 13
shocks, 196
simplex, 150, 153
    face of, 153
simplicial algorithms, 152
social networks, 258
social optimum, 291
Spatial Economics, 12
spatial mechanisms, 27

specialization, 286
Sperner's lemma, 153
spin glass, 177
spin glass theory, 177
Stackelberg dominance, 207
Stackelberg warfare, 209
staggered price setting, 232
staggered wage setting, 227-228, 232, 235
standard operating procedures, 118-119
statistical mechanics, 177
steady state, 163
stochastic, 25
stochastic model, 14, 248
stock market, 258
stock returns
    complex dynamics, 92
strange attractors, 196
strategic action, 25
Strong Law of Large Numbers, 18, 20
structure
    emergence of, 118
subdivision, 153
symmetry breaking, 11, 183
symmetry, 178
synergistic effects, 17

tâtonnement, 149
technical knowledge, 287
technologies, 10
temperature, 184
theorems
    Brouwer's fixed-point, 148, 150-151
        proof of, 154
    constant inflation-rate, 228, 235
    fixed-point index, 158
    implicit function, 158
    inverse function, 158
    Lagrange's, 237
    non-neutrality, 229, 239
    peak/average ratio, 229, 238
theorems (cont'd.)
    Sard's, 154
    turnpike, 65
theories
    central limit, 77
    chaos, 77, 86
    competitive equilibrium, 271
    complex dynamics, 77
    dynamical systems, 170
    economic, 9, 254
    equilibrium, 78, 91, 269, 278
    financial, 77, 84
    game, 205
    growth, 268
    information, 43
    international trade, 9, 11
    intertemporal duality, 83
    mean field, 189
    pattern, 268
    public choice, 258
    rational expectations, 259
    spin glass, 177
    temporal duality, 83
threshold effects, 10
time series analysis, 99
topology, 181
transfer function, 152, 157
traveling salesman problem, 129, 178
Turing's model, 143
    morphogenesis, 143
turnpike theorem, 65

ultrametricity, 181
utility, 120

vertices, 153

Walras' law, 148
Walrasian dynamics, 12
webs, 264

The Addison-Wesley **Advanced Book Program** and the SANTA FE INSTITUTE would like to offer you the opportunity to learn about our new "Studies In the Sciences of Complexity" titles and workshops in advance. To be placed on our mailing list and receive pre-publication notices and special offers, just **fill out this card completely** and return to us.

Title, Author, and Code # of this book: **Date purchased:**

____ ____________

Name ________________________________

Title ________________________________

School/Company ________________________________

Department ________________________________

Street Address ________________________________

City ____________________ State ________ Zip ________

Telephone (______) ________________________________

Where did you buy this book?

☐ Bookstore
☐ Mail Order
☐ School (Required for Class)
☐ Campus Bookstore (individual Study)
☐ Toll Free # to Publisher
☐ Professional Meeting
☐ Publisher's Representative
☐ Other ________________________________

Please define your primary professional involvement:

☐ Academic: Professor
☐ Academic: Student
☐ Academic: Researcher
☐ Industry: Administrator
☐ Industry: Researcher
☐ Industry: Technician
☐ Government: Administrator
☐ Government: Researcher
☐ Government: Technician

Check your areas of interest.

200 ☑ SFI

201 ☐ Agriculture
202 ☐ Anthropology
203 ☐ Artificial Intelligence
204 ☐ Astronomy
205 ☐ Atmospheric Sciences
206 ☐ Biological Sciences
207 ☐ Chemistry
208 ☐ Computer Sciences
209 ☐ Communication Sciences
210 ☐ Dentistry
211 ☐ Economics
212 ☐ Education
213 ☐ Engineering
214 ☐ Geology/Geography
215 ☐ History/Philosophy Science
216 ☐ Industrial Science
217 ☐ Information Sciences
218 ☐ Mathematics
219 ☐ Medical Sciences
220 ☐ Pharmaceutical Sciences
221 ☐ Physics
222 ☐ Political Sciences
223 ☐ Psychology
224 ☐ Social Sciences
225 ☐ Statistics
226 ☐ OTHER ________________________________ (please specify)

Of which professional scientific associations are you an active member?

________ ________ ________ ________ ________ ________

________ ________ ________ ________ ________ ________

Would you like to be sent information about the SANTA FE INSTITUTE and its workshops?

☐ Yes ☐ No

fold and staple